SCIENCE
and TECHNOLOGY
of SURFACE COATING

SCIENCE
and TECHNOLOGY
of SURFACE COATING

**A NATO Advanced Study Institute
on the Science and Technology of Surface Coating
held at Imperial College, University of London,
in April, 1972**

Edited by

BRIAN N. CHAPMAN

RCA David Sarnoff Research Centre
Princeton, New Jersey, U.S.A.
(on sabbatical leave from *Imperial College,*
London, England)

and

J. C. ANDERSON

Department of Electrical Engineering,
Imperial College of Science and Technology,
London, England

1974
ACADEMIC PRESS LONDON AND NEW YORK
A Subsidiary of Harcourt Brace Jovanovich, Publishers

ACADEMIC PRESS INC. (LONDON) LTD.
24/28 Oval Road,
London NW1

United States Edition published by
ACADEMIC PRESS INC.
111 Fifth Avenue
New York, New York 10003

Library of Congress Catalog Card Number: 73-9453
ISBN: 0 12 168350 8

59276 40794598

Printed in Great Britain by
William Clowes & Sons, Limited
London, Beccles and Colchester

List of Contributors

C. AMSALLEM, *Hydromecanique et Frottement, Andrezieux-Boutheon, France.*

J. C. ANDERSON, *Imperial College, London, England.*

RAGHUPATHY BOLLINI, *Dept. of Electrical Engineering, Purdue University, West Lafayette, Indiana, U.S.A.*

JOSEPH W. BOARMAN, *School of Electrical Engineering, Purdue University, West Lafayette, Indiana, U.S.A.*

ROINTAN F. BUNSHAH, *Materials Department, University of California, Los Angeles, California, U.S.A.*

DAVID S. CAMPBELL, *Department of Electrical Engineering, University of Technology, Loughborough, England.*

RICHARD CARPENTER, *Allen Clark Research Centre, Plessey Co. Ltd., Northampton, England.*

J. J. CAUBET, *Hydromecanique et Frottement, Andrezieux-Boutheon, France.*

NEWELL C. COOK, *Organic Chemistry Branch, General Electric R & D Centre, Schenectady, New York, U.S.A.*

ROGER P. CORBETT, *Wolfson Applied Electrostatic Advisory Unit, University of Southampton, England.*

DONALD A. DECKER, *School of Electrical Engineering, Purdue University, West Lafayette, Indiana, U.S.A.*

R. DIJKSTRA, *Philips Research Laboratories, Eindhoven, Netherlands.*

EUGENE F. FINKIN, *D.A.B. Industries, Inc., Detroit, Michigan, U.S.A.*

D. DE GRAAF, *Acheson Colloids, Scheemda, Holland.*

J. C. GREGORY, *Cassel Heat Treatment Service, Imperial Chemical Industries Ltd., Birmingham, England.*

M. A. K. HAMID, *Antenna Laboratory, Department of Electrical Engineering, University of Manitoba, Canada.*

K. HIEBER, *Siemens Research Laboratories, Munich, West Germany.*

L. HOLLAND, *Edwards High Vacuum International, Crawley, England.*

JOHN D. HUNTER, *Department of Electrical Engineering, University of Manitoba, Canada.*

J. DE JONGE, *Philips Research Laboratories, Eindhoven, Netherlands.*

SIMO KARTTUNEN, *Graphic Arts Laboratory, Technical Research Centre of Finland, Helsinki, Finland.*

R. P. KHERA, *Woodbridge, Connecticut, U.S.A.*

ARUN P. KULSHRESHTHA, *Middle East Technical University, Ankara, Turkey.*

S. KUT, *E. Wood Ltd., Ware, England.*

JOHN E. LAND, *Union Carbide U.K. Ltd., Swindon, England.*

R. G. LOASBY, *United Kingdom Atomic Energy Authority, Aldermaston, England.*

T. A. MADEN, *Fothergill & Harvey Ltd., Littleborough, England.*

DANIEL R. MARANTZ, *Flame-Spray Industries, Inc., Port Washington, New York, U.S.A.*

M. MARTINI, *Electronic Industries of Canada, Downsview, Ontario, Canada.*

A. R. MOSS, *Ministry of Defence, Fort Halstead, England.*

J. MUIRHEAD, *DeVilbiss Company, Bournemouth, England.*

MISS PIRKKO OITTINEN, *Graphics Arts Laboratory, Technical Research Centre of Finland, Helsinki, Finland.*

S. A. PACE, *Kynoch Press, Birmingham, England.*

ALBERTO PASSERONE, *Centre of Applied Physical Chemistry, Italian National Research Council, Genoa, Italy.*

A. POLITYCKI, *Siemens Research Laboratories, Munich, West Germany.*

STEVEN B. SAMPLE, *School of Electrical Engineering, Purdue University, West Lafayette, Indiana, U.S.A.*

M. SCHLESINGER, *University of Windsor, Ontario, Canada.*

H. SILMAN, *Oxy Metal Finishing (Great Britain) Ltd., Woking, England.*

CHRISTOPHER W. SMITH, *Special Markets Department, Metco Ltd., Chobham, England.*

ROGER G. SMITH, *Coatings Service Division, Union Carbide U.K. Ltd., Swindon, England.*

RICHARD T. SMYTH, *Electrical Engineering Department, Imperial College, London, England.*

M. J. SPARNAAY, *Philips Research Laboratories, Eindhoven, Netherlands, and Twente University, Netherlands.*

LESTER L. SPILLER, *Environmental Protection Agency of the U.S. Government, North Carolina, U.S.A.*

H.-U. STEIN, *Leybold Heraeus, Koln-Bayenthal, West Germany.*

J. C. VIGUIÉ, *Laboratoire d'Étude des Matériaux Minces, Centre d'Étude Nucléaires, Grenoble, France.*

RICHARD L. WACHTELL, *Chromalloy Research and Technology Division, Orangeburg, New York, U.S.A.*

HAROLD N. WATSON, *Hard Face Welding & Machine Co., Buffalo, New York, U.S.A.*

SHELDON WEINIG, *Materials Research Corporation, Orangeburg, New York, U.S.A.*

ARNE WITH, *Department of Surface Coatings, Teknologisk Institut, Copenhagen, Denmark.*

W. J. YOUNG, *Ministry of Defence, Fort Halstead, England.*

Introduction

THE NATO ADVANCED STUDY INSTITUTE

Surface coatings are in use in nearly all branches of science and technology. The more commonly known uses include the paint coatings applied to houses and the chromium plating on motor cars. But there are uncountable further applications ranging from the very thick copper coatings applied to printing rollers to the very thin anti-reflection coatings on camera lenses, from high temperature diffusion coatings in jet engines to vacuum deposited resistors in electrical microcircuits, from electrostatically painted golf balls to the print in the daily newspaper, and so on.

The techniques used to coat surfaces are also numerous; a wide range of techniques has been necessary to cater for the very wide range of applications. Extensive research and development has resulted in an impressive quality of coating in many cases.

However, the diversity of coating methods now available has caused its own problems. In these days of specialization, people in the coating industries tend to be familiar with just a few of the techniques and often uninformed of many others. This is because of the many diverse industries in which coating techniques have developed, and because of the many different technical backgrounds of the people involved. To a considerable extent, coating techniques have evolved in isolation from each other.

The NATO Advanced Study Institute on the Science and Technology of Surface Coatings, held at Imperial College, London, from April 9th–14th, 1972, brought together scientists and technologists representing the whole range of surface coating methods. The Institute had several aims:

(i) To gain an understanding and appreciation of a broad range of existing coating methods by realizing the interdisciplinary nature of coating problems and by promoting discussion in commonly understood technology.

(ii) To improve an existing technique, or even to formulate new processes, by making use of some of the scientific principles used in another coating technique.

(iii) To make progress with problems commonly encountered in coating technology by the exchange of ideas.

A large part of the meeting was devoted to lectures on the whole range of coating techniques, each lecture dealing with a specific technique or group of techniques and being given by an expert in that field. Some of the techniques discussed, e.g. Electrolytic Deposition, have been in very effective use for many years whereas others, such as Harmonic Electrical Spraying, are at the research stage. The lectures form the contents of this book. Much of the discussion took place in small seminar groups, each group consisting of only a dozen or so people. A brief summary is given below.

The meeting was truly international, with participants from 20 countries, both NATO members and others. The participants had an excellent opportunity to learn about the whole range of surface coating techniques, and their cumulative experience often caused fresh light to be shed on even very familiar topics.

SUMMARY OF THE LECTURES

In general, it was intended that a lecture on the basic scientific principles of a coating method be followed by a lecture on an application of the technique. However, there was some overlap in this respect, and the papers should be read with this in mind.

Some attempt at categorization of processes was made by division into conduction and diffusion processes, chemical processes, wetting processes and spraying processes. But this is only one of many means of categorization, none of which are very satisfactory. This point is discussed later in the section on seminar sessions. It was hoped, however, that there would be some advantage from trying to thread some pattern amongst the techniques.

It may be helpful to the reader, in order to gain some perspective, to review briefly the contents of the lectures:

The Surface is vital to all coating techniques, and has been the subject of extensive study. Sparnaay discusses various aspects of current understanding of surfaces, and techniques of surface analysis.

Conduction and diffusion processes

As an introduction to this section, Anderson describes how random motion of particles, in the presence of concentration gradients, leads to mass transport (diffusion), and how transport is also achieved by the action of electric fields on ions or other charged particles (conduction).

The first coating technique, that of Electrostatic Deposition, is discussed by Spiller. His paper is concerned primarily with the deposition of material in the liquid form, the solvent used then being evaporated to form a solid coating. At the source, the liquid is atomized and charged, and then can be directed onto the substrate using an electrostatic field. Although deposition is generally on conducting substrates, techniques for coating non-conductors are discussed. Kut describes how electrostatic spraying is applied to the deposition of epoxy powders, no atomization stage being required in this case; subsequent processing consolidates the powder coating into a homogeneous layer. Corbett deals with the deposition of conducting powders; he describes how charge transfer between the deposited powder and the substrate leads to repulsion of the powder, and further shows how this may be overcome.

With, in his paper on Electrophoretic Coating, shows how a coating can be deposited on a conducting substrate from a dispersion of colloidal particles. The article to be coated is immersed in an aqueous dispersion which dissociates into negatively charged colloidal particles and positive cations. An electric field is applied with the article as anode (positive electrode); the colloidal particles are transported to the anode, where they are discharged and form a film. In the case of a paint coating, this requires subsequent curing. With further shows that electrophoresis itself is not a very effective transport process, so that electrodeposition may be a better term for the coating process.

Khera deals with Electrolytic Deposition. This is closely related to electrophoresis, except that it is primarily concerned with the deposition of ions rather than of colloidal particles. Two electrodes are immersed in an electrolyte of an ionic salt which dissociates in aqueous solution into its constituent ions; positive ions are deposited onto the cathode (negative electrode). Although the electrolytic deposition of conducting materials is quite common, Dijkstra and de Jonge show how thin organic coatings can be deposited with this technique.

Anodization, described by Campbell, is closely related to electrolytic deposition. It refers, however, to a process which occurs at the anode (hence its name) for a few specific metals. The anode reacts with negative ions from the electrolyte and becomes oxidized, i.e. it forms a surface coating. Silman discusses the practical applications of both electrodeposition and anodizing.

In Diffusion Coating, material diffuses from the surface into the bulk of the substrate. There is hence no clearly defined interface or coating; the coating is *in* the substrate rather than *on* it. This leads Wachtell, who describes diffusion coating, to consider a surface coating as "a variation in chemistry or morphology of the outer layers of an object from its substrate as a whole". He has further suggested that "surface coating" is tautological! In practice, the term "diffusion coating" tends to imply the procedures used to bring the material to the surface of the substrate as well as the subsequent diffusion treatment. Wachtell describes some of these procedures and Cook, in his paper, describes a method using electrodeposition in molten fluorides, a technique which is called Metalliding. Caubet and Amsallem describe how surface treatments using diffusion are utilized in practice, and give examples of the improvements in surface qualities achieved. Stein discusses a relatively new technique called Spark-hardening, in which an arc is periodically struck between a vibrating anode and the conducting substrate (cathode); material is transferred from the anode and diffuses into the substrate.

Chemical processes

The first paper in the section on Chemical Processes illustrates the overlap between the categories of our classification. Gregory discusses the principles of Conversion and Conversion/Diffusion Coating. The basis of conversion coating is that the substrate is reacted with other substances (which may be in the form of solids, liquids, or gases) so that its surface is chemically converted into different compounds having different properties. (Anodization could probably be described as an electrochemical conversion process.) Conversion coating usually takes place at elevated temperatures and diffusion is often an essential feature.

Viguié describes the principles of Chemical Vapour Deposition (CVD), in which a chemical process takes place in the vapour phase so that a reaction product is deposited on to the substrate. Politycki and Hieber then discuss the application of a particular type of CVD known as Pyrolysis, which involves the thermal decomposition of volatile materials on the substrate. Viguié describes a novel CVD process in which the source material is transported to the heated substrate in the form of an aerosol spray.

Electroless Deposition is often described as a variety of electrolytic deposition which does not require a power source or electrodes, hence its name. But Schlesinger points out that it is really a chemical process catalysed by the growing film, so that the electroless term is somewhat of a misnomer. Politycki and Stoeger describe how both electroless and electrolytic deposition are used to fabricate interconnection structures in semiconductor technology.

Wetting processes

Wetting is the ability of a liquid to cling to a surface. Passerone's paper deals with the basic principles of Wetting Processes, and primarily the thermodynamics involved. The phenomenon of wetting is fundamental to a variety of coating processes in which material is applied in liquid form and then becomes solid by solvent evaporation or cooling. Conventional Brush Painting is a wetting process, as is Dip Coating, in which the part to be coated is literally dipped into a liquid, e.g. paint, under controlled conditions of, for example, withdrawal rate and temperature. Wetting is also an essential second stage of several coating processes in which the emphasis is usually placed on the transport process from source to substrate. De Graaf describes how brush applied graphite coatings are applied to black and white television tubes, and Watson discusses a range of coating techniques loosely described as Welding Processes, all of which rely on wetting. Welding is not often considered as a wetting process, but it is well demonstrated here that it most certainly is. Much the same can be said of the Printing Process, which also relies on wetting. This is dealt with theoretically by Karttunen and Oittinen, and practically by Pace. The ink, conventionally pigment in a solvent, is transferred to and is deposited on a paper or other substrate, usually to form a pattern; the solvent evaporates to leave the required print. Printing, although not usually considered as a surface coating process, is rather versatile, as is illustrated by Loasby: "thick film" electrical components can be screen printed, the "ink" containing, for example, an electrically resistive material and the "paper" usually being a ceramic substrate.

Spraying processes

Spraying processes can be considered in two categories:

(i) macroscopic in which the sprayed particle consists of many molecules and is usually greater than 10 μm in diameter, and

(ii) microscopic, in which the sprayed particles are predominantly single molecules or atoms.

Muirhead describes Air and Airless Spraying, the first of the macroscopic processes. When a liquid exceeds a certain critical velocity, it breaks up into small droplets, i.e. it atomizes. The atomized droplets, by virtue of their velocity (acquired from a high pressure air or airless source) can then be sprayed onto a substrate. De Graaf describes how such sprayed coatings are used in colour television tube technology.

Flame spraying is described by C. Smith. In this process, fine powder (usually of a metal) is carried in a gas stream and is passed through an intense combustion flame, where it becomes molten. The gas stream, expanding rapidly because of the heating, then sprays the molten powder onto the substrate, where it solidifies. Maden deals with an application of this process, in which it is used in the production of durable, non-stick, low slip coatings.

Detonation Coating, discussed by R. Smith, is somewhat similar to flame spraying; a measured amount of powder is injected into what is essentially a gun, along with a controlled mixture of oxygen and acetylene. The mixture is ignited, and the powder particles are heated and accelerated to high velocities with which they impinge on the substrate. The process is repeated several times a second. Land discusses the uses of detonation coating, particularly in the aircraft industry.

A limitation of flame spraying is that only those materials which will melt in the combustion flame can be successfully deposited. This limitation is removed in Arc Plasma Spraying, described by Moss. The process is very similar to flame spraying, except that the powder is now passed through an electrical plasma produced by a low voltage, high current electrical discharge. By this means, even refractory materials can be deposited. Smyth shows how arc plasma spraying is being applied to the deposition of thick film electrical circuits, in contrast to the more conventional screen printing process discussed by Loasby (q.v.).

Electric-Arc Spraying is also related to the processes described above. Marantz deals with the principles of this process, in which an electric arc is struck between two converging wires close to their intersection point. The high temperature arc melts the wire electrodes which are formed into high velocity molten particles by an atomizing gas flow; the wires are continuously fed to balance the loss. The molten particles are then deposited onto a substrate as with the other spray processes.

Harmonic Electrical Spraying is a process described by Sample, Bollini, Decker and Boarman. The material to be sprayed must be in liquid form, which will usually require heating. It is placed in a capillary tube and a large electrical field is applied to the capillary tip. It is found that by adding an a.c. perturbation to the d.c. field, a collimated beam of uniformly sized and uniformly charged particles is emitted from the tip. Since these particles are charged, they could be focused by an electric field to produce patterned deposits.

Evaporation is the first of the microscopic techniques to be discussed, and is dealt with by Martini. When a liquid is boiled, it breaks up into single atoms or molecules which are almost independent of each other, i.e. it vaporizes. When boiling takes place at atmospheric pressure, e.g. the boiling of a kettle, its vapour stays close to the parent body because it cannot penetrate the barrier formed by the surrounding environment of air molecules with which the vapour atoms collide and are reflected back. But if the boiling is carried out in vacuum, where there is almost no surrounding gas, the escaping vapour atom will travel in a straight line for some considerable distance before it collides with something, for example, the vacuum chamber walls or a substrate. This is the principle of the vacuum evaporation process. Kut describes the ways in which evaporation is used industrially to produce decorative coatings; the technique is then usually referred to as Vacuum Metallization. Bunshah deals with a development of the basic process, known as Reactive Evaporation. Small traces of an active gas are added to the vacuum chamber; the evaporating material reacts chemically with the gas so that the compound is deposited on the substrate.

Sputter Deposition, described by Holland, is also a vacuum process, but uses a different physical phenomenon to produce the microscopic spray effect. When a fast ion strikes the surface of a material, atoms of that material are ejected by a momentum transfer process. The process can be likened to the collision of the cue ball with the triangular array of balls in snooker. As with evaporation, the ejected atoms or molecules can be condensed on a substrate to form a surface coating. Some of the many applications of sputtering are discussed by Weinig.

In evaporation and sputtering, almost all of the ejected material is electrically neutral. This is not the case in Ion Plating, described by Carpenter, in which a proportion of the depositing material is deliberately ionized. Once charged in this way, the ions can be accelerated with an electric field so that the impingement energy on the substrate is greatly increased. This can result in considerable improvements in coating adhesion, and the achievement of, for example, ohmic contacts on semiconductors.

Ion Implantation is very similar to ion plating, except that now all of the depositing material is ionized, and in addition the accelerating energies are much higher. The result is that the depositing ions are able to penetrate the surface barrier of the substrate and be implanted *in* the substrate rather than *on* it. The process, described by Martini, is used mainly in the semiconductor industry.

The three remaining papers are intended to give an idea of the type of research work associated with surface coatings. Finkin discusses frictional and wear properties. Hamid and Hunter describe a microwave examination technique, and Kulshreshtha deals with semiconductor thin film properties.

Seminar sessions

Many topics were discussed in the parallel seminar sessions which took place. Of these, two particular topics were obviously of general interest:

The first was a need to establish some sort of organization chart for surface coating techniques, to enable one to see the range of techniques available and select a suitable process for a specific application. However, it was not found possible to select suitable criteria for classification. It appeared

that, at best, one might develop a classification based on one's own requirements and interests.

The second topic, of even greater interest, was that of adhesion. There
were several approaches to this topic; the "physical" and "chemical"
approaches were both based on atomistic bonding, albeit expressed in somewhat different terms; the "macroscopic" approach was based on mechanical
interlocking; and the "empirical" approach denied that adhesion had any-
thing to do with science! It seems that adhesion is a subject of major general
interest to surface coaters, but it is also little discussed because of our present
lack of knowledge on the topic.

GLOSSARY

In addition to the index, an alphabetical glossary of terms is included in this
book. Although far from comprehensive, some of the specialist terms which
are used in surface coating technology and in the articles in this volume, are
explained there.

UNITS

The International System of units known as SI (Système International
d'Unités) is now coming into worldwide use. In this volume, in addition to
the units supplied by the author, the SI equivalent has been given where
possible, even where its use might appear pedantic, for example in cases
where the SI unit is not yet in common use.

The International System of units is founded on the seven base units
listed below:

Quantity	Name of base unit	Symbol
length	metre	m
mass	kilogram	kg
time	second	s
electric current	ampere	A
thermodynamic temperature	kelvin	K
luminous intensity	candela	cd
amount of substance	mole	mol

This table is taken from "The use of SI units", PD 5686 published by the
British Standards Institution, London, 1972, which should be consulted for
further information on SI.

Equivalent values of some Imperial units in terms of SI units are given in the glossary under "Imperial units".

ACKNOWLEDGMENTS

The Advanced Study Institute and these Proceedings have been made possible only through the assistance and support of many people, whose invaluable contributions we acknowledge and thank them for. We are also indebted to NATO, Imperial College, and RCA for all of the help and facilities they have provided.

August, 1973   B. N. CHAPMAN
                                                      J. C. ANDERSON

Contents

Surfaces

M. J. SPARNAAY

Philips Research Laboratories, Eindhoven, Netherlands

I. THE NATURE OF A SURFACE

A. Adsorption. The surface tension

Through the centuries, surfaces have attracted the attention of many scientists. Newton (1686, 1717) was acquainted with the phenomenon of two flat glass plates which, after being pressed together, could hardly be separated and he was also acquainted with the effect of liquid droplets placed between the plates. The presence of a liquid made the separation much more difficult. Later Taylor (1711), Hawksbee (1711, 1712), Jurin (1719) and others (see Wolf, 1967) systematically carried out a number of experiments, notably those in which the rise of liquid columns in capillaries of varying diameter was determined. In their considerations they often used a force law of the following form, assumed to be valid between two mass units m_1 and m_2, a distance r apart:

$$F = \lambda \frac{m_1 m_2}{r^n} \tag{1}$$

where F is the force and λ a constant. The exponent n is equal to 2 in the case of gravitation, but Newton also considered cases in which $n > 2$. He carried out integrations using equation (1) in order to arrive at expressions for the law of attraction between two large bodies. These preceded similar integrations by Hamaker (1937) by about 250 years.

Although not clearly stated by him, Newton probably considered that cases in which $n > 2$ were applicable to problems of adhesion, capillary action and so on. Since a determination of n was experimentally impossible to make, the law expressed by equation (1) with $n > 2$ was forgotten in later years. Thus, Van der Waals, at the time of the preparation of his thesis (Van der Waals, 1873) had no knowledge of it.

However, it is now generally recognized that electrically neutral atoms (and molecules) attract each other according to an r^{-7} force law. The force constant is not directly related to the masses of these atoms (molecules), but to such quantities as the atomic polarizability, or in the case of polar molecules to the dipole moment. We return to the force constant in Section II.A. For the moment it is sufficient to note that the statements regarding the r^{-7} law and the force constant are compatible with the statements made by Van der Waals in his thesis. Forces of this type are therefore called Van der Waals' forces.

We now consider briefly one phenomenon in which these forces are essential—the phenomenon of physisorption. The purpose of this brief consideration is to gain an insight into the relation that exists between the number per unit area of physisorbed atoms or molecules, and the value of the surface tension of the surface of the solid or liquid where the adsorption has taken place. The surface tension is the free energy which must be spent to form one unit area of the surface in a reversible manner (reversible: the disappearance of one unit area of the surface must be accompanied by the production of the same amount of free energy). This is the definition which is used in thermodynamics. Alternatively, the surface tension can be viewed upon as a force per unit length. To see this, imagine an arbitrary line in the plane of the surface (Defay and Prigogine, 1951). At both sides of the line the surface is "pulling" at the line. The "pulling" force per unit length is the surface tension. Its direction is normal to that of the line and in the plane of the surface. This is the mechanical approach. The important point to be made in connection with adsorption is that the value of the "pulling" force depends on the adsorbed amount per unit area. What has been said here of the surface tension also holds true for the interfacial tension. In the former case solid or liquid surfaces bordering vacuum are considered, in the latter the interface between two solids, between a solid and a liquid or between a solid or liquid and a vapour is considered. It is easier to visualize surface and interfacial tensions for liquids than it is for solids, but there are no *a priori* reasons to discard the concept for solids.

When a solid or liquid surface is brought in contact with a gaseous atmosphere, gas molecules or atoms are attracted toward the surface by Van der Waals' forces. We exclude here the case of chemical interaction between the adsorbate (the gas) and the adsorbent, i.e. we exclude chemisorption and consider only physisorption. The number of physisorbed atoms or molecules per unit area, to be denoted as Γ, depends on the gas pressure p, on the temperature T, and on the physical nature of adsorbent and adsorbate. In the simplest case Γ and p are linearly proportional and we have the following equation:

$$p = \frac{p_0}{\Gamma_0} K\Gamma \qquad (2)$$

where K is a dimensionless constant and where p_0 and Γ_0 are standard values of the pressure and the adsorbed number per unit area of atoms or molecules. Suitable standard values are: $p_0 = 1$ atmosphere, and $\Gamma_0 = N_0$ where N_0 is the number of available adsorption sites per unit area of the surface. The dimensionless constant K depends strongly on the temperature. At elevated temperatures thermal movements of adsorbed and non-adsorbed gas atoms and molecules are intense and prevent the particles from being permanently adsorbed. In that case K has a small value. Curves of p vs. Γ at a given constant temperature are isotherms. Equation (2) (a linear p–Γ relationship) is the equation for a Henry-type adsorption.

Usually p is sufficiently low so that the gas obeys the ideal Boyle–Gay Lussac law and we have the well-known equation of state for an ideal gas:

$$p = NkT \qquad (3)$$

where N is the number of gas atoms or molecules in the gas per unit volume and k is Boltzmann's constant. The quantity kT is the thermal energy per atom or molecule in the gas. The pressure p is an energy per unit volume. The pressure can also be viewed as a force per unit area.

The linear proportionality between p and Γ suggests also that the physisorbed particles behave as an ideal gas. Instead of a gas pressure p we introduce the two-dimensional pressure π and write:

$$\pi = \Gamma kT \qquad (4)$$

in which π can be expressed as an energy per unit area or as a force per unit length. π has the same dimensions as the surface or interfacial tension γ. However, it has the opposite tendency. Whereas γ tends to minimize the total surface (interfacial) area, π tends to maximize it. A very good example of this tendency is provided by soap solutions. Soap molecules, introduced in water, show a strong tendency to reside at the water surface. At the same time the value of the surface tension of water decreases from about 7×10^{-2} N/m to 3×10^{-2} N/m or less. The difference, 4×10^{-2} N/m in this example, can be interpreted in terms of a surface pressure π. Indeed the example of soap solutions led Gibbs to the formulation of the famous Gibbs adsorption equation (Gibbs, 1878, 1961). A simplified yet very useful version of this equation is (at constant temperature):

$$d\gamma = -\Gamma kT \, d(\ln c) \tag{5}$$

where c is the concentration (of soap molecules in water in the given example). When the adsorption takes place from the gas phase instead of the liquid phase, c must be replaced by p, the gas pressure. Note the minus sign in equation (5). An *increase* of c or p leads to a *decrease* of γ. When Γ and c (or p) are linearly proportional, c in equation (5) can be replaced by Γ and integration leads to:

$$\Delta\gamma = -\Gamma kT \tag{6}$$

where $\Delta\gamma$ is the difference between the surface tensions before and after the adsorption has taken place. Comparison between equations (4) and (6) shows that the surface pressure π is just equal to $-\Delta\gamma$.

An interesting corollary is that, once the relation between Γ and c or p is known, in other words once the adsorption isotherm has been determined, the decrease of the surface (or interfacial) tension can be obtained. We give one more example. A Henry-type adsorption is only found for $\Gamma \ll \Gamma_0$. When this case no longer applies, i.e. when the coverage $\theta = \Gamma/\Gamma_0$ is no longer very small, the proportionality between p (or c) and Γ is lost. Saturation is observed, a large increase of p (or c) is needed to produce only a small increase of θ. A tendency toward saturation is often well represented by the Langmuir equation (Langmuir, 1916, 1917). Assuming the case of gas adsorption this equation reads:

$$p = p_0 K \frac{\theta}{1-\theta} \tag{7}$$

Introduction of this equation into equation (5) (after first replacing c by p in this equation) and integration gives:

$$\Delta\gamma = +kT\Gamma_0 \ln(1-\theta) \tag{8}$$

Since $\theta < 1$, this is again a negative quantity. Usually Γ_0 is of the order of 10^{18}–10^{19} m^{-2}. Since at room temperature kT is equal to about 4×10^{-21} J, we see that $\Delta\gamma$ typically becomes of the order of $-10^{-2}\,\mathrm{J/m^2}$ (or $-10\,\mathrm{erg/cm^2}$ or $-1\,\mathrm{N/m}$).

It is remarkable to note that in now classical texts in the field of thermodynamics, such as those of Gibbs (1961) and Van der Waals and Kohnstamm (1927), much attention was given to the surface. Van der Waals and Kohnstamm developed a macroscopic concept of the nature of a surface. They considered a transition region between a liquid and its vapour, whose width was large compared to the dimensions of an atom. In their concept the density variation in the transition region was connected with average values of Van der Waals' forces in that region. They obtained the result that these average values were dependent on the location in the transition region of the atoms or molecules and this dependence finally led to values of the surface tension (see also Bakker, 1928; Tolman, 1949; Kirkwood and Buff, 1949; Ono and Kondo, 1960).

Another approach to the nature of the surface tension is that based on the rupture of the bonds between neighbouring atoms in a solid or liquid. For a number of solids this rupture can be produced by a cleavage process. For liquids such a process is not possible and must therefore be considered as imaginary. However, imaginary or not, this process leads to a fairly satisfactory physical picture of the nature of the surface tension. Before we illustrate this, we note that a rupture of bonds is always accompanied by side effects. Thus, upon breaking bonds the structure near and at the surface becomes slightly loosened or, more generally, distorted; moreover a redistribution of electrical charge carriers (electrons, ions), which may already be present in the system, often takes place. These and similar phenomena are accounted for when a reversible process to enlarge the surface area can be carried out. Fortunately these side effects are usually unimportant and merely amount to saying that the difference between the result of a slow process during which equilibrium remains established at any small time interval (a reversible process), and the result of a fast irreversible process, which only accounts for bond breaking, is small. Bond breaking alone (formally at 0 K) leads to a value of the surface energy, the complete reversible process leads to a value of the surface tension, which is a *free* energy.

We now compare the value of the surface energy $\mathrm{H^S}$, taken per atom (molecule), with the heat of vaporization $\mathrm{H^V}$, also taken per atom or molecule. In a table given by Wolf (1967b) (see Table I), it is shown that the ratio $\mathrm{H^S/H^V}$ (this ratio is sometimes called Stefan's number) has for many different liquids a value close to $0\cdot4\pm0\cdot05$. This ratio is explained as follows: when each atom in the bulk has z_0 nearest neighbours, the number of bonds broken per atom when the liquid is evaporated, is $\frac{1}{2}z_0$ (or z_0 per *pair*). The number of nearest neighbours per *surface* atom is less than z_0 and is denoted

TABLE I

Liquid	H^S/H^V (Wolf)	Solid/melt	γ_{SL}/H^l (Turnbull)
Argon (85 K)	0·37	Ag	0·45
Oxygen (70 K)	0·40	Cu	0·44
SiCl$_4$	0·33	Fe	0·45
Hexane	0·34	Ga	0·44
Benzene	0·33	Hg	0·53
Fatty acids	0·20 ± 0·05	Ni	0·44
Water (300 K)	0·15	Sn (tetrag.)	0·42
Na (100 °C)	0·30	Water	0·32
Hg (20 °C)	0·49	Bi	0·33
Ag and Au (1000 °C)	0·17	Ge*	0·35

* According to Jaccodine (1963) the surface free energy of solid Ge [(111) planes] is 1·06 J/m², and that for solid Si (111) is 1·23 J/m². Estimates for the {100} planes led to values which were 1·8 times higher. For the {110} planes, estimates led to 1·3 times higher values. This means that the table has only a qualitative significance. γ_{SL} denotes the solid/melt interfacial tension. This quantity is expressed here in energy units per atom.

as z_S. In order to form a surface, $\frac{1}{2}(z_0 - z_S)$ bonds per atom must be broken. Thus, very roughly, H^S/H^V must be equal to $(z_0 - z_S)/z_0$. For liquids it seems surprising that the ratio H^S/H^V can be given in terms of straightforward numbers z_0 and z_S but, as Frenkel (1943, 1955) has pointed out, a liquid can often be considered as a distorted solid rather than as a compressed gas. Furthermore the table (see Table I) given by Turnbull (1950) shows that the ratio of the interfacial tension of a solid and its melt, taken per atom, and the latent heat of melting per atom (H^l) is for many substances 0·45 ± 0·05. This suggests that there is some support for a model in which a fraction $(z_0 - z_S)/z_0$ plays a role, $(z_0 - z_S)/z_0$ now representing the fraction of the bonds per surface atom which is weakened, not broken.

For solids in which there exists a strong covalent bonding such as is the case in Ge and Si (see Table I) the surface tension is of the order of 1 J/m²; for liquids such as water, the bonding forces are much weaker and the surface tension is of the order of 0·1 J/m² or less. The effect of foreign adsorbed atoms and molecules leads to a decrease $\Delta\gamma$ of about 0·01 J/m² depending on the quantity adsorbed per unit area and on the shape of the adsorption isotherm.

B. Electrical effects at surfaces and interfaces

Another effect which can lead to a decrease of γ, is the electrification of the surface or interface. A large number of experiments have been carried out with a system, sketched in Fig. 1, consisting of a mercury reservoir ending in a glass capillary, and an adjoining (aqueous) electrolyte. In this electrolyte a reversible electrode (often calomel) is present. A voltage could be applied between the Hg electrode and the reversible electrode by means of a potentiometer. The applied voltage can be varied and the variations dV are wholly found at the Hg/electrolyte interface, no electrical transport taking place across this interface. This means that excess negative charges are found on one side of the interface and positive charges on the other. We can conceive of this phenomenon as an adsorption of electrical charges, which again must

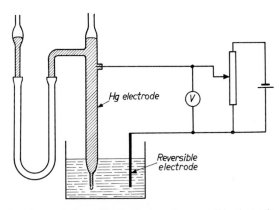

Hg electrode

V

Reversible electrode

FIG. 1. Electrolytic cell, containing a Hg electrode and a reversible electrode, both in contact with an aqueous electrolyte. The interfacial tension of the Hg/electrolyte interface (and therefore the shape of the mercury droplet) can be varied by varying the potential difference V.

result in a decrease of the interfacial tension. This is indeed found in the experiments, the decrease amounting to 0.1 J/m^2, depending of course on the value of the applied potential difference. The behaviour is described by the Lippmann equation (Lippmann, 1875, 1883):

$$\sigma = -\frac{\partial \gamma}{\partial V} \qquad (9)$$

in which σ is the surface charge density, i.e. the charge quantity per unit area accumulated on one side of the interface (the mercury side); γ is, as before, the interfacial tension; and ∂V is the change of the applied voltage. The temperature is kept constant, and no material is added to or extracted from the system. Therefore the symbol "∂" is used instead of "d" to indicate the differentiation in equation (9). Figure 2 gives an illustration. The surface charge density is a function of V. As indicated by equation (9), when σ is zero, γ shows a maximum, the electrocapillary maximum (e.c.m.). At the e.c.m. the potential drop is usually not zero, because the molecules at the interface form a dipole layer, and also because spontaneous adsorption of ions in the solution takes place. Ions of the opposite sign are attracted to the interface to restore the electrical neutrality. Since the (electrostatic) forces operating on these ions are only weak, i.e. of the same order of magnitude as the forces connected with the thermal energy kT, they are found over a fairly large distance x in the solution. This distance is of the order of 10^{-8} m. Ions distributed in this way form a Gouy layer (Gouy, 1908, 1917). The potential-distance relationship pertaining to the Gouy layer is represented in Fig. 2 as a fairly wide curve both for $\sigma = 0$ and for $\sigma > 0$.

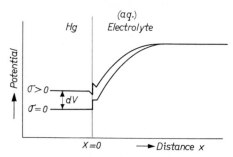

FIG. 2. Potential against distance x at the Hg/aqueous electrolyte interface. dV: change of applied potential difference; σ: surface charge at Hg side of interface; vertical parts at boundary $x = 0$: potential jump due to boundary dipole layer; curved parts: space charge or Gouy layer, extending to about $x = 10$ nm, which is a typical value for a 10^{-3} molar monovalent electrolyte. In both cases ($\sigma = 0$ and $\sigma > 0$), a layer of adsorbed ions is present at the right-hand side of the plane $x = 0$.

The Lippmann equation is a special form of the Gibbs adsorption equation. It is convenient to consider the electrons in the metal as the adsorbing component and to write $\sigma = -e\Gamma_e$, where Γ_e is the number per unit area of "adsorbed" electrons (this number may also be negative). Furthermore $-e\,dV$ of equation (9) (after introducing Γ_e instead of σ) and $kT\,d(\ln c)$ of equation (5) are special cases of variations of the chemical potential of the adsorbing components. Electrons in a metal which is part of a polarizable electrode such as the mercury electrode, form an interesting case. Their chemical potential can be varied at will (see Grahame, 1947) by manipulating the potentiometer. For electrons the name "chemical potential" may be somewhat misleading. Usually for charged components the name "electrochemical potential" is used. For electrons the name "Fermi energy" is the most common one.

Electrical effects play an important role in many surface systems. Parallel to, but almost independent of, the development of the views on the electrical double layer at the Hg/electrolyte interface, was the development of the theory of the stability of lyophobic colloids by Derjaguin and Landau (1941a and 1941b) and Verwey and Overbeek (1942, 1948). Here also the electrical double layer around the colloidal particles plays an important role. Another important field, where electrical double layers are studied, is that of soap films and soap bubbles. We have seen already that Gibbs was interested in these systems. Soap molecules form, at the surface of water, more or less orderly-arranged monolayers which are often ionized. At the water surface the ionized parts are pointing toward the water and the hydrophobic hydro-carbon tails (length about 2×10^{-9} m) are pointing outward. Small ions (diameter about 0.3×10^{-9} m) in the water phase serve for charge compensation. These ions (H^+ or Na^+ for instance) are introduced together with the soap molecules in non-ionized form. In soap bubbles such monolayers are facing each other. Soap bubbles can be considered as membranes and have served in the past as a model for biological membranes.

A most important moment for the study of electrified surfaces, and of surfaces in general, was the invention of the transistor (see Pearson and Brattain, 1955) and the discovery of the importance of surface states by Bardeen (1947). The concept of surface states may be explained as follows. We have seen that, in order to create a new surface, bonds must be broken. This has consequences for the electrons originally participating in these bonds. Sometimes it may be advantageous for the newly formed surface atoms to hold fixed excess electrons. Energy states are formed through the process of rupture and in the underlying case these energy states are called surface acceptor states. Alternatively it may be advantageous for the surface atoms to repel electrons and to donate them to the bulk of the crystal. In that case surface donor states are formed. One of the results of semiconductor surfaces research in the nineteen sixties was, that clean surfaces of pure ger-manium crystals were found to have a slight excess of negative charges (see Many et al., 1965; Frankl, 1967) which are held fixed in surface states (Fig. 3a). The compensating positive charge carriers reside in the crystal under-neath the surface in the space charge layer, which is the equivalent in semi-conductors of the Gouy layer in solutions. The space charge layer has a thickness of about one micrometre. The charge carriers in this space charge layer are mobile and consist mainly of positive holes. One method used to obtain this result was that of the field effect. In the case drawn in Fig. 3(b) (a condenser system, in which a metal plate is placed parallel to a plate of semiconducting material), the metal plate has been given a positive charge. The semiconductor plate has obtained a compensating negative charge and a transverse field exists between the two plates. The negative charge partly resides in the surface states and partly recombines with positive holes, thereby reducing the number of mobile charge carriers. Indeed a decrease of the conductance of the semiconductor sample is found in this experiment (see also Boonstra, 1967). For silicon the situation is qualitatively the same as for germanium. The surface state density is, both for germanium and for silicon, of the order of one state per surface atom, in agreement with predictions made by Tamm (1932) and by Shockley (1939). There are acceptor

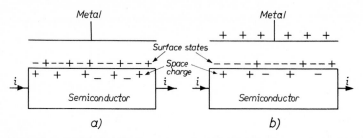

FIG. 3. Illustration of field effect. Surface states: charged donor and acceptor states are shown. Space charge: charge carriers here are mobile and determine field effect. The effect of the positive charges brought to the metal surface is a *decrease* of the surface conductance. This is due to a recombination of holes with induced electrons. i: electrical current.

states and donor states, their effects almost balancing each other. Chemically active ambients such as oxygen, tend to reduce the surface state density. Transistors have oxidized surfaces and it is therefore understandable that the surface state density in those systems is much lower, e.g. of the order of $10^{15}/m^2$.

The double layer aspects of the Hg/electrolyte interface, of colloidal solutions, of monolayers spread on water, and of semiconductor surfaces have formed the subject of a book by the present author in which their common properties have received special attention (Sparnaay, 1972).

II. Two solids in close proximity

A. *Three principles underlying their interaction*

If the cleavage process, discussed in the previous section, had been reversible, the broken bonds would have been repaired after bringing the two solids together to a distance which is less than an atomic diameter. This diameter, to be denoted as $2a$, is of the order of a few Ångström units (1 Ångström unit = 1 Å = 10^{-10} m). The implicit assumption is that there is a perfect match of the opposing atoms of the two surfaces. If so, the junction of the two solids leads to a decrease of the total free energy, which is numerically the same as the increase which takes place when the surfaces are created.

This in turn is the same as the value of the surface tension: about 2×10^3 erg/ cm^2 or 2 mJ/m^2. However, this is an idealized situation. In practical cases foreign molecules, originating from the atmosphere, contaminate the surfaces after separation and this contamination prevents a perfect match being preserved. When the two surfaces are formed independently and belong to different species, their junction becomes even more difficult. Only when a chemical interaction (or alloying) between the solids is possible, can there be a free energy decrease of the above mentioned magnitude. It is, moreover, assumed here that the flatness of the surfaces is perfect. However, in practical cases there are only a few places of contact (i.e. places where the distance is about $2a$), surface roughness and surface contamination providing for an average surface-to-surface distance of 10 nm or more. In those cases an external force must be exerted to bring the average distance down to a few Å. The combination of surface roughness and external force may make the final area of contact larger than the geometrical area and this promotes the chemical interaction or the alloying. Alternatively a third substance may be brought between the two others. The silicon nitride procedure for obtaining glass-to-metal seals may serve as an example (Stoller *et al.*, 1970). The metal, for instance copper, tantalum, or gold, is exposed at 700 $^\circ$C to NH$_3$ and SiH$_4$ vapours. A layer of about 1000 Å (100 nm) of Si$_3$N$_4$ is formed. This serves as an intermediate layer for a metal-to-glass seal. The seals are generally better than those made without Si$_3$N$_4$. Although there is no direct proof, it seems probable that there is a chemical interaction of Si$_3$N$_4$ with both the glass and the metal.

A second, more physical way to understand the process of joining of two solids, makes use of equation (1). London (1930, 1931) and Eisenschitz and London (1930) showed that this equation is applicable to the case of two atoms a distance r apart ($r > 2a$), and that the exponent $n = 7$. Usually the energy equation is presented instead of a force equation. We have:

$$v_L = -\frac{\lambda_L}{r^6} \tag{10}$$

where v_L is the interatomic energy of interaction and λ_L is a constant for a particular species, which depends on the physical nature of the atoms. For a simple model, London wrote:

$$\lambda_L = \tfrac{3}{4} h v \alpha^2 \tag{11}$$

In this model the atoms were represented as identical three-dimensional harmonic oscillators, the oscillating mass (an electron) having a frequency v, and the oscillator polarizability being α. Finally h is Planck's constant. The energy hv is comparable to the ionization energy which is often of the order of 10^{-18} J. Since α is typically of the order of 10^{-24} cm^3, λ becomes about 10^{-59} erg cm^6 or 10^{-78} J m^6. For two different atoms 1 and 2, London found that λ becomes $\frac{3}{4}h\sqrt{v_1 v_2}\,\alpha_1\alpha_2$.

Although the atoms were assumed electrically neutral Coulomb's law is the basis of the derivation of equations (10) and (11). It is important to note that the electrical charges inside each atom are in different positions. The further they can move apart inside each atom, the larger becomes the attraction (at constant r). The tendency to move apart is expressed in the polarizability. In the oscillator model this quantity can be viewed as the volume in which the outer electron moves around the centre (the nucleus plus the inner shell electrons). For polar molecules instead of atoms, the constant λ_L has been shown to be $\frac{1}{3}p_1 p_2$ $(1/kT)$ by Keesom (1912), p_1 and p_2 being the dipole moments of the two polar molecules. The interaction between a polar molecule and an atom represented as an oscillator, leads to $\lambda_L = \alpha p_2$ (Falkenhagen, 1922; Debye, 1920; Sparnaay, 1959). In all these cases $n = 7$.

There is a characteristic difference between the "chemical" and the "physical" approach. There is no chemical interaction beyond a distance $r = 2a$ but, as can be inferred from equation (10) with $\lambda_L = 10^{-78}$ J m^6, the London attraction is still appreciable at a much larger distance. In order to obtain the energy of attraction U_A between two solids, integrations must be carried out, based on the interaction energy v_A between two atoms. For two solids, a distance l apart, and both having a thickness large compared to l, one has

$$U_A = -\frac{A}{12\pi l^2} \quad \text{or} \quad F_A = -\frac{dU_A}{dl} = -\frac{A}{6\pi l^3} \tag{12}$$

where the energy U_A is expressed in energy units per unit area and the force F_A in force units per unit area; the constant $A = \pi^2 N_1 N_2 \lambda_L$, N_1 and N_2 being the numbers of atoms per unit volume of the two solids (N_1 and N_2 are usually of the order of 10^{22}–10^{23} cm^{-3}). For $l = 5$ Å (0·5 nm) it is seen that $U_A = -10$ to -100 erg/cm^2 (-10 to -100 mJ/m^2) or more than an order of magnitude smaller than the free energy loss obtained after "chemical" bonding. The force F_A is large, of the order of 10 MN/m^2 (about 100 kgf/cm^2). Again, lack of flatness may prevent the solids from having an overall separation of 0·5 nm, but again an external force will be helpful. A third principle must be mentioned. This is, unlike the "chemical" and the "physical" principles, based on the conductivity of the solids, on the electrical contacts between them and, above all, on the difference $\Delta\phi$ of the work functions of the two solid surfaces. Physically a difference of work functions can be understood as a difference of the affinities for electrons of the two surfaces. The affinity can be modified by small traces of foreign molecules.

It is a well known fact that even two apparently identical metal electrodes, with surfaces which are cleaned in the same way, may show work function differences up to one electron volt ($1·6 \times 10^{-19}$ J).

According to Kelvin (1870) there is an attraction between the two plates obeying the conditions enumerated above. This attraction can be represented by the following force law:

$$F_K = -\frac{(\Delta\phi)^2}{8\pi l^2} \tag{13}$$

which, of course, can also be written as an energy law. In equation (13) F_K is the force of attraction in force units per unit area. This law shows an even slower decay with increasing distance than did the one derived from London's law. With $\Delta\phi = 1$ electron volt ($1·6 \times 10^{-19}$ J), and $1 = 10$ Å (1 nm), F_K becomes $4·1 \times 10^2$ kgf/cm^2 or 40 MN/m^2. Since small amounts of impurities have a large effect upon $\Delta\phi$, the work function will not be uniform along the surface. However, even when $\Delta\phi$ changes sign locally, the force F_K represents an attraction.

In the next subsection a principle will be described which, although not dealing with an attraction between two macroscopic bodies, is relevant because it describes an expansion of their interface, in this way tending again to a unification of the two macroscopic bodies.

B. Interfaces and the effect of electrical charges

When a droplet of an electrolyte is placed on the surface of a conducting substance, a complete wetting of the surface can be obtained when a potential difference is applied between droplet and conducting substance. For an explanation of this effect consider the force equilibrium, which determines the contact angle θ. The equilibrium equation is:

$$\gamma_S = \gamma_{SL} + \gamma_L \cos\theta \tag{14}$$

which is Young's equation (Young, 1805). In this equation γ_S is the surface tension of the solid surface, γ_L that of the liquid surface, and γ_{SL} is the interfacial tension of the solid/liquid interface (Fig. 4). Let us assume that before the application of a potential difference no interfacial charge is present. After this application the value of γ_{SL} decreases by an amount $\Delta\gamma_{SL} = -\int_0^{\psi_0} \sigma \, d\psi_0$, where ψ_0 is that part of the potential difference which is present in the region including the solid/electrolyte interface and extending to the bulk of the adjoining phases. As γ_{SL} decreases, $\cos\theta$ must increase, γ_S and γ_L remaining constant. In this way wetting is enforced (Sparnaay, 1964).

The application of a potential difference leads here to a displacement of matter in the droplet. There is no difficulty because one of the phases is a liquid. However, essentially the same phenomenon must happen when both phases are solids. Of course equation (14) is no longer applicable, but one of the features remains: the application of a potential difference leads to a decrease of the interfacial tension, and this decrease introduces a tendency towards an increase of the solid/solid interfacial area. The value of this

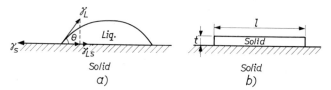

FIG. 4. Illustration of the effect of the interfacial tension upon the mechanical strength of a solid/solid system as proposed in Section II.B. (a) γ_{LS}, γ_S, γ_L: liquid/solid interfacial tension and surface tensions of solid and liquid respectively; θ: contact angle. (b) A thin solid layer forming an interface with a second solid.

increase depends on the elastic characteristics of the solids. Let us consider the following example: a bar, width a, thickness t and length l ($l \gg t$ and a) is subject to a stress in the direction of l. This stress produces a strain δl and an elastic force tries to restore the original situation. For $\delta l \ll l$ this force, to be denoted as F_{elast}, is proportional to the strain δl. We introduce a relative elongation $\Delta = \delta l / l$. The proportionality condition $\Delta \ll 1$ is usually obeyed. Many crystalline solids (metals, semiconductors) cannot support Δ values larger than about 10^{-2}–10^{-3}. Rupture takes place which begins at the weakest spots in the material. For F_{elast} we write, in accordance with general practice:

$$F_{elast} = - Eta\,\Delta \qquad (15)$$

where E is Young's modulus. A typical value is 10^{10} N/m² (or 10^{10} J/m³ or 10^{11} erg/cm³).

The elastic energy U_{elast}, necessary to produce the strain δl is:

$$U_{elast} = E(la)t\,\frac{\Delta^2}{2} \qquad (16)$$

Here la is the surface area of one of the sides of the bar. Assume now (Fig. 4b) that this side is part of the interface with a second solid. In such a case the energy U_{elast} can be provided by a change of the interfacial energy (or, similarly by a change of the interfacial tension). We have seen that such a change can be brought about in a number of ways, one of them being the electrification of the interface. With $E = 10^{10}$ J/m^3, $t = 10^{-7}$ m, $la = 1$ m^2, and $\Delta = 10^{-2}$, the elastic energy is 50 mJ/m^2 (50 erg/cm^2) which can easily be produced by a change of the interfacial tension. A larger change will lead to a destruction of the interface, because, as we have seen, Δ values larger than 10^{-2} can rarely be supported by most materials.

Now equation (16) is only valid for a very special model: a bar $(l \times t \times a)$ which is subject to a force applied to the whole $t \times a$ plane. However, it can be shown (Sparnaay, to be published) that an equation similar to equation (16) holds true for systems in which we are directly interested, i.e. for thin solid layers of any shape forming a large interface with a solid substrate. In those cases a numerical factor must be added to the right-hand side of equation (16) which is of the order of 0·1 depending on the shape of the thin layer.

Electrical effects and their consequences, as described here, are of special importance in semiconductor technology, where potential differences over layered systems are often applied. We mentioned already SiO$_2$ layers and Si$_3$N$_4$ layers. Si$_3$N$_4$ layers on Si devices are popular because the diffusion coefficient of foreign atoms and molecules, which may affect the electrical characteristics of the device, is extremely low (McMillan and Misra, 1970). Young's modulus for SiO$_2$ is 9×10^{10} N/m^2 and it can support a Δ value of $1\cdot5 \times 10^{-4}$. For Si$_3$N$_4$ these values are $1\cdot8 \times 10^{11}$ N/m^2 and 3×10^{-4}. These values indicate that in applications of semiconductor devices some care is needed, because double layers may build up at the Si/SiO$_2$ interface, which may lead to such an alteration of the interfacial tension that the adhering oxide layer is damaged.

III. MODERN METHODS OF SURFACE ANALYSIS

Ten years ago surface problems could be summarized in three unanswerable questions:

(1) Is the surface clean, i.e. free from foreign atoms?
(2) What is the exact position of the surface atoms?
(3) What is the physical nature of surface states?

Now there are powerful experimental techniques capable of giving an answer to the first two questions and theory is beginning to become sufficiently advanced to attack the third. It must be borne in mind, however, that it serves no purpose to attempt to answer the third question if no adequate knowledge exists concerning the state of cleanliness of the surface and the location of the surface atoms. Therefore we will briefly consider some of the experimental techniques providing us with this knowledge.

A. Low energy electron diffraction (LEED)

Historically the first technique is that of low energy electron diffraction (LEED). In this technique (which was used by Davisson and Germer, 1927) a beam of electrons with a kinetic energy (T)* of the order of 100 eV is directed onto the surface of a crystal. The electrons hardly penetrate the crystal, but a number of processes take place when they hit the surface. We will consider two of these processes. The first is a diffraction (T being conserved). When the surface is crystalline, the diffraction takes place at distinct angles, which are specific for the crystal structure at the surface. The diffracted beams can be observed as light spots on a fluorescent screen. The final interpretation still offers many problems, but the starting point is that the interatomic distance d at the surface is of the same order as the wavelength λ assigned to the electrons. This wavelength is found by using the de Broglie equation, which for free electrons reads: $\lambda = \sqrt{150/T}$ Å. Here the kinetic energy T is expressed in eV. The interatomic distance at the surface, d, which is a repeat distance in the case of a crystalline solid, is the basis of each attempt toward an understanding of a LEED pattern. The crystal structure at the surface is often quite different from that in the bulk (for a review see Estrup and McRae, 1971). A well known example is that of the Si (111) plane (Schlier and Farnsworth, 1959; Lander et al., 1963) where the periodicity at the surface is 7 times that in the bulk, provided heating at 700 °C in vacuum has taken place.

LEED patterns can only be obtained when the atoms at the surface are arranged in a periodic structure. This condition is not necessary in the second process—the Auger process (Auger, 1925)—which takes place when an electron hits the surface.

B. Auger electron spectroscopy (AES)

The impact of a free electron may result in the liberation of an inner shell electron of the target atom. This is the starting situation for an Auger process. Figure 5 gives an illustration. Here the impact has resulted in a vacancy (open circle) in the K-shell. A second electron of the same atom, in a higher energy level (L_I in Fig. 5) occupies the emptied low-lying K level. By this process the atom gains energy and this energy can be used to remove an electron from a higher level (L_{III} in Fig. 5). This is a $K L_I L_{III}$ transition. Alternatively the atom may emit an X-ray photon, but for light elements

---

* There should be no confusion between the symbol T used here for kinetic energy, and the same symbol used elsewhere in this paper for absolute temperature.

the emission of Auger electrons dominates. Auger electrons have well-defined energies, which can be measured. These energies are independent of the energy of the primary, incident, electrons. They are characteristic for the target atoms. In this way [i.e. in Auguer Electron Spectroscopy (AES)] foreign atoms can be identified and also "counted". In addition a shift of the energy of Auger electrons may be observed depending on the binding state of the target atom. This is still a matter of investigation.

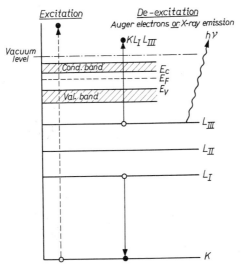

FIG. 5. Schematic energy diagram illustrating excitation and de-excitation process. $\bigcirc$: empty electron states; $\bullet$: occupied electron states; E_F: Fermi energy (or: electrochemical potential of the electrons); E_C: lower limit of conduction band; E_V: upper limit of valence band; E_C–E_V: forbidden energy gap. The E_V, E_C, E_F band scheme applies to a semiconductor.

C. Ellipsometry

A sensitive non-destructive optical method may be mentioned, that of ellipsometry. Here a light beam of a known state of polarization (i.e. of known ellipticity) is reflected against a surface. The state of polarization of the reflected beam is, usually, different from that of the incident beam. By varying the ellipticity of the incident beam, the reflected beam can be made linearly polarized (i.e. the polarization ellipse becomes a straight line) and this can accurately be analysed. Used in conjunction with gas adsorption measurements, the optical method can be applied to study the effect of a given quantity of adsorbed atoms per unit area. Used in conjunction with AES it may be a helpful method to clarify quantitatively the Auger process

(Vrakking and Meyer, 1971). Here ellipsometry has fulfilled the role of an auxiliary method. However, it can also be applied as an independent method. Thus, by using ellipsometry it could be shown (Bootsma and Meyer, 1969; Meyer, 1971) that the optical absorption for visible light near the surface of Si, Ge [and GaP (Morgan, 1973)] was larger than the optical absorption in the bulk of these crystals. This absorption difference was abolished after chemisorption (with, for example, O_2, NH_3, H_2S) had taken place. The interpretation of this effect was that a surface layer of about 1 nm thickness was somewhat disordered owing to the broken bonds at the surface. Chemisorption restored the "bulk" situation. The phenomenon is linked to the disappearance of surface states (Meyer, to be published).

D. Ion beam techniques

During recent years ion beam techniques have become available. Two applications may be mentioned. Firstly, for ion beams (mostly noble gas ions are used) with energies larger than 3 KeV, the impact may result in the liberation of a surface atom. The mass of this atom and its yield can be measured in a mass spectrometer (Benninghoven, 1970 and 1973; Benninghoven and Storp, 1970; Castaing et al., 1960, 1962). Also a surface radical consisting of a cluster of atoms, may be liberated. The identification of such a cluster in a mass spectrometer leads to information concerning the chemical properties of the surface. Thus Benninghoven (1973) identified SiO_3 clusters which were liberated from an Si (111) surface after traces of oxygen had been introduced. This indicates that already at a low oxygen surface concentration, at least two O_2 molecules interact with one Si atom. This technique is known as Secondary Ion Mass Spectrometry (SIMS).

Secondly, for energies smaller than about 1 KeV, scattering of the impinging ions (mass m_1) can be studied (Smith, 1971). When such an atom hits a target atom t (mass m_t) at the surface, it changes its direction and also its kinetic energy. Denoting the energy of the incident ions as T_0, and that of those ions which are observed at a scattering angle of 90° as T_{90}, the ratio T_{90}/T_0 is equal to $(m_t - m_1)/(m_t + m_1)$. Knowing mass m_1 and energy T_0, the method provides m_t when T_{90} is measured. Using this technique it can be checked (Brongersma and Mul, 1973) that the (111) plane of a ZnS crystal has only Zn atoms at the surface, whereas at the opposite side of the crystal (the $(\overline{1}\overline{1}\overline{1})$ plane) only S atoms are at the surface. The technique can, of course, also be applied to detect impurity atoms. Another aspect of the method may be stressed here (Brongersma and Mul, 1972): assume that atom t is a foreign atom, adsorbed on the surface under investigation. What is its position with respect to the plane of the surface? An answer to this

question can be given by changing the position of the plane of the surface
with respect to the direction of the incident beam. When the foreign atoms
are placed on top of the plane of the surface and not between the atoms of
the surface proper then, at grazing incidence, the effect which target atom t
(mass m_t) has upon the scattering of the incident atom (mass m_1) is larger
than at normal incidence. The atom t has a "shadow" effect, which becomes
larger when the angle of incidence becomes smaller. This is illustrated in
Fig. 6. Brongersma and Mul (1972) carried out the experiment for the
adsorption of bromine molecules on the Si (111) plane, the incident beam
consisting of 1 KeV Ne^+ ions. They found indeed a shadow effect.

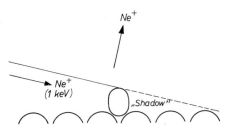

FIG. 6. Illustration of "shadow" effect. The oval represents an adsorbed atom. There is, at the
left of this oval, a second "shadow" region (not shown in this Figure).

Most of the methods mentioned in this section permit the detection of
10^{16}–10^{17} foreign atoms per m^2 of the surface or less, depending on such
parameters as the efficiency of the processes involved. The efficiency of the
Auger process reaches a maximum when the energy of the incident electrons
is around 3 KeV, and is of the order of 10^{-5}. The efficiency with which, in
SIMS, secondary ions are liberated varies strongly from case to case. In
those (rare) cases in which it approaches unity, the sensitivity of SIMS may
be such that 10^{14} atoms per m^2 or less can be detected.

Surface conductance measurements, such as those dealt with in Section
I.B allow the detection of 10^{13} mobile charge carriers (conduction electrons
or holes) per m^2 of the geometrical surface. As pointed out, these mobile
charge carriers are found in a space charge layer underneath the surface of
a semiconductor crystal. The sensitivity of these electrical measurements is
rather more sensitive than that of the methods described in this section, but
they are rather non-specific.

REFERENCES

Auger, P. (1925). *J. Phys. Radium* **6**, 205.
Bakker, G. (1928). *In* "Handbuch der Experimentalphysik VI". Leipzig.
Bardeen, J. (1947). *Phys. Rev.* **71**, 717
Benninghoven, A. (1970). *Z. Physik* **230**, 31.

Benninghoven, A. (1973). *Surface Sci.* **35**, 427.
Benninghoven, A. and Storp, S. (1970). *Z. Angew. Physik* **31**, 171.
Bootsma, G. A. and Meyer, F. (1969). *Surface Sci.* **14**, 52.
Brongersma, H. H. and Mul, P. M. (1972). *Chem. Phys. Lett.* **14**, 380.
Brongersma, H. H. and Mul, P. M. (1973). *Surface Sci.* **35**, 393.
Brongersma, H. H. and Mul, P. M. (1973). *Chem. Phys. Lett.* **19**.
Castaing, R., Jouttry, R. and Slodzian, G. (1960). *Compt. Rend.* (*Paris*) **251**, 1010.
Castaing, R. and Slodzian, G. (1962). *J. Microscopie* **1**, 395.
Davisson, C. J. and Germer, L. H. (1927). *Phys. Rev.* **30**, 705.
Debye, P. (1920). *Z. Physik.* **21**, 178.
Defay, R. and Prigogine, I. (1951). "Tension Superficielle et adsorption", page 1. Liège.
Derjaguin, B. V. and Landau L. (1941a). *Acta Physiocochim. U.R.S.S.* **14**, 633.
Derjaguin, B. V. and Landau, L. (1941b). *J. Exp. Theor. Phys.* (*Russ.*) **11**, 802.
Eisenschitz, R. and London, F. (1930). *Z. Physik* **60**, 491.
Estrup, P. and McRae, E. G. (1971). *Surface Sci.* **25**, 1.
Falkenhagen, H. (1922). *Z. Physik.* **23**, 87.
Frankl, D. R. (1967). "Electrical Properties of Semiconductor Surfaces". Pergamon Press, Oxford.
Frenkel, J. (1955). "Kinetic Theory of Liquids". Dover (first Russ. edition 1943).
Gibbs, J. W. (1961). "The Scientific Papers", Vol. I. Dover. (The part "Equilibrium of heterogeneous substances", which deals with thermodynamics, was published 1878 and preceding years.)
Gouy, G. (1908). *Ann. chim. Phys* (8), **8**, 294.
Gouy, G. (1917). *Ann. Phys.* (9), **9**, 129 (a review).
Grahame, D. C. (1947). *Chem. Revs.* **41**, 441.
Hamaker, H. C. (1937). *Physica* **4**, 1058.
Hawksbee, F. (1711). *Phil. Trans.* **27**, 395, 473, 539.
Hawksbee, F. (1712). *Phil. Trans.* **28**, 151.
Jaccodine, R. J. (1963). *J. Electrochem.* **110**, 524.
Jurin, J. (1719). *Phil. Trans.* **30**, 1083.
Keesom, W. H. (1912). *Versl. Koninkl. Nederland Akad. Wetenschap* **20**, part 2, 1414 (in Dutch).
Keesom, W. H. (1921). *Z. Physik.* **22**, 129.
Kirkwood, J. G. and Buff, F. P. (1949). *J. Chem. Phys.* **17**, 338.
Lander, J. J., Gobeli, G. W. and Morrison, J. (1963). *J. Appl. Phys.* **34**, 2298.
Langmuir, I. (1916). *J. Am. Chem. Soc.* **38**, 2221.
Langmuir, I. (1917). *J. Am. Chem. Soc.* **39**, 1848.
Lippmann, G. (1875). *Ann. Chim. Phys.* (5), **5**, 494.
Lippmann, G. (1883). *J. Phys. Radium* (2), **2**, 116.
London, F. (1930). *Z. Physik* **63**, 245.
London, F. (1931). *Z. physik. Chem.* (*B*) **XI**, 222.
Many, A., Goldstein, Y. and Grover, N. B. (1965). "Semiconductor Surfaces". Amsterdam.
McMillan, R. E. and Misra, R. P. (1970). *IEEE Transactions of Electrical Insulation* **EI5**, 10.
Morgan, A. E. (1973). To be published.
Newton, I. (1686). "Principia", London, Prop. LXXXV, Theorema XXXV; Prop. LXXXVI.

Newton, I. (1717). "Optics", London, Query 31.

Ono, S. and Kondo, S. (1960). "Encyclopedia of Physics", Vol. X, p. 134. Springer Verlag, Berlin.

Pearson, G. L. and Brattain, W. H. (1955). *Proc. I.R.E.* **43**, 1794 (History of Semiconductor Research).

Schlier, R. E. and Farnsworth, H. E. (1959). *J. Chem. Phys.* **30**, 917.

Shockley, W. (1939). *Phys. Rev.* **56**, 317.

Smith, D. P. (1971). *Surface Sci.* **25**, 171.

Sparnaay, M. J. (1959). *Physica* **25**, 444.

Sparnaay, M. J. (1964). *Surface Sci.* **1**, 213.

Sparnaay, M. J. (1972). "The Electrical Double Layer". Pergamon Press, Oxford.

Stoller, A. I., Schlip, W. and Benbenek, J. (1970). *RCA Review*, June, 443.

Tamm, I. (1932). *Physik. Z. Sowjetunion* **1**, 733.

Taylor, B. (1711). *Phil. Trans.* **27**, 538.

Thomson, W. (later Lord Kelvin) (1870). *Nature Lnd.* **2**, 56.

Turnbull, D. (1950). *J. Appl. Phys.* **21**, 1022.

Van der Waals, J. D. (1873). Thesis, Amsterdam.

Van der Waals, J. D. and Kohnstamm, Ph. (1927). "Lehrbuch der Thermostatik", Vol. I. Leipzig.

Verwey, E. J. W. and Overbeek, J. Th. G. (1948). "Theory of the Stability of Lyophobic Colloids". Elsevier, Amsterdam.

Verwey, E. J. W. (1942). *Chem. Weekblad* **39**, 563 (in Dutch).

Vrakking, J. J. and Meyer, F. (1971). *Appl. Phys. Lett.* **18**, 226.

Wolf, K. L. (1967). "Physik und Chemie der Grenzflächen", Vol. I. Springer Verlag, Berlin.

Young, T. (1805). *Phil. Trans.* **95**, 65, 82.

Conduction and diffusion processes

J. C. ANDERSON

Department of Electrical Engineering, Imperial College, London SW7, England

I. CONDUCTION

A. Charge, field and potential

Experimentally it is observed that a force exists between electrostatic charges. This may be expressed mathematically as

$$\mathbf{F} = \frac{q_1 q_2}{4\pi\varepsilon r^2}\, \mathbf{a} \tag{1}$$

where $\mathbf{F}$ is the force and $\mathbf{a}$ is a unit vector pointing from one charge along the line joining the charges (separation r), away from the other charge, as

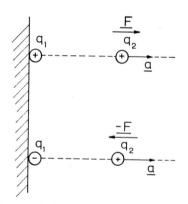

FIG. 1. The electrostatic forces F acting between charges q_1 and q_2.

shown in Fig. 1 (i.e. like charges repel), and ε is a property of the medium, known as the permittivity. We define an electric field of strength $\mathbf{E}$ as the force per unit charge on a positive test charge placed in the field. Thus the force on a charge $+q$ in a field $\mathbf{E}$ is given by

$$\mathbf{F} = q\mathbf{E} \qquad (2)$$

and the direction of the electric field is the direction of the force. For a negative charge the direction is opposite to that of the field.

If a field exists in a medium and a unit point charge is placed in it, the work done on unit charge in moving between two points, A and B, is defined as the potential difference between A and B, i.e.

$$V_A - V_B = -\int_A^B \mathbf{E}.\mathbf{dl} = -\int_B^A E \cos \alpha \, dl \qquad (3)$$

where $\mathbf{dl}$ is an element of path length as illustrated in Fig. 2.

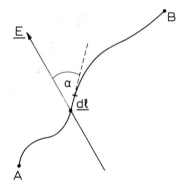

FIG. 2. $\mathbf{E}$ is the electric field acting at an angle α to the tangent as a unit charge moves along $\mathbf{dl}$.

Usually we take the potential at ∞ as zero (by definition) so that the potential at a point P is given by

$$V_P = -\int_\infty^P \mathbf{E}.\mathbf{dl}$$

Suppose a charged particle of unit charge starts from zero potential and moves along the field lines to a point where the potential is V. Since the charge moves along the field line, $\mathbf{E}$ and $\mathbf{l}$ are everywhere parallel so that

$$V_0 - V_1 = -\int_0^1 E \, dl \qquad (4)$$

where distance, l, is measured from the point at which potential is zero, i.e.

$$V_1 = V = El \quad \text{or} \quad E = \frac{V}{l} \qquad (5)$$

If the particle carries a charge q, the work done, from the above, is qV. If it is lifted vertically by the field against the force of gravity, then

$$qV = mgl \quad \text{or} \quad q\frac{V}{l} = mg \quad \text{i.e. } qE = mg \tag{6}$$

In a parallel plate capacitor $Q = CV$ from Coulomb's law of capacitance and, in S.I. units, $C = \varepsilon A/d$

$$Q = \frac{\varepsilon A V}{d} = \varepsilon A E \quad \text{i.e. } E = \frac{Q}{\varepsilon A} \tag{7}$$

where $C = $ capacitance in farads, d is the distance between the plates of area A and ε is the permittivity of the medium between the plates.
In general $dV = -\mathbf{E}.\mathbf{dl}$ where dV is the potential difference between two points dl apart. Now

$$\mathbf{dl} = dx.\mathbf{i} + dy.\mathbf{j} + dz.\mathbf{k} \tag{8}$$

where $\mathbf{i}$, $\mathbf{j}$ and $\mathbf{k}$ are unit vectors, and

$$\mathbf{E} = E_x.\mathbf{i} + E_y.\mathbf{j} + E_z.\mathbf{k} \tag{9}$$

$$\therefore \ dV = -(E_x.\mathbf{i} + E_y.\mathbf{j} + E_z.\mathbf{k}).(dx.\mathbf{i} + dy.\mathbf{j} + dz.\mathbf{k}) \tag{10}$$

i.e.

$$dV = (E_x\,dx + E_y\,dy + E_z\,dz) \tag{11}$$

By partial differentiation

$$dV = \frac{\partial V}{\partial x}\,dx + \frac{\partial V}{\partial y}\,dy + \frac{\partial V}{\partial z}\,dz$$

$$\therefore \ \mathbf{E} = -\left(\frac{\partial V}{\partial x}\cdot\mathbf{i} + \frac{\partial V}{\partial y}\cdot\mathbf{j} + \frac{\partial V}{\partial z}\cdot\mathbf{k}\right) \tag{12}$$

or

$$\mathbf{E} = -\text{grad } V = -\nabla V \tag{13}$$

This means that the electric field lines are everywhere perpendicular to the lines of constant potential (equipotential lines) as shown in Fig. 3.
In one dimension $E = -\partial V/\partial x$ where E is in the x-direction. It can also be shown that, where there is a distributed charge density ρ per unit volume

$$\nabla^2 V = -\rho/\varepsilon \quad \text{(Poisson's equation)} \tag{14}$$

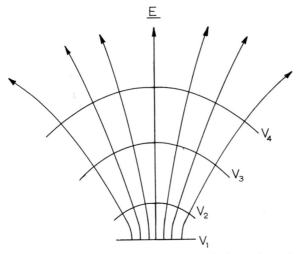

FIG. 3. The electric field lines **E** are everywhere perpendicular to the equipotential lines.

B. Electronic conduction

In a metal, electrons behave as if they are "nearly-free", i.e. more or less independent of the lattice. Thus if a potential V is applied between the ends of a bar of metal, length l, an electron in the metal experiences a force due to the electric field E given by $-eV/l$, where $E = V/l$ from equation (5), and it is accelerated.

The rate of gain of velocity is given by Newton's law, so that

$$\left(\frac{dv_{\mathrm{d}}}{dt}\right)_{\mathrm{gain}} = \frac{-eE}{m} \tag{15}$$

where m is the mass of the electron and v_{d} is its drift velocity in the field. But electrons collide with the lattice after a mean time τ and lose the memory of their acceleration, so that, in the steady state

$$\left(\frac{dv_{\mathrm{d}}}{dt}\right)_{\mathrm{gain}} + \left(\frac{dv_{\mathrm{d}}}{dt}\right)_{\mathrm{loss}} = 0$$

thus

$$\frac{-eE}{m} + \frac{v_{\mathrm{d}}}{\tau} = 0 \quad \text{or} \quad v_{\mathrm{d}} = \frac{eE\tau}{m} \tag{16}$$

Current density J (amp/m^2) is defined by rate of flow of charge per unit area, i.e.

$$J = nev_d \qquad (17)$$

where n is the number of charges per unit volume.

$$\therefore \; J = \frac{ne^2\tau E}{m} \qquad (18)$$

Thus conductivity σ is given by

$$\sigma = \frac{J}{E} = \frac{ne^2\tau}{m} \qquad (19)$$

and the resistivity $\rho = 1/\sigma$.
Mobility μ is defined as drift velocity per unit field, i.e.

$$\mu = \frac{v_d}{E} = \frac{e\tau}{m} \; \text{m}^2/\text{Vs} \qquad (20)$$

Typical electron mobilities are in the range 10–50 cm^2/Vs for metals and 50–100,000 cm^2/Vs for semiconductors.

C. Ionic conduction

Where ions are in a liquid to which a field is applied, they migrate in a direction determined by their charge. The current density is given by

$$J = (n_1 z_1 \mu_1 + n_2 z_2 \mu_2)eE \qquad (21)$$

where n_1, n_2 are the numbers of positive and negative ions per unit volume, z_1, z_2 are their valencies and μ_1, μ_2 their mobilities. Because the ions are relatively massive, the mobilities are lower than those of electrons in a solid. For weak electrolytes, mobility is constant. For strong electrolytes, where dissociation is complete at all concentrations, μ falls at high concentrations because each ion attracts an "atmosphere" of ions of opposite charge which retards its progress through the lattice. For very weak electrolytes, since μ is constant, conductivity is proportional to concentration (i.e. to n_1 and n_2), but increases less rapidly at higher concentrations.

In solids, ions can only move under the influence of a field if there is room for them to move. Two possibilities arise: ions in interstitial positions can jump with the assistance of the field to adjacent interstitial positions, or vacancies can migrate by ions jumping into them with the assistance of the field. Currents are low and mobilities are very low. Both can be increased with temperature since this increases the equilibrium number of interstitials and vacancies.

II. DIFFUSION

Ionic conductivity always involves mass transport. But we do not necessarily need a field to produce it, since a concentration gradient will cause mass transport—of ions *or* neutral atoms—by diffusion.

The space–time diffusion equation is Fick's Law

$$\frac{dc}{dt} = -\operatorname{div} D \operatorname{grad} c \tag{22}$$

where c is the concentration, in number per unit volume of diffusing atoms or ions, and D is the diffusion coefficient whose units are area per second. In an isotropic medium, D is the same for all directions, but in a single crystal it varies with crystallographic direction. It is generally a function of concentration and of temperature and its S.I. units are m^2/s. For fixed concentration, generally

$$D = D_0 \exp\left(\frac{-E}{kT}\right) \tag{23}$$

where D_0 is typically $0.02 \ cm^2/sec$ and E the activation energy for diffusion. In a solid E ranges from 0.75–3.0 eV.

In an isotropic medium the flux of atoms or ions, J_N, crossing unit area in one second is, from equation (22)

$$J_N = D \frac{\partial c}{\partial x} \tag{24}$$

The relation between diffusion coefficient and mobility is given by the Einstein relationship

$$kT\mu = qD \tag{25}$$

where q is the charge on the ion.

The basic principles of electrostatic coating

L. L. SPILLER

Environmental Protection Agency of the U.S. Government, Research Triangle Park, North Carolina, U.S.A.

I. INTRODUCTION

Surface coatings have the general function of physically preventing certain surface areas from coming into contact with destructive atmospheres or materials. Obviously, if the coatings are to protect the covered surfaces from the effects of exposure, they too must withstand these effects. They may also serve to enhance the beauty of the surface. Many coating materials have been developed to accomplish this dual protective–decorative function simultaneously.

The application of these coatings has led to the development of many techniques. There is still a very active programme concerned with the improvement of the quality of applied materials and with the advancement of the technology of their application. The chief aim of any of these developments is to get the surface coatings in place with minimum effort but with maximum efficiency. This means the development of methods which best apply the materials at maximum solids, maximum speed and with minimum waste.

II. ELECTROSTATIC LIQUID COATING

Urged on by this need for improved efficiency, coating engineers have developed the electrostatic method of coating articles with liquid coatings. At present, two basic systems, air-electrostatic and No. 2 electrostatic processes, are well established and widely used for the coating of metals.

A. Air-electrostatic processes

Two modifications of air-electrostatic processes have been developed and are being used which employ air forces to atomize the liquid material and then charge and deposit these particles using electrostatic forces. A short description of the equipment used with these two modifications will aid in understanding the details of this discussion.

In one of these modifications the articles to be coated are supported on a grounded conveyor which carries them into a spraying enclosure (Fig. 1).

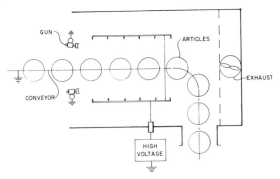

FIG. 1. Schematic plan view of air-electrostatic process with independent electrode system.

In the enclosure an electrode system is positioned along both sides of the path of the articles. The electrodes are about 12 in. (30 cm) away from the surface of the articles and are made of a series of fine wires. The electrode is connected to the negative terminal of a 100,000 volt d.c. supply capable of delivering 5 milliamperes of direct current. The charged electrodes create an abundance of negatively charged air particles. A low pressure air atomizer is arranged to spray finely divided paint into the space parallel to the conveyor. By collision with the charged air particles, the paint droplets pick up a negative charge by ion bombardment. The paint, once charged, will move toward the grounded articles and deposit there. This attraction to the articles results in a marked increase in the paint deposition efficiency.

In the second air-electrostatic modification, the droplets are formed by air and then charged electrically by a gun-like device that is shown schematically in Fig. 2. The charging electrode is mounted on the front of the spray gun. The ions are produced from the sharpened point of this electrode and charge the paint particles as they are dispersed by the atomizing air stream. The electrical properties of the coating material have an important influence on the operation of these units if the charging electrode is so mounted as to be in intimate contact with the coating fluid.

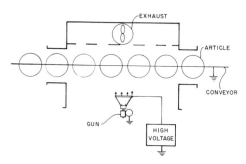

FIG. 2. Schematic plan view of air-electrostatic process with electrodes supported on gun.

B. No. 2 electrostatic process

This process also has two basic modifications which are referred to as *bell* type and *disc* type equipment. The basic electric operational principles of the two are the same.

Figure 3 is a schematic arrangement of a typical bell modification installation. In this arrangement the articles are supported on the grounded con-

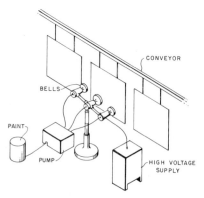

FIG. 3. Schematic view of a bell type electrostatic spraying installation.

veyor. A paint atomizer of special three head design is located opposite the surface of the articles. A high voltage is applied to the atomizers with respect to the articles on the conveyor. The liquid coating material is introduced to the rotating atomizer and then is formed by the rotation into a thin film

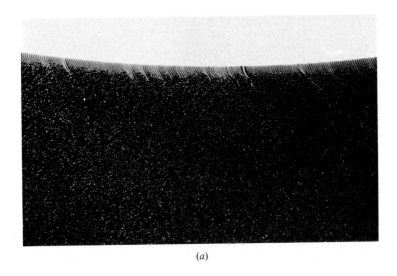

(a)

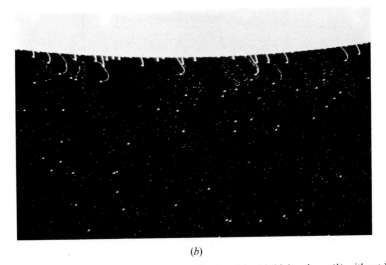

(b)

FIG. 4. High speed photographs of liquid at bell edge: (a) with high voltage (b) without high voltage applied.

flowing across the inner bell surface to its exposed edge. At the edge the electrical forces form the liquid paint film into a series of closely spaced liquid cusps or extensions. These extend out into the space ahead of the atomizer directed toward the article. Since the liquid has some electrical conductivity, the electric charge on the atomizer and thus also on the material concentrates at the tips of these extensions. When the electrical repulsion acting on the liquid at the tip equals the surface tension force holding the liquid together, the column of liquid in the extension breaks and a highly charged, liquid body is formed. The aggregate of all these particles forms a spray or mist. This transition can be seen from observation of Figs. 4a and 4b, which are high speed photographs of the edge phenomena with and without voltage. The mist travels to the work under action of the forces of the electrical field. It deposits on the article with practically 100% painting efficiency.

In Figure 5 the disc modification of the No. 2 electrostatic spray process is shown schematically. Here the atomizer is a disc member located in the

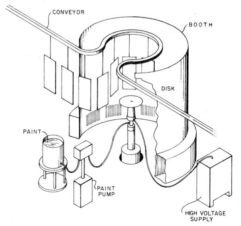

FIG. 5. Schematic of disc type electrostatic spraying installation.

centre of a loop in the conveyor. Liquid paint delivered to the centre of the disc flows across the disc surface to its outer edge. The disc is charged to a high voltage and this electrostatically atomizes and charges the paint particles which are then attracted to the work. Here again the efficiency is about 100%.

Both the disc and the bell are charged with respect to the work parts by being connected to a 100,000 volt d.c. supply. Normal operating current is of the order of 0·2 to 0·5 milliamperes.

III. Electrostatic Liquid Coating of Non-conductive Materials

The advantages to be gained in metal finishing by electrostatic application have been established beyond doubt. These are high deposition efficiency, high speed conveyorized automation capability, manpower savings and better pollution control.

The successful application of electrostatic methods to metals can be accomplished readily because the metal surface is electrically conducting and can easily be established as a collecting electrode. Surfaces which are not electrically conductive present problems.

The attractive benefits of using electrostatic methods in the metal coating field, however, make it desirable to extend these same benefits to the coating of non-conductive articles. This has become increasingly important since increased quantities of coating materials are being used on non-conductive items. The rapidly growing use of plastics as a replacement for metals and the need for most of these materials to be coated has further increased this desire.

As described earlier, electrostatic spraying is carried out by using one or the other of two basic systems. In one of these, the air-electrostatic process, the liquid coating material is pneumatically atomized, then these atomized spray particles are charged and deposited in an electrostatic field. In the other, the No. 2 Process, the material is atomized under the influence of an electrostatic field which forms, charges and then deposits the spray particles. In all these systems electric charges are conveyed by the paint particles to the object being sprayed. If the sprayed article is conductive and connected to the earth, a charge transport to ground can be detected. However, if the sprayed surface is non-conductive, the charge transported by the paint particles accumulates but cannot be dissipated to ground.

When attempts are made to electrostatically spray non-conductive surfaces, this charge accumulation on the parts lowers the voltage drop between the atomizing device and the surface to be coated; this causes less effective deposition in the first method and less effective atomization and deposition in the second. The charge built up on the non-conductive surface by deposited particles, repels the like charged coating particles as they approach the surface. This causes waste of coating material and lack of uniformity in the coating.

Through the years there have been developed a number of techniques by which non-conductive articles can be rendered adequately conductive so they can be coated electrostatically. These methods have met with varying degrees of success. In discussing them, these methods will be classified into two categories—physical and chemical.

A. Physical methods of obtaining conductivity

The several methods of physically modifying a surface to obtain adequate conductivity for electrostatic spraying are:

1. Wetting the surface with water or solvents

If water is used to wet the surface, the moisture serves as an electrode to produce the needed attraction. This method is successful for water soluble coatings. Difficulties are encountered when the coating is an oil base material because of the incompatibility of these materials with the moist surface. Low boiling organic solvents have been proposed for use as the wetting agent. Such materials are not completely successful since a non-homogenous surface, such as wood, absorbs the conductive solvent to varying degrees. Various surface areas of the article then exhibit variable surface conductivities. This condition of different surface conductivities results in a non-uniform coating deposition.

Moisture—water or solvent—has been employed, however, as a means of establishing conductivity in special cases. We will examine the case of finishing wood.

Kiln-dried wood has a resistivity of the order of 10^{15}–10^{16} Ωm at room temperature. The resistivity decreases sharply as the moisture content is increased. A linear relationship exists between logarithm of resistivity and the moisture content, up to about 7% moisture. The break in the linearity of the log resistivity versus moisture content corresponds with the transition zone from monomolecular absorption of water by wood to multimolecular absorption. This is the fibre saturation point where the path available for electric conduction gradually shifts from the cell walls to the cell cavities of the wood structure allowing the conductivity to take place along the continuous water columns. The electrical conductivity is ionic. The migration of ions through the bulk of the wood under the influence of an applied field has been demonstrated by radioactive tracer techniques.

Since conductivity varies with grain direction, it becomes an additional variable to be considered when coating untreated wood. Across-the-grain resistivity ranges from 2·3 to 5·5 times higher than that along-the-grain for soft woods and from 2·5 to 8·0 times higher for hard wood. This means that cross-grain areas may accumulate a different coating than along-the-grain areas.

Normally any wood which has less than 10% moisture does not have sufficient conductivity to allow it to be consistently coated by electrostatic methods. On the other hand, if the wood has moisture content above 10%, it can be coated by these methods without further trouble. The dimensional stability of such materials after they are coated, however, may well prevent their having this high moisture content.

2. *Heating the article surface*

The second physical method of creating conductivity to be considered is the heating of the non-conductive article. It has long been known that dielectric materials change their insulating properties when they are heated.

The degree of this variation depends upon the material. Generally conductivity characteristics change in parallel fashion with increased temperatures. The mobility of the surface atoms of solid dielectrics under normal temperature conditions is very low; the charge carriers remain stationary in more or less irregular positions. When heat is applied, some activity develops. The first activity is in the form of an increasing lateral mobility of the surface atoms, then it changes to a diffusional traffic of an irregular rotational horizontal pattern, which increases as the melting point is approached. The conductors which are normally bound become more free and mobile as the temperature is raised.

This phenomena is utilized to obtain effective electrostatic deposition on glass surfaces (Fig. 6). Many glass articles are coated to improve their

Fig. 6. Disc electrostatic sprayer as used to coat glass bottles.

appearance or better their surface characteristics. Scratch resistance can be improved by the deposition of appropriate materials on the surface. Container glass at 300–350 °F (149–177 °C) has sufficient electrical conductivity to allow it to act as the collecting electrode during the electrostatic process. Surface coatings can be electrostatically applied under these conditions. If the temperature is raised to about 1200 °F (650 °C) quite good conductivity exists and metal organic salt solutions can be applied. These solutions decompose as they strike the surface and reduce to their elementary metal. The coatings thus formed present a unique reflecting coloured decorative surface.

Heat as a means of reducing resistivity can also be used in the coating of rubber covered automobile steering wheels which acquire enough conductivity to be coated electrostatically when heated to about 230 °F (110 °C).

3. *Backing the article surface with a suitable electrode*

The third means of physically reducing the effects of high resistivity is effected by backing the non-conductive surface with a suitable conductive electrode. This is applicable to thin non-conductive materials such as paper, textiles and plastic in sheet form. An electrode is placed next to the sheet on the side opposite to that to be coated. The non-conducting dielectric material then accepts the charged coating material particles and the charge can leak off to the backing electrode through the sheet.

4. *Altering the electrical character of the surface*

The fourth method of inducing conductivity is to use, where possible, the conductivity inducing capabilities of carbon black or extremely fine metal particles dispersed in a suitable matrix. Materials which will not respond acceptably to any of the methods discussed earlier can be coated with a primer or sealer coat of carbon black dispersed in a resin binder. After conventional application and drying, such coatings produce on the surface a layer which will provide the desired range of conductivity. These types of conductive primers, although not transparent, serve as fillers and provide good smooth primed surfaces. Items such as moulded, sawdust-phenolic toilet seats are coated by this method. Metal powders can be used to replace the carbon black in some instances.

B. *Chemical methods of obtaining conductivity*

Because of the fact that mechanical methods of producing conductivity were extremely limited in the breadth of their applications, a new approach was needed. Some new method was desired which would be generally applicable on many types of surface, provide conductivity while still not covering the inherent surface features, have extreme ease of application or simplicity of use, and not modify the characteristics of the applied cover coat materials. Several chemical methods were developed to produce conductivity on dielectric surfaces. These approaches, which generally can be described as chemically absorbed gel layer methods, surface sulphonation methods, or methods utilizing a layer of photoconductive pigments, will be described in detail.

1. *Chemically absorbed gel layer*

The first truly chemical method of treating non-conductive articles was introduced in 1966 by the author of this paper and utilizes a material consisting of a quaternary ammonium compound dissolved in a relatively fast evaporating solvent. This solution is applied to the surface of the non-conductive article; the carrying solvent evaporates and leaves the chemical on the surface of the article, where it exists as a very thin transparent conductive gel layer swollen with a polar liquid of high dielectric constant. The gel layer contains ionizable organic compounds, which are chemisorbed by the insulating surface. This layer is approximately 1000 Å (100 nm) thick and is oriented on the surface so it chemically associates with the ions which form part of the coating material. This chemical association occurs through either hydrogen bonding or ionic bonding. Normally the surface of an insulating material contains enough functional groups capable of forming the necessary hydrogen bonds. The quaternary ammonium compound contains a halide, preferably a chloride, and a single substituent carbon chain. The straight chain groups are without unsaturation. The structural formula is:

$$\left[\begin{array}{c} CH_3 \\ R-N-CH_3 \\ CH_3 \end{array}\right]^{+} Cl-$$

This material as a dilute (1 to 5% by volume) solution in a fast solvent is applied over the surface to be coated in a relatively humid atmosphere. The solvent evaporates leaving the compound distributed over the surface. This material absorbs water and becomes swollen. The swollen gel layer which is formed on the article surface is electrically conductive and forms the needed surface conductivity. The layer is relatively stable and can stand long storage times of the order of several months, but is sensistive to temperature. An extended exposure at over 140 °F (60 °C) can destroy the gel structure and bring about loss of conductivity. If thermal degradation of the film has not occurred, conductivity may be restored by exposing the layer to a humid atmosphere.

Figure 7 is a picture of the No. 2 bell units being used to finish plastic picture frames that have been rendered conductive by this chemical treatment.

Fig. 7. Electrostatic bell as used to coat conditioned plastic picture frames.

2. Surface sulphonation

Sulphonating of surfaces is the second true chemical method to be discussed. The usefulness of sulphonates as wetting and emulsifying agents was realized in the early stages of surface chemistry. These sulphonates contain alkyl substituents and sulphonate groups, i.e. SO_3 or SO_3M. "M" is an ammonium or an alkali metal radical. The first sulphonating method was performed in liquid phase during condensation using sulphuric acid as the sulphonating agent in the presence of an inert solvent. A subsequent development used a gaseous sulphonating agent such as sulphur trioxide in the absence of a liquid media for the sulphonation of solid pieces of resinous polymer. Thus gas was employed instead of liquid. This method is an important one, because a solid having many surfaces can be uniformly and homogeneously sulphonated. To be capable of being sulphonated, a polymer should have in its molecular structure at least one aromatic nucleus such as styrene. In some instances the sulphur dioxide vapour is diluted with an inert gas such as nitrogen, air, or carbon dioxide. The extent to which the sulphonation reaction is allowed to proceed can be varied to obtain appreciable changes of the polymer surface. Exposure of the surface to 0·1 sulphonic acid group per molecule of the monomer tends to make the polymer more hydrophilic (wettable with water). The principal sulphonation reaction is the formation of sulphonic acid groups on the polymer molecules. A secondary effect is a cross linking reaction involving formation of a sulphone linkage between polymer molecules, resulting in ionic polymer.

These sulphonated polymers have the unusual properties of being able to form a "microgel" with water. The individual sulphonated resin aggregates of approximately 10 μm size are swellable by water, but are not disintegrated by it. These gel structures create on the dielectric surface a conductive path and they reduce the surface resistivity to below 10^{12} ohms per square. Thus they make the surface electrostatically coatable. The preferred degree of sulphonation is in the range from 0·1 to 1·0 milligram of SO_3 in the form of the sulphonic acid groups, per square centimetre of surface.

The dielectric surface containing sulphonic acid groups can be neutralized to prevent any tendency to chemically degrade the coatings and pigments subsequently used upon the surface. Additional precautions must be exercised to diminish the water sensitivity of the sulphonic acid layer by proper paint formulation techniques. It must also be realized that working with toxic gases is hazardous and proper precautions must be taken.

The sulphonation process has much promise if the development work being done at present continues to solve the presently existing difficulties which are encountered when this technique is applied in a production line.

3. Utilization of photoconductive pigments

A very new method, photosensitive coating, is the third one to be considered. In the past many attempts have been made to replace the carbon black prime coat used to produce conductivity on non-conductive articles. Such carbon black primers, although giving good conductivity, are difficult to cover. "Strike-through" and "bleeding" also often result when carbon black is used in the undercoat for lighter colours. The sought-for ideal primer-sealer is one which is conductive, is light coloured, and has good hiding. These properties are especially essential for wood and wood moulded products which need a sealer coat to fill the porosity of the surface. Preferably the primer should not lose its conductivity during oven baking and should not be sensitive to humidity changes.

The electrophotographic industry in its reproduction work uses a layer of photoconductive material dispersed in an insulating resinous binder as a recording element. The most popular of these materials is a group of specially manufactured zinc oxide pigments. These types of pigments are manufactured by the so called French or Direct Process in which metallic zinc is vaporized and then the vapours are burned in a reducing gas. Only the zinc oxides prepared by the French Process are photosensitive.

It was decided to experiment with primers in which French Process zinc oxide materials were used as pigments. Since this pigment, when exposed to light, was successfully used as a conductor in the electrophotographic industry, it was assumed that the same condition would be exhibited when such primers were used on item surfaces.

This special zinc oxide pigment is dispersed in a non-conductive binder resin and a solvent mix, and is applied to a non-conductive surface such as wood or plastic by any one of the conventional methods. After the coating is properly dried, the article covered with this coating is exposed for 2–3 seconds to a light source emitting a visible wavelength peaking at 3800–5500 Å (380–550 nm). After this exposure, the coated surface shows a resistivity below the threshold value of 2.5×10^8 ohms per square. Such zinc oxide coated articles, if properly grounded and irradiated, can be successfully coated electrostatically.

To simplify the light source problem it was necessary to improve the light sensitivity of zinc oxide coatings. This was done by sensitizing the zinc oxide through adding dyes such as Rose Bengal, Cyanine, etc. The optical sensitization by dyestuffs is explained by an electron transfer mechanism. The electron transfers from the dye molecule to the conduction band of the zinc oxide lattice. The semiconducting ZnO in turn replenishes the emptied ground level of the dye molecule.

When a primer is formulated using appropriate quantities of resin builder, French Process zinc oxide and a suitable sensitizing dye, a very effective, light coloured, good sealing primer is obtained. Such a primer when applied to a non-conducting surface, dried and then irradiated, will exhibit and retain an electrical conductivity sufficient to allow the surface to be electrostatically coated.

C. Conclusions

As late as 1964, the electrostatic coating of non-conductors had very little significance, despite the fact that electrostatic coating processes had already been widely accepted in the metal finishing field. Wood finishing was a highly specialized field not readily adapted to automation and the volume of nonconductor articles other than wood used by the industry was very small. With the growing use of plastics, attention was turned to coating these non-conductive surfaces. Simple mechanical methods and techniques whereby non-conductors can be given a conductive surface layer have been developed and used. More recently chemical methods of making these items conductive have been developed.

By using the discussed methods which employ chemically absorbed gel layers, surface sulphonation, and photoconductive pigments, practical coating operations have been developed which now allow the electrostatic finishing process to be used on all types of non-conducting items. The great economic benefits of these application techniques have thus been extended to the wood, glass and plastics field.

IV. Electrostatic powder coating

A relatively new coating technology employs powders. Although many articles have been written and many discussions have been held about this technique, a limited number of companies use it to date.

The current popularity of powder coatings has stimulated an interest in coating nonconductive articles. Practical electrostatic powder deposition requires electrically conductive surfaces as does liquid electrostatic coating. The range of surface resistivity of the article must be at the same level as for liquid: surface resistivity not less than 10^7 ohms per square.

The practicality of this technology is proven by the fact that for decorative purposes glass articles are coated very satisfactorily with epoxy powders. The important consideration is the fusion temperature of the deposited powders. The industrial powders available at this time are being fused at 400–500 °F (205–260 °C). Substrates such as wood, plastics, etc., cannot endure this temperature. Current research projects are aimed at trying to overcome this difficulty so that powder coating techniques, surface preparation methods, and electrostatic deposition means can be combined in coating non-conductors.

BIBLIOGRAPHY

Arhart, G. (1961). U.S. Pat. 2 992 139.
Braun, J., Davidson, W. and Skaar, Ch. (1953). *Forest Prod. J.* **10**, 445.
Bulgin, D. and Owen, G. (1951). Brit. Pat. 663 885.
Eichorn, J. and Steimetz, J. (1960). U.S. Pat. 2 945 842.
Elder, J., Bieling, R. and Blakeslee, A. (1962). U.S. Pat. 3 060 134.
von Gottberg, H. (1966). *Metall-Oberflache*, 304.
Helmuth, R. E. (1957). U.S. Pat. 2 662 833.

Lin, R. T. (1967). *Forest Prod. J.* **6**, 54.
Miller, E. P. (1959). U.S. Pat. 2 888 362.
Spiller, L. and Bobalek, E. (1971). U.S. Pat. 3 236 639.
Stainhauer, A. (1958). U.S. Pat. 2 854 477.
Stewart, P. H. (1961). Brit. Pat. 885 715.
Taft, D. and Heidecker, S. (1967). TAPPI 18 Coatings Conference, 1966.
Walles, W. and Bird, A. (1971). U.S. Pat. 3 578 484.

Product finishing with electrostatically sprayed powder coatings

S. KUT

E. Wood Limited, Talbot Works, Stanstead Abbotts, Hertfordshire, England

I. POWDER COATINGS

Application of plastic powders to metal surfaces is by no means new. Some thirty years ago preheated metal components were dipped in pulverized polyethylene, which adhered by melting, the article then being post-heated to form, by fusing, a continuous plastic film. This procedure only became economic and reliable following the introduction in 1953 of fluidized beds by Edwin Gemmer. Polyamides were first applied, and then later other thermoplastics such as polyethylene, cellulose acetabutyrate, polyethers, P.V.C. (polyvinylchloride) and polyacrylate.

A major advance in recent years has been the introduction of thermosetting epoxy powders, which can be economically applied by electrostatic spraying.

II. ADVANTAGES OF POWDER COATING

The more important advantages may be summarized as follows:

(1) No solvent employed, resulting in:
    (a) No costly wastage of solvents which are essentially only carriers for the coating resins in solvent based paints;
    (b) No inflammable solvents—cost savings in statutory safety features in plant, and reduced insurance premiums;
    (c) No pollution from solvents;
    (d) Reduced health hazards to operatives, with freedom from solvent odour, and reduced contamination of the atmosphere in the work area.
    (e) No flash-off period required—can place coated article directly in oven—saving space and time;
    (f) Economies in oven usage—no energy wasted in driving off solvents, and in their evacuation from the oven;
    (g) No possibility of retained solvent adversely affecting the resistance properties of the cured coating as can arise with solvent-bearing paints;
    (h) Dust free films are more readily obtained;
    (i) Economies in ventilation, and consequently heating of the work area, due to absence of solvents.
(2) Approaching 100% effective use of the powder coating can be attained. It is essential to use a powder recovery unit, re-using overspray powder, when 98–99 +% effective powder utilization is possible.
(3) A fairly uniform and higher film thickness than with a solvent-bearing one-coat baking enamel—in *one* application—highly decorative *and* protective.
(4) Sagging, bridging, poor coverage at edges and other surface defects liable to occur with solvent-bearing paints do not arise.

Other advantages have previously been detailed (Kut, 1971).

III. Pollution aspects

It is probably correct to say that, in Europe, acceptance of powder has been more on economic and performance grounds, although due reference has been made to pollution aspects. However, in the U.S.A. the reverse appears to be the case in so far as the attention being focused on ecology is directing attention to powder coatings as one alternative. Water based coatings are, of course, another, and even with solvent based coatings, pollution can be countered by using appropriate incinerators or scrubbers.

IV. Plastic powders—generally

Thermoplastic powder coatings have distinct disadvantages by definition, softening at elevated temperatures. Under conditions of fluctuating temperature, with the different coefficients of expansion and contraction of substrate and powder, haircracks can appear in some instances with reduced adhesion. In contrast, the adhesion of epoxy powders is of a very high order, and superior to that of some thermoplastic powders, which after ageing may tend to form essentially an envelope around the coated article.

Epoxy powders also exhibit superior hardness, chip, impact and scratch resistance, and are being increasingly employed in industry, where the hazards of physical damage and environmental conditions demand a coating superior to the lower cost thermoplastics.

V. Application of powder coatings

A. Fluidized bed coating

A fluidized bed essentially consists of a purpose designed tank, divided by a porous plate into two chambers, with the powdered plastic on top of the porous plate, and a main air chamber under the plate. On air pressure being applied, the air moves through the porous plate and powder, the finely divided plastic powder becoming "fluidized" and behaving like a liquid. Instead of air, nitrogen or an inert gas can be used.

Metal parts to be coated are preheated to a temperature above the melting point of the specific powder used, and then dipped into the fluidized bed. The plastic powder melts, fusing to the part which is withdrawn and unsintered powder shaken or blown off. Thermoplastic powders are mostly post-heated to achieve a smooth coating (Kut, 1971).

B. Epoxy powders—fluidized bed

The average film thickness obtained with a typical epoxy powder is about 200 μm although special formulations can be prepared to increase this figure, as distinct from other controlling factors noted above. A high film thickness is, however, not necessarily of advantage.

C. Electrostatic spray coating

In 1962 the electrostatic spraying of plastic powders was introduced, a vital technological advance in powder coating, utilizing the principle that opposite-charged particles attract each other, as successfully used for many years in spraying solvent-based paints.

Powder is fed from a reservoir to a special spray gun, where a high voltage, low amperage electrostatic charge is applied—the powder picking up the charge from a transfer of electrons from the gun to the powder. This transfer takes place (Pascoe, 1967) both through contact with the highly charged (70–90 kilovolts) gun nozzle, and through the surrounding ionized air. The part to be coated has an opposite charge, and it must be electrically conductive and grounded. On the charged powder approaching the earthed part, it is attracted, wrapped round the surface and clings until heat-fused into a uniform coating.

D. Surface preparation

The same requirements apply irrespective of the method used for applying the epoxy powder. It is essential that the substrate to be coated is free from grease and rust. Any residual grease will melt, and though initially satisfactory results may be obtained, it will impair the long term adhesion. Grease can also cause surface defects such as cissing. Conventional methods of degreasing are employed for mild steel—a trichloroethylene or perchloroethylene vapour bath is preferred. Appropriate washing is also required if contaminated with drawing fluids.

Steel surfaces must be freed from scale and rust—preferably by abrasive blasting for long term durability, though acid cleaning may be employed—ensuring there is no rusting prior to coating. Service requirements of the article also influence selection of the method of surface preparation.

Pretreatments, such as iron or zinc phosphating, are also employed, and confer additional protection to the steel (particularly with a zinc coating) if the coating is damaged. Such damage is, of course, less likely than with a conventional paint. Some care in selection must be exercised as epoxy powder coatings do not adhere equally well to all such pretreatments. Adhesion is at the optimum direct on the steel, without any such intermediate layer, though perfectly good results are being obtained with such pretreatments when their use is required—as for exposure to exterior or to humid environment.

Stiefel (1971) advises that the thickness of the phosphating layer should be below 4 μm.

E. Powder recovery

Reference has already been made to the economic necessity for reclaiming non-adhering and overspray epoxy powder. Specific plant is available for reclaiming the powder, fulfilling two main functions. Firstly, to collect the overspray powder, and to sieve it to remove dirt particles. Secondly, to reduce powder dust in the working area.

VI. CURING OF EPOXY POWDERS

Epoxy powders are thermosetting, and sufficient time and temperature must therefore be allowed to fully cure the specific formulation. In contrast, thermoplastic powders require only melting to form a continuous film.

Epoxy powders melt at 90–95 °C and are cured at 140–220 °C, depending on the time and specific powder formulation.

The typical curing schedule for a now "conventional" slow curing epoxy powder is 20 minutes at 180 °C, 10 minutes at 200 °C, 5–8 minutes at 220 °C.

A. Fast curing epoxy powders

Constant technical advances are being made, particularly in shortening the curing cycle and lowering the minimum temperature, whilst still maintaining adequate storage life. Typical production cycles that can now be achieved with a "fast curing" epoxy powder are 35 minutes at 140 °C, 25 minutes at 150 °C, 10–12 minutes at 180 °C, 5 minutes at 220 °C.

Employment of lower curing temperatures such as 130–140 °C requires particular care, as this is a threshold area and if—as is not infrequently the case—the oven temperature is not maintained overall, inadequate curing will result. In practice short curing schedules at 150 °C and 180 °C are widely used.

B. Curing schedules

Curing times and temperatures as generally quoted are for full curing of the powder, after the component has reached the specified temperature. The time for the article to reach the appropriate temperature in the oven employed must, therefore, be established. Significant oven temperature fluctuations can give misleading results. For components of varying thickness, the controlling factor is the thickest section.

C. *Adequate curing essential for optimum performance*

During the heating of epoxy powders, a chemical reaction takes place between the epoxy resin and hardener, transforming the powder from a brittle to an elastic film. Incomplete curing may give adhesion, but leads to poor elasticity and a brittle film, not withstanding even light impact—a useful practical indication of inadequate curing.

VII. FILM THICKNESS

Generally, if epoxy powders are applied at a film thickness of less than about 50 μm, the film will tend to be inadequately opaque. It is normal practice in electrostatic spraying to apply 50–75 μm, to cold surfaces.

For general anti-corrosive and chemical resistance service, 75–100 μm is applied, and for exterior pipe coatings about 250 μm.

VIII. PARTICLE SIZE OF EPOXY POWDERS

The particle size range of the epoxy powder used depends on the method of application. For electrostatic spraying the maximum particle size of commercial powders is 90–120 μm, with the average being 30–60 μm. With such powders, pore-free films of about 40 μm can be applied. For fluidized bed, the maximum is of the order of 300 μm. Finer powder is required for electrostatic spraying than for fluidized bed application. For electrostatic spraying, a high proportion of fine particle size is undesirable as it tends to lead to poorer powder recovery.

IX. PROPERTIES OF EPOXY POWDERS

Excellent adhesion, tough, mar and abrasion resistant. Decorative, glossy and matt coloured powders, including metallics, as well as textured finishes are available. The cured coating is non-toxic.

A. *Chemical and solvent resistance*

High order of resistance—superior to liquid epoxy coatings in some respects, particularly to acids.

B. *Electrical resistance*

Epoxy powders are outstanding as electric insulators, resistant to temperature changes.

C. Weathering

Similar to conventional epoxy paints, the epoxy powders on exposure lose their gloss, and surface chalk. The rate of chalking is, however, extremely slow—and, if at all measurable, of the order of 1–2 μm per year. Full film integrity and protection is maintained. At a film thickness of 60 μm this is equivalent to a 3% annual loss.

X. Position today

Epoxy and other powders for electrostatic spraying are today competitive product finishes and can be justified on economic grounds not only where their specific properties and application technique are of advantage, but also in competition to many conventional baking systems. In some fields, epoxy powders will replace solvent-based paints, particularly when considering the installation of new or replacement plant.

Epoxy powders are dual purpose—high order of anti-corrosive resistance and as a decorative and durable finish for a wide variety of industrial and consumer articles.

From the point of view of costs, epoxy powder coating merits serious consideration where currently using two coats of baking enamels. In some instances, even instead of one coat, depending on cost, or if the epoxy coating can dispense with or simplify a specific pretreatment. Some cost aspects are discussed, and these are considered as quite distinct from the superior film properties obtained with epoxy powders.

A. Industrial uses for epoxy powder

Typical outlets for epoxy powder coatings, particularly by electrostatic spraying are for general purpose decoration and protection, and also for high performance protection. Some examples are detailed below.

B. General and consumer applications

Steel furniture, and particularly all manner of tubular articles and wire-work, such as chairs, table legs, fencing and radiators, lend themselves particularly well to electrostatic "wrap round". In contrast to conventional paints, there is minimal damage of the coating in transport and in usage. Garden equipment, domestic appliances, such as washing machines and drums, refrigerator shelves, dishwasher racks, wire screens and meshes. Shelving; grilles and roller blinds; luggage racks. Supermarket trolleys and baskets; display goods. Curtain walling and window frames. Hospital and office equipment.

C. Industrial uses

Protection of steel structures and cables in aggressive atmospheres—marine, chemicals or solvents; processing plant; interior and exterior coating of pipes; pumps, valves, couplings, risers and other fittings. Plating racks, drum lining; scaffolding. Electrical insulation for electric motor starters, armatures, capacitors and resistors. Sheet metal which may be postformed. Marine equipment, such as sonar units. Machinery housings. Fire extinguishers. Coils and springs. Various automotive uses such as wheel rims, tie rods, interior trim, springs and suspension arms.

XI. ACRYLICS

Acrylic powders have now been on the market for some time, and as compared to epoxies their particular virtues are resistance to weathering (with gloss retention) and to discoloration at elevated temperatures. These powders give well adherent coatings of good decorative appearance.

Acrylic powders are cured in the range of 12 minutes at 220 °C; 20 minutes at 200 °C or 45 minutes at 180 °C. Newer developments visualize the use of 30 minutes at 160 °C, 15 minutes at 180 °C and 7 minutes at 200 °C.

The high order of weather resistance of acrylic powder has, for example, lead to its use during the last $1\frac{1}{2}$ years on aluminium window frames. Another major application is its use as a coating on heating units and oven casings on a mass production line.

XII. POLYESTER POWDERS

These can be considered as somewhat intermediate between epoxy and acrylic in terms of anticipated performance and probably, in due course, in terms of price. Chemical and solvent resistance will, of course, not be of the same high order as epoxy powders, and acrylics are likely to have superior weathering resistance.

A typical curing schedule varies from 45 minutes at 180 °C to 15 minutes at 220 °C and 30 minutes at 200 °C. With another formulation basis a schedule of 7 minutes at 200 °C is possible. Some latitude in formulation can be visualized and considerable activity is going on in this field, particularly to improve mechanical properties as well as other resistance characteristics.

XIII. POWDER RECOVERY PLANTS

Ofner (1970) has summarized the essential requirements of an efficient recovery unit as: (1) Filter powder from the air stream to an efficiency of at least 99–99·9%; (2) Store the powder in a loose, non-compacted form; (3) If necessary, dry the powder; (4) Mix the powder with dry gas and maintain it fluid for removal; (5) Sieve out impurities from the fluidized powder.

With modern plant, above 99% powder usage can be achieved. A filter is required to supplement a cyclone for removal of the fine powder.

XIV. EXPLOSIVE HAZARDS

Some qualification is required of the hitherto frequently repeated claim that there are no fire or explosion hazards with powder coating. Firstly, all organic powders are inflammable. Secondly, at a critical powder–air concentration a dust explosion can occur. According to tests reported by one manufacturer drawing attention to this vital aspect, the critical concentration is in the region of about 53 g/m^3 and the ignition temperature at about 460 °C.

Any spark with a temperature above 460 °C—with a poorly earthed article—could lead to a dust explosion. In order to prevent this, the powder concentration in the booth must under no circumstances be higher than half of the above critical level.

Furthermore, the booth should be designed in such a manner that the powder does not settle in it, but is immediately drawn into the recovery unit.

Adherence to these and other related safety precautions is recommended, although epoxy powders—in view of their high ignition point and higher minimum ignition energy—are quite clearly markedly less hazardous than solvent-air mixtures.

This aspect has been recently fully dealt with by the author (Kut, 1970).

REFERENCES

Kut, S. (1970). *In* "Proc. 2nd. Internat. Conf. on Plastic Powder Coating". Industrial Finishing and Surface Coatings, London.
Kut, S. (1971). *Product Finishing* **24**, 10, 31–37; **24**, 11, 28–30; **24**, 12, 32–38.
Ofner, J. (1970). *Industrie-Lackier-Betrieb* **38**, 6, 276.
Pascoe, W. R. (1967). *Trans. Tech. Paper Reg. Tech. Conf. Soc. of Plast. Eng., Akron Sect.*, 39–49.
Stiefel, M. (1971). Private communication.

The electrostatic deposition of conducting and semiconducting powders

R. P. CORBETT

Wolfson Applied Electrostatic Advisory Unit, Electrical Engineering Department, University of Southampton, Southampton, England

I. INTRODUCTION

A. Background

The electrostatic powder coating process has become widely accepted for the application of insulating materials such as epoxy resins and thermoplastic materials. Attempts to extend the range of materials that can be applied in dry powder form by electrostatic techniques have in the past proved to be plagued with a surprising number of difficulties. The reasons for these difficulties, which are related to the resistivity of the deposited powder layer, will be discussed after a brief description of the physical mechanism of the coating process.

There are two basic configurations used in electrostatic powder coating systems; namely the electrostatic powder spray gun and the electrostatic fluidized bed. Historically the first was developed from the simple liquid paint spray gun and the latter was developed from the hot dip process (in which heated objects to be coated are dipped into a bed containing fluidized powder). Although electrostatic fluidized bed systems have been used to great advantage in the coating of continuous strip wire, and also for the bulk coating of multiple objects of small vertical height, electrostatic powder gun systems have been adopted in the majority of installations. In a typical automated installation, some four or five powder guns are employed, some being mounted on vertical reciprocators to give general coverage and some being fixed to give penetration into particular regions of the object.

In this paper the mechanisms relevant to systems of the powder gun type will be described, but the physics of the systems of the electrostatic fluidized bed type differs only in detail.

B. Particle charging and transport

In electrostatic powder spray guns, particles are charged in the region immediately in front of the gun nozzle. Except in a few cases where contact charging is deliberately used, the particles are charged by being bombarded by gas ions which originate from a corona discharge. This is produced on a sharp point that is maintained at a potential of the order of 100 kV, mounted at the spray gun nozzle. The charge produced on the particles, which to a considerable extent determines the degree of overspray, wrap around and material utilization, is described by the following equation (Pauthenier, 1960) over the particle size range employed in most powder coating systems (Corbett, 1970):

$$q = 12\pi\varepsilon_0\left(\frac{\varepsilon}{\varepsilon+2}\right)E_c r^2 \qquad (1)$$

where q = particle charge (C)
  ε_0 = permittivity of free space (F/m)
  ε = relative permittivity of powder
  E_c = electric field in the charging region (V/m)
  r = mean particle radius (m).

The first important feature of this equation is that the particle charge is independent of the discharge current flowing in the charging region. This is a simplification in that sufficient current is necessary to achieve the limiting charge value given by equation (1). However, increasing the current beyond this minimum value does not further improve particle charging.

This charge is, however, directly dependent on E_c, the electric field in the charging region, which has been measured by the author (Corbett, 1971) to be of the order of 10^6 V/m. Substituting into equation (1) for a typical particle size of 50 μm, yields a particle charge of 10^{-13} C, assuming that the relative permittivity ε is of the order of 2.

The particles are transported to the object to be coated under the action of three main forces:

1. The airflow from the spray gun which was used to transport the powder from the fluidized bed hopper to the gun nozzle.
2. The electric force resulting from the interaction of the particle charge with the electric field created between gun and the earthed object to be coated.
3. Gravitational forces.

1. Airflow

The first of these, the airflow, results in a component of velocity directed towards the object to be coated. Although it is claimed by some plant designers that this can be used to help penetrate re-entrant parts of the object where only feeble electrostatic forces are assumed to operate, it is thought that when insulating materials are being applied, adequate penetration should be obtainable by means of pure electrostatics without the help of the airflow. Excessive airflow results in a considerable proportion of the material being oversprayed. Unlike liquid paint, powder paints can be collected and reused. However, this can be a costly and cumbersome process when several alternate colours are being applied.

2. Electric forces

The electric force F_E driving the particle towards the object to be coated is given by the following simple equation:

$$F_E = qE \tag{2}$$

E is the local electric field between gun and earthed object and has been measured to be of the order of 10^5 V/m. The drag on the particle, F_d, can be described with adequate precision by Stokes law (Davies, 1966)

$$F_d = 6\pi\eta r v_E \tag{3}$$

Here v_E (m/s) is the velocity relative to the airstream, of the particle.

$$v_E = \frac{qE}{6\pi\eta r} \tag{4}$$

and η is the viscosity of air in M.K.S. units.

For the 50 μm particles considered above in the region between gun and earthed object, equation (4) yields a velocity $v_E = 1 \cdot 5$ m/s. This shows the order of particle transport velocity that can be achieved by typical electrostatic forces provided that the particle is efficiently charged to the Pauthenier limit described in equation (1).

3. *Gravity*

To put the velocity derived above in perspective, the gravitational settling velocity v_g for the same particle may be determined simply by equation (5), also assuming drag forces to be described by Stokes law:

$$v_g = \frac{m_g}{6\pi\eta r} \qquad (5)$$

where m = particle mass (kg)
       g = acceleration due to gravity (m/s^2)

Assuming a specific gravity of ~ 2 for the particle considered above, the gravitational settling velocity is ~ 0.1 m/s, which is an order of magnitude lower than the particle velocity due to electric forces.

Although these calculations are considerably oversimplified in the interest of space, they at least serve to show the order of magnitude of the forces acting on a particle being transported to the object to be coated during the coating operation.

C. *Electrode phenomena*

At first sight the picture given above of the coating operation might be thought to be complete. The particles are charged and transported by electro-static forces to the object to be coated. When a layer has been built up, the object is removed to a fusing or curing oven where the layer is transformed to an adherent, non-porous and decorative coating.

The story is, however, incomplete as the mechanism of the growth of the powder layer and its adhesion up to the time when it is heat-treated, has not been considered. This is a very critical phase in the process and will be con-sidered in detail later.

D. *Powder resistivity*

The behaviour of a particular powder when being applied by electrostatic techniques has been found in recent study (Corbett, 1972) to be critically dependent on the resistivity of the deposited powder layer. It is important that this resistivity is distinguished clearly from the bulk resistivity of the material from which the particles are constituted. Current is carried through the powder layer along conduction paths over the surface of the particles (Masuda, 1962). The resistivity is thus dependent on the surface resistivity of the particles and the area of contact of the particles. Both these factors are critically dependent on the amount of water adsorbed onto the surface of the particles and thus the resistivity of a powder layer is extremely sensitive to ambient humidity and in many cases to a lesser extent to the ambient temperature.

This factor accounts for the lack of reproducibility encountered when studying electrostatic powder coating processes, particularly those employing the more conducting materials.

II. GROWTH OF THE POWDER LAYER

A. Insulating powders

These may be defined as powders in which the particles retain their charge for the duration of the coating operation. This results in a very important advantage to the coating process in that under suitable conditions the coating thickness becomes self-limiting. The effect of this is that a uniform coating

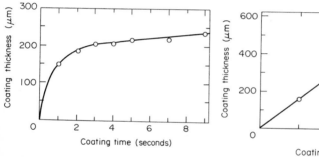

FIG. 1. Coating thickness as a function of time for an epoxy resin powder of high resistivity.

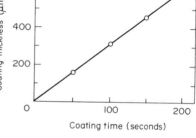

FIG. 2. Coating thickness as a function of time for vitreous enamel powder of resistivity 10^{11} Ω m.

is produced over the surface of the object with a thickness that is largely independent of coating time and the cloud density created by the spray gun. Figure 1 shows a typical coating thickness relationship as a function of coating time for an epoxy resin (Bright *et al.*, 1970). Provided that the coating time is arranged to be in the region where the thickness is insensitive to coating time, the precise conditions are not critical. Furthermore, the limiting thickness is also largely independent of gun-to-surface spacing.

Several attempts have been made to account for this self-limiting phenomenon theoretically. The original idea that the accumulation of charged particles in the powder layer repels further incident charged particles (Bright *et al.*, 1970), does not give a complete quantitative description of the process. More detailed treatments (Makin *et al.*, 1971; Corbett, 1972) describe the behaviour of the potential of the surface of the layer as the thickness increases. These yield quantitative agreement with measured limiting coating thickness values and also show the main features of the process; namely a coating thickness which is largely independent of cloud density, exposure time and gun-to-surface spacing.

The resistivities of materials which exhibit this behaviour have been measured to be typically 10^{12} Ωm or greater. This resistivity range includes important materials such as epoxy resins and many thermoplastic powders such as PTFE (polytetrafluoroethylene) and polythene.

B. Semiconducting powders

In this intermediate resistivity range (10^9 to 10^{12} Ω m), the charge on the particles leaks away during the time of the coating operation (typically 20 seconds). However, the resistivity is not so low that the problems typical of conducting materials are encountered. As would be expected, the powder particles may be applied readily by electrostatic techniques, but the coating is not self-limiting. Thus, given sufficient coating time, layers up to several millimetres thick may be produced.

The coating thickness in this resistivity range has been found to be proportional to time as is shown in Fig. 2. These measurements were made with vitreous enamel powder of resistivity 10^{11} Ω m, and should be contrasted with the self-limiting characteristics exhibited by epoxy resin powder shown in Fig. 1.

This sensitivity of the coating thickness to exposure time is a drawback to the use of semiconducting materials. A further drawback is the sensitivity to the density of the cloud produced by the powder gun and also to the gun-to-object spacing. However, by careful control of these parameters, systems have been designed to operate successfully with materials in this resistivity range, which includes many important ceramic materials such as vitreous enamels and ferrites.

C. Conducting powders

In addition to the disadvantages exhibited by semiconducting materials described above, conducting powders of resistivity less than approximately 10^9 Ω m have a further important drawback. Unless the conditions of particle size and electric field are carefully controlled, the particles are removed from the surface immediately after deposition.

In such cases, the observer normally notes that the powder was not deposited. This is not the case; the powder is deposited by electrostatic forces, but is continuously removed from the surface. This leads to the impression that no deposition has taken place. Although this difference may appear to be somewhat subtle, its observation held the clue as to how this problem could be overcome. The origin of this particle removal phenomenon was subsequently investigated systematically at Southampton University and will be described briefly below.

A conducting particle, which will be assumed to be charged negatively, is driven onto the earthed surface by the applied electric field immediately above the surface. However, being conducting, it immediately loses its charge and is now held on the surface only by mechanical adhesive forces, such as Van der Waals forces. Immediately after this, however, a positive charge is induced onto the conducting particle resting on the surface, the magnitude of which has been calculated by Felici (1966). The interaction of this positive induced charge with the field above the surface, produces a considerable force tending to remove the particle from the surface. When this force exceeds the mechanical adhesive force, particle removal results. For conducting particles the charge transfer time is extremely brief and particle removal occurs almost immediately after deposition.

By studying these phenomena experimentally and theoretically, it has been possible to define ranges of particle size and electric field where adhesion and thus successful operation are possible for conducting powders. The problem has also been overcome in some industrial plants by wetting the surface of the object to be coated prior to the coating operation, thus increasing the mechanical force of adhesion. However, the coating thickness is then dependent on the wetting of the surface being uniform, and dry surface systems are, in principle, preferable.

There are considerable economic incentives to apply conducting materials by electrostatic techniques. Such processes as the upgrading of steel by the application of powdered aluminium or stainless steel, with a subsequent pressure operation to consolidate the layer, are extremely attractive.

III. CONCLUSIONS

It has been possible to give only a very brief description of the physics of the particle charging process and transport of the particle to the surface to be coated. Using very simple basic ideas, the component of particle velocity given by the electric forces in a typical electrostatic powder gun system was derived and shown to be about 10 times the gravitational settling velocity.

The processes that occur in the growth of the powder layer were related to the resistivity of the powder layer. For insulating powders, the charge retention on the particles results in a uniform coating being produced over the surface of the object. The thickness of such coatings can be arranged to be insensitive to coating time, cloud density and gun-to-object spacing.

For semiconducting materials the advantage of a self-limiting thickness is not present. However, electrostatic techniques offer considerable savings in terms of material wastage and labour over conventional spray systems. It has been found that provided the operating parameters are carefully controlled, important semiconducting materials, such as vitreous enamels, may be applied efficiently.

Finally, the problems encountered when conducting materials such as powdered metals are applied by electrostatic systems have been described qualitatively. In this case, efficient operation is only possible in narrow ranges of the two major parameters of the system; namely particle size and the field above the surface to be coated.

REFERENCES

Bright, A. W., Makin, B. and Corbett, R. P. (1970). *In* "Electrical Methods of Machining, Forming and Coating", pp. 119–127. *I.E.E. conf. pub.* **61**, London.
Corbett, R. P. (1970). ibid., pp. 111–118.
Corbett, R. P. (1971). *In* "Static Electrification", pp. 307–319 *Inst. of Phys. Conf. publication*, **11**, London.
Corbett, R. P. (1972). Ph.D. Thesis, University of Southampton.
Davies, C. N. (1966). "Aerosol Science" Academic Press, London and New York.
Felici, N. J. (1966). *Revue General de l'Electricité* **75**, 1145.
Makin, B., Corbett, R. P. and Bright, A. W. (1971). Newcastle Electrochemical of Engineering Symposium. Inst. of Chem. Engrs.
Masuda, S. (1962). *Electro Technical Journal of Japan* **7**, 108.
Pauthenier, M. (1960). "La Phys. des Forces Electrostatiques", 255. CNRS conf. Grenoble.

The basic principles of electrophoretic coating—a study

Arne With

The Technological Institute, Department of Surface Coatings, Copenhagen, Denmark

I. Definition and description

Electrophoretic coating, or, as it is better described, electrodeposition, is a process by which a film is deposited upon a conductive surface from a dispersion of colloidal particles (or ions), dissociated in water or other suitable liquid, and charged with a charge opposite to that of the conductive surface.

Electrodeposition is primarily used as an industrial paint application method. Articles of steel or other metals are immersed in an aqueous dispersion of resin and pigments, which are dissociated into negatively charged colloidal particles and cations consisting of ammonium or amines. The dispersion is stabilized at a pH of, for example, 8. When an electric field is applied using the article as anode and the tank as cathode (or separate surrounding cathodes), the colloidal particles will migrate to the surface of the anode, where they will be discharged and coagulate, thus forming a film. The paint film is then rinsed with tap water and deionized water and generally cured in an oven.

The method is very suitable for a large production of uniform articles, where the whole of the surface to be protected by the paint is accessible to the electric field. The advantages are:

1. An excellent coverage of sharp edges and high points.
2. The film is formed nearly simultaneously on all areas.
3. The film has a uniform thickness.
4. No fire risk.
5. No runs and sags.
6. No solvent popping in the oven.
7. Normally the film is free from pores, thus giving good corrosion resistance.

The disadvantages are:

1. Very costly installation.
2. Demand for homogeneous production of suitable articles.
3. The "faults" of the surface are faithfully reproduced, even enlarged.
4. Usually only the first coat (primer) of a system can be applied in this way, as the formed film acts as an insulator.

Other coatings such as rubber and plastic can be applied by electrodeposition on metals. The "Elphal" Process of the British Iron and Steel Research Association is an example of cathodic electrodeposition, which gives a continuous coating of aluminium on to steel (Yeates, 1966).

II. Discussion of the Mechanisms of Film Formation and of Those Variables Taking Part in Each Mechanism

The following five mechanisms take part in the film formation: A. electrophoresis; B. electrolysis; C. electrocoagulation; D. electroosmosis; E. mechanical transport of the charged colloidal particles. The physical and electrical variables governing the film formation are: F. the substrate; G. the resistance of the film and of the dispersion; H. the applied electric field and the current density; I. the throwing power of the system; J. the temperature of the system; K. the remaining physical and electrical properties and the purity of the system; L. the flow of the dispersion; M. the time.

The mechanisms and the variables are closely interlinked, but we are able to consider "working models" for some of the mechanisms if we regard the interface adjoining the surface of the substrate as "mechanically frozen". We will regard in the same way the interface of liquid surrounding each colloidal particle.

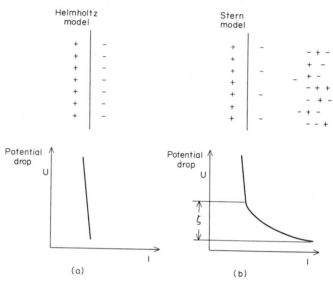

FIG. 1. Potential drop as function of distance l from the electrical double layer. (a) Helmholtz model; (b) Stern model.

Let us consider the interface as a simple Helmholtz electrical double layer of thickness l (Fig. 1a), forming a capacitor with a capacitance C per m² of (Moore, 1957):

$$C = \frac{\varepsilon}{4\pi l} \tag{1}$$

where ε is the absolute permittivity of the dispersion liquid. The potential difference across the capacitor is then, with a surface charge density of σ (c/m²):

$$\Delta V = \frac{4\pi\sigma l}{\varepsilon} \tag{2}$$

Now the Helmholtz double layer is insufficient, as Stern (Moore, 1957) has shown the existence of a diffuse double layer in the liquid outside the surface, and only this layer of charge is free to move. The Stern layer (Fig. 1*b*) causes a particular drop in the potential called the electrokinetic, ζ or zeta, potential.

A. Electrophoresis

If the particle, of charge Q and surface area A, has a constant velocity, the resultant force will be zero. This means that the electrical force is equal to and opposed to the frictional resistance:

$$EQ = \eta \frac{Av}{l} \tag{3}$$

or per m²

$$E\sigma = \eta \frac{v}{l} \tag{4}$$

where E is the applied electric field (V/m), η is the viscosity of the liquid ($kg/m\ s$), and v is the velocity (m/s), σ is taken from equation (2), where ΔV is replaced by ζ, the electrokinetic potential:

$$\sigma = \frac{\zeta \varepsilon}{4\pi l} \tag{5}$$

$$E \frac{\zeta \varepsilon}{4\pi l} = \eta \frac{v}{l} \tag{6}$$

$$v = \frac{E\zeta \varepsilon}{4\pi \eta} \tag{7}$$

The same equation is found by Henry (1925) and Hückel (1931) but based on a far more extensive mathematical basis.

However, equation (7) is only roughly true. The factor 4 in the denominator is substituted by 6 (from Stokes law) by Weigel (1957).

Moreover, the electrokinetic potential ζ is extremely difficult to measure, but it is considered to be of a magnitude of 40–50 mV (Moore, 1957). Weigel (1957) has found the velocity of the resin particle to be of a magnitude of $v = 3 \times 10^{-8}$ m/s at a field strength of 1 V/m. With a voltage of 120 V and a distance of 0·3 m between the electrodes, i.e. a field strength of 400 V/m, the velocity will be $v = 1·2 \times 10^{-5}$ m/s or 12 µm/s. And with a processing time of 3 minutes the covered distance will be $s = 2·16 \times 10^{-3}$ m. This means that the electrophoretic transport is of small importance to the total process!

B. Electrolysis

This is defined as the transport of ions through the electrolyte of an electric cell. If the voltage is high enough, generation of oxygen can be seen at the anode (and at the same time, of hydrogen at the cathode). In order to obtain a non-porous film, the gas development at the anode should be kept as low as possible. At the same time the nascent oxygen may cause unwanted oxidation of the resin in the dispersion.

On the other side the concentration of anions, especially protons but also metal ions, is of the highest importance. Normally the pH of the dispersion is from 7·5 to 8·2 (Weigel, 1957), but the pH in the interface near the anode surface is found to be as low as 3·5 (Tatton and Drew, 1971).

According to Weigel (1957) there is a generation of H^+ and O_2 at the anode:

$$H_2O \rightarrow 2H^+ + 2e^- + \tfrac{1}{2}O_2 \tag{8}$$

which explains the acidity of the interface. At the same time an anodic corrosion of the substrate will give metal ions:

$$M \rightarrow M^{n+} + ne^- \tag{9}$$

Both the protons and the metal ions are of great importance to the electrocoagulation described later.

Working models for these processes have been applied but with poor results.

The pH of the dispersion is naturally of great importance to the pH of the interface. During the use of the bath, the dispersion will "loose" resin and the cations formed by the dissociation of the resin will slowly raise the pH. Therefore it is necessary to adjust the bath in different ways in order to keep the pH constant.

C. Electrocoagulation

This is defined as the precipitation of the colloidal particles out of the dispersion upon the substrate surface, due to neutralizing of the surface charge of the particles. The dispersion of resin particles is stable only in a comparatively narrow pH range, specific for the type of resin. The more we move towards the acidic end, the less stable the dispersion will be. In the interface, protons and metal ions will discharge the colloidal particles (lower the ζ) and these will precipitate out upon the surface of the anode. Initially the discharge happens on the surface itself, but as soon as film formation has started, the protons and metal ions coming through the pores of the film will take over. This mechanism is then the most important of the five for the film formation. The velocity of the electrocoagulation can be found by measuring the thickness of the film, governed by variables such as time, voltage, current density, pH, temperature, solid content, electrical conductivity and permittivity of the dispersion, etc. It is very important to ensure that the coagulation happens very close to the surface. If the current density is too high, the protons will migrate too far, resulting in a loose, spongy film. This is known, with a not too well chosen expression, as the "rupture voltage condition".

D. Electroosmosis

This is defined as the osmotic dehydration of the deposited film, effected by the electric field. It is the opposite mechanism of electrophoresis: the liquid will migrate as a function of the electric field, through the coagulated film. In exactly the same way as was discussed for electrophoresis, we can find the velocity of the liquid (Moore, 1957):

$$v = \frac{E\zeta\varepsilon}{4\pi\eta} \tag{7}$$

and we are then able to calculate the flow through each pore. No figures for this flow are available, but the electroosmosis is very important to the dehydration of the film, which is found, after the rinse, to contain a very small percentage of water. Thus the film will be stable and firm without runs and sags and there will be no "solvent wash", and no solvent popping giving pores and craters in the film, during the subsequent oven curing. It is mentioned that the electroosmosis to some extent washes the salts out of the film (Weigel, 1957).

E. Mechanical transport of the charged colloidal particles

As shown before, the velocity of the particles due to the electrophoresis is of a very small order of magnitude. It is therefore necessary to maintain the concentration of the colloidal particles in the interface by convection, and usually this is done by circulating the dispersion. The rate of circulation is primarily governed by the need to prevent precipitation of resin and resin-covered pigment particles, but it should not be high enough to remove the acidic interface from the surface of the substrate, thus causing a loose, spongy or no film formation. Circulation rates of 40 times an hour are mentioned (Tatton and Drew, 1971).

Weigel (1957) mentions that diffusion also takes part as a film forming factor, but no discussion of this has been found.

F. The substrate

The metallic surface should be free from any trace of oil, grease, salts and corrosion products, except suitable conversion coatings. If the substrate is steel to be phosphated, a hard microcrystalline zinc phosphating, giving approximately 1·5 g/m^2 or less, is preferred. Heavier coatings will yield considerable amounts of zinc ions at the interface due to acidic attack, and these might have undesired effects on the film formation. Chromate rinse as a final step of the phosphating process is not used, as the chromate ions will contaminate the dispersion (Tatton and Drew, 1971).

With other metals, it may be necessary to adjust the pH of the dispersion in order to keep the anodic corrosion down to a minimum.

G. The resistances of the film and of the dispersion

Weigel (1957) has discussed these, but laboratory experiments have shown the resistances to be of non-ohmic character.

H. The applied electric field and the current density

Usually one of these is kept constant; as a rule the voltage is kept constant in commercial plants, giving an initial current density of approximately 50 A/m^2 from 90–400 V (Tatton and Drew, 1971). Figure 2 shows a typical relationship between film thickness and time for different current densities; and Fig. 3 the relationship between film thickness and time for different constant voltages.

The field lines will concentrate on sharp edges and corners, but will move on to other areas when the first are covered by the insulating film. In the same way the lines will later concentrate through the pores, so as to fill them out.

I. The throwing power of the system

This is defined as the ability to coat recessed and shielded areas (Tatton and Drew, 1971) and is dependent on the resistance of the film and the conductivity of the dispersion, and of the voltage and current density.

J. The temperature of the system

A considerable part of the consumed electrical energy is transformed into heat. It has been found that the optimum temperature for film formation is

27–32 °C (Tatton and Drew, 1971), and cooling of the dispersion is normally necessary.

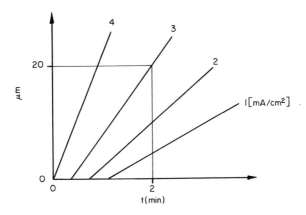

FIG. 2. Film thickness as a function of time at constant current.

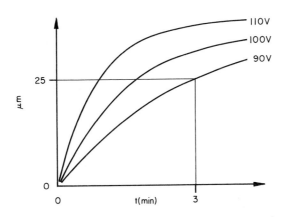

FIG. 3. Film thickness as a function of time at constant voltage.

K. The remaining physical and electrical properties of the dispersion

and

L. The flow of the dispersion
will not be discussed here.

M. The time

This is not the least interesting factor and I want to point out that the time also governs the quality, that is the dehydration and the number of pores in the film.

III. CONCLUSION

Electrodeposition has shown promising qualities which are now used to apply primer coats on suitable metal subjects.

REFERENCES

Henry, D. C. (1931). *Proc. R. Soc.* **A133**, 106.
Hückel, E. (1925). *Phys. Z.* **25**, 204.
Moore, W. J. (1957). "Physical Chemistry" 3rd edn. Longman, Green and Co., London–New York–Toronto.
Tatton, W. H. and Drew, E. W. (1971). "Industrial Paint Application", 2nd edn. Newnes–Butterworths, London.
Weigel, K. (1957). "Electrophorese Lacke". Wissenschaftliche Verlagsgesellschaft M.b.H. Stuttgart.
Yeates, R. L. (1966). "Electroplating". Robert Draper, Teddington.

The basic principles of electrolytic deposition

R. P. KHERA

*Semi-Conductor Products Division, Motorola Inc., McDowell Road, Phoenix,
Arizona, U.S.A.*

I. INTRODUCTION

Electroplating or electrodeposition may be defined as the production of metal coatings on solid substrates by the action of an electric current. In contrast to various other processes of applying coatings, electroplated coatings are applied to improve appearance, corrosion resistance and physico-chemical properties of the surfaces (hardness, electrical and thermal conductivity, solderability, reflectivity etc.). Some of the advantages of electroplated coatings over the other methods of applying coatings are:

(i) Absence of an intermediate layer between the coatings and the substrate metal as in the case of hot dip and diffusion processes;

(ii) Fine structure and often very valuable physical properties mentioned above;

(iii) Easy control of the coating thickness to fractions of a micrometre;
(iv) Most convenient method of applying coatings of metals with high melting points as for example copper, nickel, chromium, iron, silver, gold and platinum.

Electrodeposition of metals from aqueous solutions is mainly limited by the decomposition potentials of the metals to be deposited, their hydrogen overvoltages and various other polarization phenomena. Also the reproducibility of plated parts with regards to adhesion and other physical properties of the plated metals, maintenance of uniform conditions in the electroplating baths, and depositing an equal thickness of the metal over the entire surface of the plated parts, pose some problems.

Developments in the electroplating industry have resulted almost entirely from empirical investigations, mainly undertaken to solve practical problems. With the rapid industrial evolution of large mechanized electroplating plants and stringent requirements of coatings for good adhesion, protection against corrosion, pollution control and other specific purposes, an empirical approach to the problem is no longer sufficient.

A scientific treatment of electroplating processes is complicated since most electroplating baths are rather complex. Effects of various constituents added to confer specific properties to the deposited coatings, cannot be clearly ascertained. The numerous superimposed processes, such as incorporation of non-metallic foreign substances into the deposited coatings, codeposition of metallic impurities, hydrogen evolution and several other oxidation-reduction processes, make it almost impossible to distinguish clearly the individual steps in metal deposition. Laboratory investigations under controlled conditions may clarify some of our hypotheses concerning electron transfer processes, but yield no definite conclusions concerning the overall situation occurring in actual plating practice.

II. CHEMICAL AND ELECTROCHEMICAL PRINCIPLES

A. Ions and electrolysis

An electroplating system consists essentially of a plating bath, a d.c. source and two electrodes. The electrode connected to the positive pole of the d.c. source is called the anode and the one connected to the negative pole is called the cathode. The plating baths contain a conducting solution called the electrolyte whose principal component is a salt or other compound of the metal to be plated.

Figure 1 shows a simple electroplating system for the deposition of copper from copper sulphate solution. When dissolved in water, copper sulphate dissociates into positively charged copper ions (cations) and negatively charged sulphate ions (anions).

$$CuSO_4 \rightleftharpoons Cu^{2+} + SO_4^{2-} \tag{1}$$

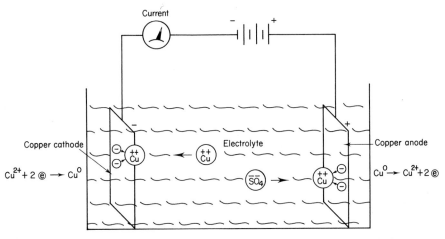

FIG. 1. Electrolytic cell for the deposition of copper from copper sulphate solution.

Under the influence of the impressed current, reactions occur at the electrodes. The cations migrate to the cathode where they are discharged and deposited as metallic copper.

$$Cu^2 + 2e \rightleftharpoons Cu(metal) \tag{2}$$

Copper from the anode dissolves to replace the copper ions removed at the cathode thus maintaining electrical neutrality

$$Cu \rightleftharpoons Cu^{2+} + 2e \tag{3}$$

The overall process is known as electrolysis. Should the anode be a non-corrodible metal such as platinum or some other noble metal, the overall reaction at the anode is the oxidation of water

$$2H_2O \rightleftharpoons 4H^+ + O_2 + 4e \tag{4}$$

The sulphate ions remain unchanged in quantity during the electrolysis under this latter condition.

In the latter case reactions (2) and (4) show that the concentration of Cu^{2+} ions will decrease and that of H^+ ions will increase with time. To maintain uniform conditions, copper ions must be added as copper sulphate from time to time and the hydrogen ions must be removed by neutralization with alkali or by buffering the solution. A buffer is a substance which resists change in the hydrogen ion concentration of a solution by combining with hydrogen or hydroxyl ions produced to form an almost undissociated compound.

It should be noted that the ions responsible for the current transport in the electrolyte may or may not participate in the electrode reactions, as observed for sulphate ions above. Conversely the neutral molecules which do not migrate under the influence of the applied voltage but can arrive at the surface by diffusion and convection, may appreciably influence the electrode reactions. The interactions of the chemicals in the plating bath and their chemical reactions on the electrode surface, can greatly influence the electro-chemical processes at the cathode.

B. Faraday's laws

The quantity of metal deposited is given by Faraday's laws, which can be simply stated as

$$Q = F.E_q \tag{5}$$

where E_q is the number of gram equivalent weights of metal deposited, F is the Faraday constant (96,490 coulombs per gram equivalent) and Q the quantity of electricity in coulombs. The value of Q can be determined from the total amount of electricity which has flowed during electrolysis.

$$Q = \int_0^t i \, dt \tag{6}$$

where i is the current in amperes and t is the time in seconds. If the electrolysis current is maintained constant, Q is obtained simply by the equation

$$Q = i \, \Delta t \tag{7}$$

C. The Nernst equation

The Nernst equation is the most fundamental equation of electrochemistry. For an electroplating process, it can be stated as

$$E = E_0 + \frac{RT}{zF} \ln a_{M^{z+}} \tag{8}$$

where E is the electrical potential at the electrode surface (the electrode potential), R the gas constant, T the absolute temperature, and z is the valency of the metal ion M^{z+} being deposited as metal. At unit activity of the depositing metal ions ($a_{M^{z+}} = 1$), the electrode potential E from equation

(8) is equal to E_0, which is known as the standard potential, normal potential or ground potential.

Converting the natural logarithm in equation (8) to a common logarithm, the Nernst equation can be expressed as

$$E = E_0 + 2 \cdot 3 \frac{RT}{zF} \log a_{M^{z+}} \qquad (9)$$

The Nernst equation is derived from thermodynamic laws, from the considerations of free energy change for the reaction. The practical application of this relationship is limited only to the thermodynamics of a reaction. Catalytic effects, kinetic processes, overvoltages, reaction inhibitions etc., can greatly alter the behaviour of an electrode from that expected on the basis of the Nernst equation.

III. OVERVOLTAGE AND POLARIZATION

The Nernst equation (9) can be used to determine electrode potentials for processes at reversible equilibrium, i.e. the condition in which no net reaction takes place. Under practical conditions nearly every electrode reaction is irreversible to some extent. This irreversibility causes the potential of the anode to become more noble and the cathode potential less noble than their respective static potentials calculated from the Nernst equation. The overvoltage is a measure of the degree of this irreversibility and the electrode is said to be "polarized" or to exhibit "overpotential".

IV. ELECTRODE POTENTIAL

Under dynamic conditions, the electrode potential is given by

$$E = E_{rev} + \eta \qquad (10)$$

where E_{rev} is the electrode potential calculated from the Nernst equation and η is the total polarization. The total polarization η is the sum of the concentration polarization (η_c), the activation polarization (η_a), and the ohmic polarization η_o

$$\eta = \eta_c + \eta_a + \eta_o \qquad (11)$$

The various types of polarizations in equation (11) have further subdivisions to denote specific inhibitions during metal deposition. These polarizations are usually superimposed to varying degrees during the electrodeposition of

metals. To understand the nature of these polarizations, various steps involved in the electrodeposition of metals and the structure of the electrical
double layer at the cathode must be understood.

V. ELECTRICAL DOUBLE LAYER

Under the influence of an applied potential, there is a rearrangement of ions
in the vicinity of the electrode surface resulting in the formation of an
electrical double layer called the Helmholtz double layer. This leads to
further disturbance in the electrolyte so that there is formed overall a diffused
double layer referred to as the Gouy-Chapman diffuse layer, which consists
of the Helmholtz fixed double layer and the diffusion layer, as shown in
Fig. 2.

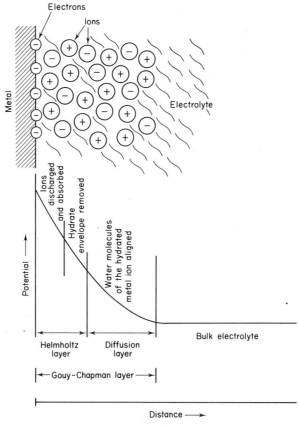

FIG. 2. Electrical double layer.

VI. Kinetics of Cathodic Deposition of Metals

The various steps involved in the cathodic deposition of metals may be summarized as follows:

(i) *Migration:* The hydrated metal ion in the bulk of the electrolyte migrates towards the cathode under the influence of impressed current as well as by diffusion and convection.

(ii) *Electron transfer:* At the cathode surface the hydrated metal ion enters the diffused double layer (Fig. 2) where the water molecules of the hydrated ion are aligned by the weak field strength in this layer. The metal ion then enters the fixed (Helmholtz) double layer where, on account of the high field strength, it is deprived of its hydrate envelope.

(iia) The dehydrated ion is neutralized and adsorbed on the cathode surface.

(The electron transfer process for the complex ions are quite complicated and not always clearly understood. It is generally agreed that the deposition of metal from complex ions is the result of discharge of simple ions which are formed as a result of the dissociation of complex ions in a previous reaction. The dissociation of the complex ions is believed to be so fast that the kinetics of the overall discharge process are not affected by this reaction.)

(iii) The adsorbed atom then migrates or diffuses to the growth point on the cathode surface and is incorporated in the crystal lattice.

The slowest step in the above mentioned processes is the rate determining step. If migration is the rate determining step, then the overpotential developed is referred to as the concentration polarization.

Quantitatively, concentration polarization is given by

$$\eta_c = \frac{2 \cdot 3RT}{zF} \log \frac{j_L}{j_L - j} \tag{12}$$

where j_L is the limiting current density, and j the applied current density. At the limiting current density, the concentration of the ionic species at the electrode surface reaches zero so that no further reaction can occur by further increasing the current density. Equation (12) shows that when j is equal to j_L, η_c tends to infinite values.

j_L can be determined from the relationship

$$j_L = \frac{DzFa}{(1 - t)\delta} \tag{13}$$

Where D is the diffusion coefficient, z the valency, a the activity, t the transport number of the reacting ion, and δ is the thickness of the diffusion layer. The value of δ varies between 10 μm for well stirred solutions to 500 μm for unstirred solutions. Most electrodeposition processes, especially those involving complex ions, occur under conditions of limiting current density; so that the consideration of concentration polarization is quite important to electroplaters.

Equations (12) and (13) show that η_c can be decreased by increasing temperature, increasing concentration and increasing agitation of the bath.

The polarization arising from steps (ii), (iia) or (iii) is referred to as the activation polarization. The activation polarization is also termed the penetration polarization, since it is caused by the inhibition of the passage of the potential-determining ions through the electrical double layer.

The value of activation polarization is given by the well known Tafel equation:

$$\eta_a = \frac{2 \cdot 3RT}{\alpha zF} \log \frac{j}{j_o} \tag{14}$$

where α is the fraction of the overpotential which assists in the deposition of metal. It is called the transfer coefficient. Its value varies between 0·2 to 0·8 depending upon the plating system. j_o is the current density at the equilibrium potential (where $\eta = 0$) and is known as the exchange current density.

If it is established that step (iii) is the rate determining step (as opposed to the situation in the previous paragraph), then the polarization is also referred to as the crystallization polarization. Other terms such as lattice-incorporation polarization, poisoning polarization, blocking polarization etc., represent further subdivisions depending upon which one is the inhibiting process.

A knowledge of various types of polarizations in a practical plating system is quite useful in eliminating problems. Thus, if a metal is depositing at a high concentration polarization, defects in the appearance and crystal structure are usually observed because of the slow supply of the depositing ions to the cathode surface. Providing some relaxation time, in the form of periodic reverse current electrolysis, d.c. superposed with a.c., or using pulsing current, helps eliminate the defects in crystal structure. Such techniques will have little beneficial effect when the deposition is entirely activation controlled.

VII. ELECTRODEPOSITION OF ALLOYS

Co-deposition of two or more metals is possible under suitable conditions of potential and polarization. The necessary condition for the simultaneous deposition of two or more metals is that the cathode potential versus current density curves (polarization curves) should be similar and close together. As an illustration let us suppose that curves A and B in Fig. 3 are the polarization curves of metal A and metal B respectively. In the absence of any polarization effects, at potential V_1, the deposition rates of metal A and metal B are given by the current densities j_1 and j_2 respectively. These rates at potential V_2 are given by current densities j_3 and j_4 respectively.

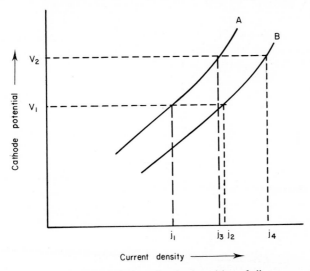

Fig. 3. Polarization curves for the deposition of alloys.

From the current densities, the weight of the metals deposited during a given interval of time can be determined from equation (5) (Faraday's laws).

It is observed that in the above illustration the plating bath is getting depleted of metal B ions faster than of metal A ions. To keep the conditions uniform the metal ions must be replenished in proportion to their rates of deposition. Evidently the most desirable situation would be for the polarization curves of the component metals during their codeposition, to be identical. Unfortunately it is extremely difficult to realize this condition in practice.

In the case of codeposition of metals with their standard potentials wide apart, the deposition potentials are brought together by complexing the more noble metal ions as illustrated below for the codeposition of copper and zinc as brass.

The reversible potential of copper in unit activity of copper ions is $+0.34$ V and that of zinc in unit activity of zinc ions is -0.77 V. If the solution contains unit activities of copper and zinc ions, zinc will not codeposit with copper unless the overpotential for copper deposition is high enough to compensate for the large difference in deposition potentials.

The deposition potentials of these metals can, however, be brought together by the adjustment of their ionic concentrations. Thus, if potassium cyanide is added to the solution of salts of these metals, it binds the copper and zinc ions as stable cuprocyanide anion $\{Cu(CN)_4\}^{3-}$ and zinc cyanide anion $\{Zn(CN)_4\}^{2-}$ respectively. In solution the copper complex dissociates to cuprous and cyanide ions according to the equilibrium

$$\{Cu(CN)_4\}^{3-} = Cu^+ + 4(CN)^- \tag{15}$$

The value of the instability constant, i.e.

$$\frac{[Cu^+][(CN)^-]^4}{[\{Cu(CN)_4\}^{3-}]},$$

is $10^{-27.3}$ at 25 °C, so that at unit concentrations of $\{Cu(CN)_4\}^{3-}$ and $(CN)^-$ ions, the equilibrium concentration of $[Cu^+]$ is $10^{-27.3}$ g molecules/litre ($10^{-24.3}$ mol/m^3). The instability constant of $\{Zn(CN)_4\}^{3-}$ is 10^{-18} so that the concentration of (Zn^{2+}) under the above conditions will be 10^{-18} g molecules/litre. With these ionic concentrations the deposition potentials of copper and zinc in the absence of any polarization, are calculated from the Nernst equation (9) to be approximately -1.30 V for each metal.

The individual polarization curves for the metals are nearly always modified as a result of interactions resulting from the codeposition. If the alloy deposition occurs at low polarization, the nobler metal will be preferentially deposited. All factors which increase the polarization during the electrodeposition, e.g. use of high current density, low temperature and quiescent solution—factors which increase concentration polarization noted above—will favour the deposition of the less noble metal.

The evolution of hydrogen during the electrodeposition of an alloy, has a significant effect on the polarization and composition of the alloy deposited. If a considerable amount of hydrogen is evolved, the potential of the cathode during the alloy deposition may be almost entirely determined by the hydrogen evolution reaction. If, as is usually the case, the overpotential for hydrogen evolution is high in the above instance, the currents corresponding to the individual metals would be close to limiting values. Under these circumstances, any increase in current will increase the amount of hydrogen evolved resulting in a poor efficiency for alloy deposition with a minor change in the composition of the alloy deposited. This is graphically represented in Fig. 4.

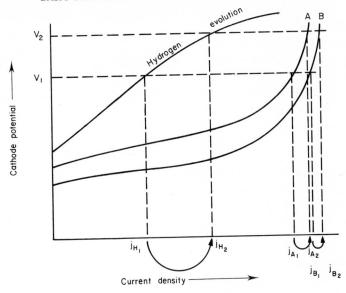

Fıɢ. 4. Alloy deposition with hydrogen evolution.

VIII. Some Practical Aspects of Electroplating

A. Plating on corrodible metals

Electrodeposition of metals, especially the reactive and more electro-negative metals is not always possible from aqueous solutions. Thus, for example, aluminium, molybdenum, tungsten, titanium, tantalum, zirconium and niobium have not yet been electroplated from their aqueous solution, to the best of the author's information. Even iridium has not been success-fully deposited from its aqueous solution, though processes for the electro-deposition of other platinum group metals (platinum, rhodium, ruthenium) are known.

Electrodeposition on metals in corrosive electrolytes or electrodeposition of noble metals such as gold and silver on a base metal substrate such as copper, must be carried out carefully so that chemical attack or the electro-chemical displacement of the base metal by the noble metal does not take place. Should this happen, the adhesion of the plated metal will be poor and, in addition, the plating bath will be contaminated.

To overcome the above-mentioned difficulties, two methods are generally used. In the case of silver plating of copper, for example, a thin undercoat of mercury formed by the electrochemical displacement of copper in a solution of mercury cyanide can first be deposited. The thin film of mercury will satisfactorily resist the corrosive action of the silver plating bath during electrodeposition. This process is known as "quicking". The second method known as "striking" consists of coating the base metal with the plating metal very quickly by initially operating the bath for a short time at a high current density. The part to be plated in this process must be made cathodic immediately on immersion in the bath. Nickel plating of zinc castings and steel is carried out by this procedure.

B. Plating on aluminium and titanium

The plating of metals on aluminium and titanium poses a problem of a different nature. These metals are always immediately covered with a protective oxide film when exposed to air. They form oxide films even in solutions containing dissolved air. Electrodeposits adhere very poorly on this film. In the case of aluminium, the problem is overcome by dipping in a solution of sodium zincate, or by a short cathodic treatment in alkaline zinc cyanide. A thin film of zinc deposits on the metal during this process. The zinc covered surface is then given a copper strike before plating nickel and chromium.

An entirely satisfactory process for plating on titanium is not yet known. The obvious solution to this problem would appear to be to find a suitable plating bath which can reduce the oxide film before plating or to form on the titanium a labile film which is reduced during electrodeposition. This has not yet been successfully achieved.

C. Plating on plastics

Before plastics can be electroplated, they must be made electrically conducting by a process known as electroless plating. The process consists in etching the plastic surfaces with chromic-sulphuric acid solutions followed by activation by immersion in tin and palladium solutions, or a colloidal solution containing these ions. The etching process is known as "conditioning" of the plastic surface, and is probably the most important step in the electroless plating of the plastic surface. The purpose of "conditioning" is to improve the wettability of the plastic surface and to give it a suitable structure for good adhesion of the plated metal. Usually an organic solvent treatment prior to conditioning is helpful in improving adhesion of the deposit on plastics. Commercially successful electroless nickel, cobalt and copper plating baths are known. The activation process deposits nuclei of tin and palladium which act as the starting materials for chemical deposition of nickel, cobalt or copper by the reducing action of hypophosphite or other ions such as borohydrides or boranes. The process after the deposition of the first layer of electroless metal is autocatalytic. A thickness of the electroless metal of 2 to 5 μm can be deposited in 15 to 30 minutes. The metal coated plastic can then electroplated with the required metal under specific conditions.

IX. PROPERTIES OF ELECTRODEPOSITS

A. Adhesion

Good adhesion of the plated metal to the substrate is one of the most important requirements. It is mostly dependent upon the substrate. For proper adhesion, the substrate must be thoroughly clean and free of any surface films.

For optimum adhesion, the substrate and the deposited metal should interdiffuse with interlocking grains to give a continuous interfacial region. A porous interface or separation between the deposited metal and the substrate observed under the microscope indicates poor adhesion. Alloy formation by the interdiffusion of the substrate and deposited metals is desirable; an intermetallic compound formation is not. The latter behave like inorganic salts and give poor adhesion, as in the case of copper and aluminium.

Etched or electroplated surfaces give better adhesion than mechanically polished, sand blasted or abraded surfaces on account of the presence of inclusions and oxide films in the latter case.

When the adhesion strength is optimum, failure during test takes place within one of the metals with lower tensile strength, but not at the interface.

B. Hardness

The hardness of the electrolytic deposit is of special interest to the coating industry. It can be easily measured with the Vickers or Knoop diamond tester. When a microhardness tester is used, the measurement can be carried out even on thin films, provided the film thickness exceeds seven times the penetration depth.

Hardness is influenced by the incorporation of impurities in the deposits. Thus, the high hardness value of chromium coatings is due to the trivalent chromium oxide inclusions. Deposits with impurity incorporation are often fine-grained. This fact has led to the observation that fine-grained deposits are harder than coarse-grained ones. However, fine-grain structure cannot be considered the reason for greater hardness. The hardness, in fact, results from the lattice strain produced by the incorporation of impurities. Alloy deposits have high hardness values because of the lattice strains produced by the lattice defects.

Maximum hardness is not desirable for coatings which are subjected to wear during use, since in that case embrittlement and internal stress are also at a maximum.

C. Brightness

Since Schlötter first developed a brightening electrolyte for nickel coatings in 1935, the field of bright coatings has progressed tremendously. However, a theoretical study of the subject has been difficult because of the complex nature of the problem. Added to it is the fact that the additives used as brighteners are usually technical products containing impurities. Even if pure compounds are used, they undergo electrochemical reactions involving decomposition, polymerization or condensation, and it is hardly possible to determine which compound is actually responsible for the brightening effect.

Bright coatings can be deposited on matt surfaces by using brightening additives. These are mostly organic compounds such as dextrose, saccharine, lactose, formaline, citrates, tartarates, etc. However, most good brighteners are sulphur compounds, especially thiourea and its derivatives and organic sulphonic acids. The brighteners are surface active agents and increase the polarization for the metal deposition process. However, some brightening agents, especially those containing sulphur compounds, and especially at low concentrations, quite often have a depolarizing effect on metal deposition.

Brightening agents must be regarded as foreign inclusions in the deposit. Overdosage of these additives can cause brittleness and lead to cracks and peeling off of the deposit from the base metal.

X. DISTRIBUTION OF THE DEPOSIT ON THE CATHODE SURFACE

The characteristic of the system which denotes its ability to deposit a relatively uniform coating over an irregularly shaped cathode, is known as its "throwing power". There is no fully accepted measure of throwing power. The percentage ratio of the smallest and greatest thicknesses of the coating on the object may be used as a measure of the throwing power.

The ability of a plating solution or a specified set of plating conditions to deposit metal in pores or scratches is known as the "microthrowing power" of the system.

Unequal deposit distribution results from an unequal current density distribution, which is a function of the geometry of the cell. Thus the areas on the cathode which are close to the anode will have a higher current density for the same voltage. In the absence of any extraneous reactions, such areas will receive a higher thickness of deposit. This type of current density is based on Ohm's and Kirchhoff's laws and is often termed the "primary current density distribution".

The current distribution is also affected by polarization and therefore by all factors which influence polarization such as concentration, velocity of transport of electrolyte and current density. Since, as observed above, the primary current distribution is different for different elements on the electrode surface, it causes polarizations of various elements to varying extents. Thus, the elements receiving higher current densities will also show higher polarizations and vice versa. These higher polarizations at the high current density areas will tend to decrease the current density. Similarly, the lower polarizations at the low current density areas will tend to increase the current density. This will set up a secondary current density distribution, the function of which is to make the current density distribution uniform, thus causing the metals to deposit uniformly. This effect will be more noticeable for systems which deposit metals at high polarizations, so that such systems will have a better throwing power.

The range of microthrowing power, by definition, is restricted to profiles in which the primary current densities are the same everywhere. Thus, the systems which deposit metals at low polarizations will tend to have better microthrowing power. Thus, also, plating baths containing a high concentration of dischargeable ions, and operated at low current densities with well stirred electrolytes, will have good microthrowing power. The reverse is true for macrothrowing power.

As observed above, micro- and macrothrowing powers change in opposite directions. All the factors which improve one kind of throwing power give rise to worsening of the other. For example, cyanide copper baths which operate at a high concentration polarization have good macrothrowing power, but poor microthrowing power. On the other hand, acid copper baths which contain a high concentration of dischargeable ions and operate at low concentration polarization, have good micro- but poor macrothrowing power. In practical applications, this means that when a uniform copper deposit is required over an irregularly shaped cathode surface, such as a car bumper (prior to nickel and chromium plating), a cyanide copper bath is used. However, for uniform deposits in holes or pores as for printed circuit boards, an acid copper bath is used.

BIBLIOGRAPHY

Brenner, A. (1963). "Electrodeposition of Alloys". Academic Press, New York and London.

Kortum, G. (1965). "Treatise on Electrochemistry". Elsevier Publishing Co., Amsterdam.

Lowenheim, F. A. (Ed.) (1971). "Modern Electroplating". John Wiley and Sons, New York.

Potter, E. C. (1961). "Electrochemistry". Cleaver-Hume Press, London.

Raub, E. and Muller, K. (1967). "Fundamentals of Metal Deposition". Elsevier Publishing Co., Amsterdam.

Schlotter, M. (1935). *Z. Metall.* **26**, 236.

Vetter, K. J. (1967). "Electrochemical Kinetics". Academic Press, New York and London.

Thin organic coatings by electrolysis of phenol

R. Dijkstra and J. De Jonge

Philips Research Laboratories, Eindhoven, The Netherlands

I. General remarks on the procedure

The well-known compound phenol, C_6H_5OH, and its alkyl derivatives, are very weak acids with pK values of about 10 (in water). The electrochemical oxidation of these phenolic compounds can lead to products of high molecular weight especially if the procedure is carried out in a non-aqueous system. The electrolytic bath contains the (water-free) phenol in the liquid state or a solution of the phenol in a polar solvent such as acetonitrile or dichlorobenzene. About 0·25 mole of a base such as triethylamine, $(C_2H_5)_3N$, per mole of phenol is added in order to promote ionization and thus electrical conductivity. The main species in such a system are accordingly phenol molecules ROH, negatively charged phenoxide ions RO^- and positively charged triethylammonium ions $(C_2H_5)_3NH^+$.

If an electric current is passed through the electrolyte, the ions $(C_2H_5)_3NH^+$ are discharged at the cathode giving rise to the evolution of hydrogen gas. At the anode the phenoxide ions RO^- are discharged and converted into phenoxy radicals $RO^\cdot$. The latter react to give large molecules with a polyphenylene-ether structure like

$$-R'-O-R'-O-R'-O-R.$$

Depending on the particular phenolic compound under investigation, the high molecular weight product may be soluble or insoluble in the electrolytic bath. In the first case, e.g. when using p-methylphenol (p-cresol) or 2,6-dimethylphenol, one observes a gradual increase in viscosity of the electrolyte during the experiment. The products, after isolation and purification, are thermoplastic polymers with softening points between 200 to 350 °C. The polymer molecules will have a linear or branched structure.

II. COATINGS

If phenol or 2-methylphenol (*o*-cresol) is used to make up the electrolyte, insoluble products are formed at the anode. The polymer will have a cross-linked structure (network structure). Sometimes it does not adhere to the anode and can be found at the bottom of the vessel in the form of a fluffy powder. This is often the case when the anode is made of platinum. With many other metals, however, the polymer is obtained in the form of a very thin continuous coating covering the anode. For good results, oxygen (air) and especially water should be excluded. It is advisable to limit the current density during the formation of the coating to about 1 mA/cm^2. If one wants to grow the coating with a constant current, it is necessary to increase the voltage linearly with time during the formation. If the voltage is no longer increased the current decreases gradually to very low values. The maximum voltage used determines the thickness of the coating (20 Å per V). Any thickness from 0·05 μm to about 0·6 μm can be obtained; in the latter case a maximum voltage of 300 V is applied.

Various kinds of objects with an electrical conducting surface can be provided with a very thin film of an organic polymer if the objects are made the anode in the electrolytic process.

Examples of electrically conducting materials that have been coated successfully are: aluminium, zinc, iron, cobalt, nickel, copper, gold, steel, brass, silicon and transparent layers of tin oxide on glass. The organic films adhere to the surface of the anode which may have an intricate shape.

The coatings are free from pinholes and form a good electrically insulating layer. A capacitor was constructed by evaporating a layer of aluminium on a microscope slide, depositing a 1500 Å (150 nm) thick organic coating using the electrochemical procedure and evaporating a second aluminium layer on top of this. The capacitance amounted to 55,000 pF/cm^2 and the breakdown voltage was about 60 V. The dielectric constant of the organic dielectric material was 4·5 and tan $\delta = 50 \times 10^{-4}$ at 1000 Hz.

The coating material is thermally stable up to 240 °C and is insoluble in common organic solvents. Non-oxidizing mineral acids do not attack the coating; it is destroyed, however, by strongly alkaline solutions. The coatings are often decorative due to Newtonian colours. If a piece of ordinary steel is used as the anode in the phenol bath, the resulting polymeric film of, for example, 0·6 μm thick, inhibits corrosion. Electroless deposited nickel or cobalt can be protected in a similar way. Although the electrolytically prepared organic coating is rather hard, it is also very thin and it does not give much mechanical protection.

More details can be found in a forthcoming publication by the same authors in *J. Electrochem. Soc.*

The basic principles of anodization

D. S. CAMPBELL

Department of Electronic and Electrical Engineering, University of Technology, Loughborough, England

I. SUMMARY

This paper gives a broad review of the preparation and properties of films prepared by anodization. In the first instance the basic concepts are considered where a comparison is made with the growth of films by thermal methods. The effect of applied voltage on the growth and thickness of the film is then examined. It is shown that growth stops when the electric field across the film is insufficient to drive the ions through the layer. Growth can be effected in practice under "constant current" or "constant voltage" conditions and both of these are examined.

Secondly, the effect of the substrate is examined. The limitations of the materials that may be anodized are discussed and the effect of purity of surface etc. of the materials that may be anodized is examined. Specific examples of surfaces that are used are mentioned, including etched aluminium and sintered tantalum.

Finally, the structure of the oxide obtained is considered. The majority of these films are amorphous, and porous oxides may be obtained by a suitable choice of electrolyte in the anodization bath. The properties of the oxide in dielectric terms are examined, particularly with regard to the effect of amorphicity on the electrical behaviour.

Allied systems related to anodization are mentioned, including plasma oxidation and gaseous oxidation.

II. Basic considerations

Many different classifications are possible for characterizing the deposition of thin films onto substrates. One type of classification (Campbell, 1970a) that may be used is:

(i) those layers that deposit independently of the substrate;
(ii) those layers that deposit because of reaction with the substrate.

The former grouping covers such techniques as electroplating, vapour phase deposition, evaporation, painting etc., while the latter group covers such techniques as thermal growth, anodization, gaseous anodization and plasma oxidation.

Anodization is thus a technique that depends on the presence of a substrate on which a film can be grown from the substrate. What is required for this substrate-dependent growth is first of all a suitable substrate, secondly conditions for the correct chemical activity for the growth to occur, and finally coherence in the growing film. If this final condition is not obtained the chemical attack on the substrate will continue indefinitely and the process is known as corrosion.

To examine the situation with regard to anodization in any detail, it is

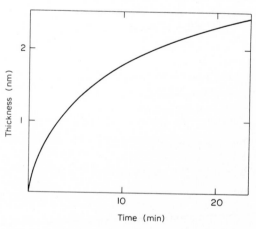

Fig. 1. Thickness versus time curve for thermal growth of Al_2O_3 on aluminium in air at 20 °C.

necessary first to briefly consider the situation that obtains with thermal growth. A large number of metals will form films if heated in gases of the required type (Evans, 1958). For example, heating in oxygen will give layers of oxide; in nitrogen, layers of nitride; and in carbon monoxide, layers of carbide. If the films obtained are coherent, the main characteristic is that growth slows down as the thickness increases, in fact, an exponential relationship with time is obtained (see Fig. 1). A limiting thickness will eventually be reached [e.g. in the case of aluminium in air at 20 °C, a limiting thickness of 40 Å (4 nm) will be obtained after a considerable period of time, certainly a matter of hours]. Coherent oxides are formed on a variety of metals,

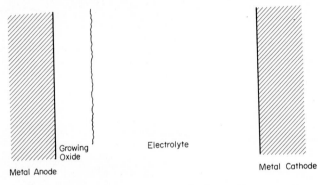

Metal Anode · Growing Oxide · Electrolyte · Metal Cathode

FIG. 2. Diagram of an anodization system.

including aluminium, tantalum and silicon. The growth phenomena can be described in electrochemical terms. It is found that the surface of the metal is positive relative to the surroundings and hence negative ions are attracted to it. Growth of the layer will stop when the potential across the film is no longer large enough to drive these negative ions through the film.

Anodization (Campbell, 1970a; Young, 1961; Dell'Oca *et al.*, 1971) is an assisted form of thermal growth. The metal to be anodized is made the anode in a bath of electrolyte (Fig. 2) and growth then occurs in the same way as with thermal growth. However, in the case of anodization, the final thickness will depend on the voltage applied and growth curves of the type shown in Fig. 3 will be obtained.

The reactions that occur at the anode and cathode can be represented by the following equations:

$$M + nH_2O \rightarrow MO_n + 2nH^+ + 2ne \qquad \text{anode} \quad (1)$$

$$2ne + 2nH_2O \rightarrow nH_2\uparrow + 2nOH^- \qquad \text{cathode} \quad (2)$$

The above equations express the fact that oxide will grow on the metal anode surface and hydrogen will be evolved at the cathode. The equations imply the presence of water and anodization usually occurs in aqueous electrolytes (e.g. a solution of phosphoric acid). However, only minute traces of water appear to be necessary so that it is possible to use other electrolyte media such as certain pure alcohols or molten inorganic salts. The pH of the electrolyte is important in obtaining a coherent film. If the electrolyte is too acid or too alkaline, the film will dissolve as it grows, and porous oxide structures will result in certain cases (Wood *et al.*, 1968).

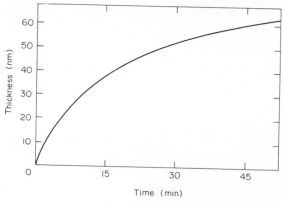

FIG. 3. Thickness versus time curve for anodization of aluminium at "constant voltage" (30 V applied, 85 °C).

Given the correct acidity of the electrolyte, coherent films can be grown by anodization on a variety of metals. Some of the more important are aluminium, tantalum, niobium, silicon, titanium and zirconium. The final thickness obtained will depend on the metal, on the voltage applied to the anode, on the temperature of the bath and on the time that the metal is immersed in the electrolyte. If the metal is kept in the anodization bath for a sufficient time, a reproducible thickness will be obtained, dependent on the applied voltage. The quantity known as the anodization constant is this thickness for 1 volt applied and in the case of aluminium is 13·6 Å (1·36 nm), tantalum 16·0 Å (1·60 nm), silicon 3·5 Å (0·35 nm), niobium 43 Å (4·3 nm). However, in practice, keeping the metal in the anodization bath for sufficient time to obtain thicknesses corresponding to those expected from the anodization constant is time consuming. Therefore, the metal is often only kept in the bath for a limited time so that the thickness obtained is not that given by the anodization constant, but is somewhat less, usually between 70–80% of the maximum that could be obtained. Under these conditions anodization is only necessary for a period of up to ten minutes.

The growth of the oxide film under "constant voltage" conditions has been illustrated in Fig. 3, and this is a technique of growth which is widely used. The disadvantage is that in the initial stages of growth very high current densities are required and for high voltage anodization, e.g. 450 volts and above, this means that anodization must be performed in a series of baths so that the anode can be immersed at successively higher voltages. One way round this difficulty is to anodize at "constant current". The film is then grown until an amount of charge corresponding to the desired thickness has been passed through the electrolyte metal interface. Growth will be as illustrated in Fig. 4. The limitation in this case will be that the voltage across the

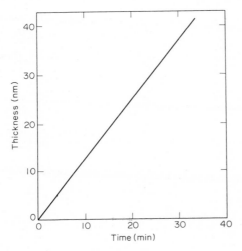

FIG. 4. Thickness versus time curve for anodization of aluminium at "constant current" (2 mA/cm^2).

film must increase as the thickness of oxide increases and eventually this will lead to breakdown of the film. In the case of aluminium, this occurs at a limiting thickness of 1·5 μm and with tantalum 1·1 μm. This limit is a function of the purity of the substrate and of the purity of the electrolyte, amongst other factors, and the use of impure materials will cause breakdown to occur at lower voltages.

In the above discussion it has been mentioned that the oxygen ions are driven through the oxide by the influence of the applied field so that it would be assumed that the anodization reaction occurs at the oxide-metal interface. This is true in certain cases (e.g. zirconium, Whitton, 1968), but with other

D. S. CAMPBELL

metals both the negative oxygen and positive metal ions can move (Davies *et al.*, 1965) so that the reaction can be at the oxide-electrolyte interface as well.

III. Substrate

The oxide film will grow on the substrate in a non-nucleated manner as continuous layers of amorphous material. Under these circumstances it is found that the oxide replicates the surface features underneath. An example of this can be seen in Fig. 5, which shows the oxide surface of an anodized zirconium metal with the grain boundaries of the underlying metal clearly replicated (Jackson, 1972).

Fig. 5. Oxide-electrolyte interface for anodic film on zirconium showing grain boundaries. (By permission of the Plessey Company Ltd.)

If the surface of the metal contains impurities then these can cause defects such as holes in the growing oxide. Even if the metal impurities are not large enough to cause holes to occur in the oxide, during anodization high current flow will occur at the weak points in the oxide produced by the imperfections in the surface and this will heat up the oxide and cause crystallization. This is illustrated in Fig. 6 in the case of a tantalum surface which has not been

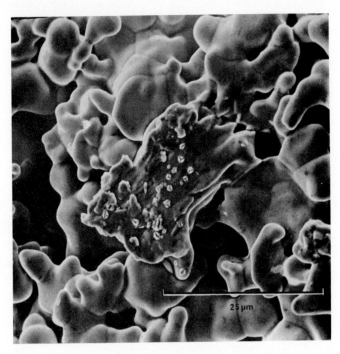

FIG. 6. Effect of impurities in tantalum surface on growth of tantalum oxide. (By permission of the Plessey Company Ltd.)

sufficiently purified prior to anodization. The crystallization of the oxide during anodization can be clearly seen (Jackson, 1972). Thus the metal substrate used must be pure; in the case of aluminium foil 99·99% purity material is used; in the case of tantalum a pure surface is often obtained by heating the surface to a high temperature in a vacuum.

The type of substrate used is often flat, but etched aluminium, etched tantalum and sintered tantalum blocks are widely used in order to increase the surface area and hence reduce the material cost of devices made with these materials. In the case of etched aluminium and tantalum, the etching is done electrochemically and Fig. 7 shows an example of etched aluminium. Porous tantalum powder compacts can be prepared by pressing the tantalum powder into suitable shapes and sintering at 2000 °C. The sintering has the second advantage of purifying the surface as noted above, and an example of a

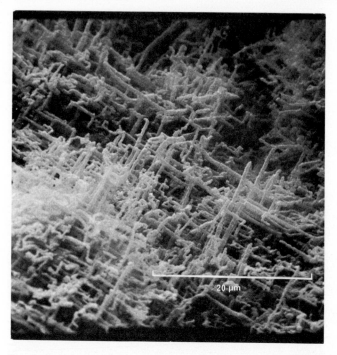

FIG. 7. Etched aluminium surface as used in an aluminium electrolytic capacitor. Replica obtained by anodization subsequent to etching followed by chemical removal of remaining aluminium. (By permission of the Plessey Company Ltd.)

sintered block is shown in Fig. 8. Figure 9 shows a similar block but in this case the tantalum has been anodized and then the compact has been broken open to show the oxide on the metal particles.

The fact that etched aluminium and sintered tantalum can be successfully and evenly anodized illustrates one of the major advantages of anodization; the surfaces can be attacked and an oxide layer deposited even if the surfaces are highly convoluted. This is sometimes described by saying that the "throwing power" of an electrolyte used for anodization is extremely high.

IV. OXIDE

As has already been briefly mentioned under Section III, the oxide obtained is amorphous; there is no long-range order in the atomic structure (e.g. studies on Al_2O_3 by Stirland and Bicknell, 1959). The electron diffraction patterns obtained will be very diffuse showing no definite lines. The most satisfactory structure studies of this type of material are usually performed

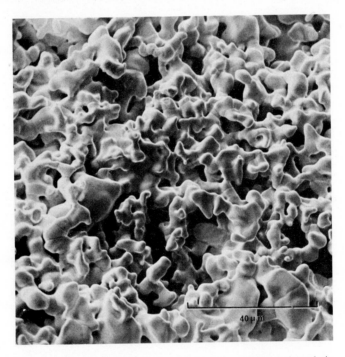

FIG. 8. Sintered tantalum block as used in a tantalum electrolytic capacitor. (By permission of the Plessey Company Ltd.)

in terms of radial distribution analyses which express the probability of another atom being at a given distance from a reference point (c.f. the work performed on amorphous silicon, Coleman and Thomas, 1967).

Thicknesses of oxide obtained will usually be in the range which shows pronounced interference colours and these are especially vivid on materials such as tantalum.

Electrical conduction through amorphous oxide films at low fields will be by an electron hopping mechanism (Jonscher, 1967), and impurities in the film are found to be relatively unimportant as the electron trapping sites for

conduction are due to the amorphicity of the structure. The electronic structure of such films has been examined in detail in terms of features such as the
forbidden band gap (Harrop and Campbell, 1968, 1970), the energy band
bending at the oxide-metal interface (Simmons, 1970) and the temperature
behaviour (Cockbain and Harrop, 1968; Harrop and Campbell, 1970). The
electrical break-down has also been examined (Klein and Gafni, 1966;
Harrop and Campbell, 1968) by several authors. Information on the electrical

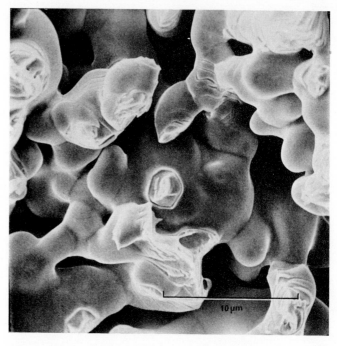

FIG. 9. Sintered tantalum block showing oxide prepared by anodization. (By permission of the Plessey Company Ltd.)

properties of anodized films is contained in several recent reviews (Harrop
and Campbell, 1970; Simmons, 1970; Campbell and Morley, 1971; Dell'Oca
et al., 1971). Typical electrical results on alumina are that the permittivity is
8, the band gap is 12 eV, the loss under low field (< 100 kV/cm) conditions
is 2%, temperature coefficient of permittivity is approximately + 1000 p.p.m.
and the breakdown field around 5 MV/cm.

The mechanical properties of anodic films assume importance particularly
with regard to the stresses obtained in the film during growth (Young, 1961;
Campbell, 1970b). These stresses define the stability of the layer. Since it is

found that the oxide usually occupies a larger volume than the parent metal, a compressive stress would generally be expected in a film. Normally such a film will be stable. However, if the compressive stress is too high, buckling will occur with the film being torn from the substrate. Tensile stresses are sometimes obtained and films with such stresses will very often crack or craze either during preparation or use.

V. Uses of Anodized Materials

There are three main uses of coating prepared by anodization. Firstly, they may be used as protective coatings. For such an application coherent films are required so that subsequent corrosion is difficult. Secondly, they may be used for decorative purposes. If porous structures are prepared (Wood et al., 1968), then the pores can be filled with dyes of various colours to give pleasing effects. The porous structure can be sealed to prevent continuous corrosion, by boiling the system in water (O'Sullivan and Wood, 1969). Finally, oxide films prepared by anodization can be used as electrical insulating layers. As the dielectric layer in electrolytic capacitors, oxide films on aluminium or tantalum obtained by anodization have obtained considerable importance (Campbell, 1971). Such layers are also used in microcircuit applications, two examples being the use of anodized layers in tantalum thin film circuits (Gerstenberg, 1970) and the anodization of silicon in integrated silicon circuitry (Schmidt et al., 1968). In the former example, tantalum films are anodized to give capacitors but it should also be noted that tantalum nitride resistors are trimmed to value using anodization techniques (Schwartz and Berry, 1964a). The anodization of the resistors can also provide excellent protection against moisture (Schwartz and Berry, 1964b).

VI. Other Related Growth Techniques

A. Gaseous anodization
(Jackson, 1967; Campbell, 1970a; Dell'Oca et al., 1971)

It has been found that it is possible to replace the liquid of the electrolyte by a low pressure gas discharge (see Fig. 10). This arrangement, which is usually run at a pressure of approximately 10^{-2} torr (1·3 Pa), has been found to give anodized layers on metals. A non-reactive discharge anode must be used in the experimental set-up so that all the voltage in the anodizing circuit is dropped across the oxide and not on any oxide formed on the discharge anode. The anodization constant is usually found to be higher than with conventional wet anodization and this is thought to be due to the higher temperature of the anode in the discharge conditions.

B. *Plasma oxidation* (Ligenza, 1965; Campbell, 1970a)

Ions created by an r.f. discharge in an oxygen pressure of between 0·1 torr (13·3 Pa) and 1 torr (133 Pa) can be used to cause anodization. The metal is made the anode of a 50 V system so that the negatively charged oxygen ions bombard the surface. Whether such a system gives true anodization is, however, still under debate.

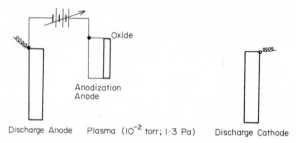

Fig. 10. Schematic diagram of the system used for gaseous anodization.

VII. Conclusions

This paper has attempted to give a broad review of the preparation and properties of films prepared by anodization. The basic concepts have been considered and this has been followed by a summary of the effect of the substrate on the growth of films. The structure of the oxide obtained by anodization has then been briefly examined and this has been followed by a summary of the uses of anodized materials. Finally, other related growth techniques have been considered, namely gaseous anodization and plasma oxidation.

REFERENCES

Campbell, D. S. (1970a). *In* "Handbook of Thin Film Technology" (L. I. Maissel and R. Glang, eds.) pp. 5.1–5.25. McGraw-Hill, New York.
Campbell, D. S. (1970b). ibid., pp. 12.35.
Campbell, D. S. (1971). *Rad. and Elec. Eng.* **41**, 5–16.
Campbell, D. S. and Morley, A. R. (1971). *Rep. Progr. Phy.* **34**, 283–368.
Cockbain, A. G. and Harrop, P. J. (1968). *Brit. J. App. Phy.* **18**, 1109–1115.
Coleman, M. V. and Thomas, D. J. D. (1967). *Phy. Stat. Solid* **22**, 593–602.

Davies, J. A., Dameif, B., Pringle, J. P. S., and Brown, F. (1965). *J. Electrochem Soc.* **112**, 675.

Dell'Oca, C. J., Pulfey, D. L. and Young, L. (1971). *In* "Physics of Thin Films" (M. H. Francombe and R. W. Hoffman, eds.) pp. 1–79. Academic Press, New York and London.

Evans, U. R. (1958). "An Introduction to Metallic Corrosion". Edward Arnold, London.

Gerstenberg, D. (1970). *In* "Handbook of Thin Film Technology" (L. I. Maissel and R. Glang, eds.) pp. 19.1–19.36. McGraw-Hill, New York.

Harrop, P. J. and Campbell, D. S. (1968). *Thin Solid Films* **2**, 273–292.

Harrop, P. J. and Campbell, D. S. (1970). *In* "Handbook of Thin Film Technology" (L. I. Maissel and R. Glang, eds.) pp. 16.1–16.36. McGraw-Hill, New York.

Jackson, N. F. (1967). *J. Mat. Sci.* **2**, 12–16.

Jackson, N. F. (1972). Private communication.

Jonscher, A. K. (1967). *Thin Solid Films* **1**, 213–234.

Klein, N. and Gatri, H. (1966). *I.E.E.E. Trans. Elec. Devices* **ED13**, 281–289.

Ligenza, J. R. (1965). *J. App. Phy.* **36**, 2703–2707.

O'Sullivan, J. P. and Wood, G. C. (1969). *Trans. Inst. Metal Finishing* **47**, 142.

Schmidt, P. F., O'Keefe, T. W., Oroschink, J. and Owen, A. E. (1968). *J. Electrochem. Soc.* **112**, 800.

Schwartz, N. and Berry, R. W. (1964a). *In* "Physics of Thin Films" (G. Hass and R. E. Thun, eds.) Vol. 2, pp. 398. Academic Press, New York and London.

Schwartz, N. and Berry, R. W. (1964b). ibid., p. 397.

Simmons, J. G. (1970). *In* "Handbook of Thin Film Technology" (L. I. Maissel and R. Glang, eds.) pp. 14.1–14.50. McGraw-Hill, New York.

Stirland, D. J. and Bicknell, R. W. (1959). *J. Electrochem. Soc.* **106**, 481–485.

Whitton, J. L. (1968). *J. Electrochem. Soc.* **115**, 58.

Wood, G. C., O'Sullivan, J. P. and Vaszko, B. (1968). *J. Electrochem. Soc.* **115**, 618.

Young, L. (1961). "Anodic Oxide Films". Academic Press, New York and London.

Applications of electrodeposition and anodizing

H. SILMAN

Oxy Metal Finishing (Great Britain) Ltd., Sheerwater, Woking, Surrey, England

|---|---|---|
I.	Introduction	100
II.	Continuous coating of strip	100
III.	Chromium plating	102
IV.	Zinc, cadmium, lead and copper	103
V.	Electrophoretic painting	104
VI.	Anodizing	104

I. INTRODUCTION

The applications of electrodeposition can be divided into four main groups:

(1) Decoration: improving the appearance of base metals by coating them with a more expensive one. This includes articles such as imitation jewellery, furniture fittings, builders' hardware and tableware.
(2) Protection: applying a corrosion-resisting deposit to increase the life of another metal. Chromium plating of automobile parts and domestic appliances, zinc and cadmium plating of nuts, screws and electrical components are included in this category.
(3) Wear resistance: using a coating to make products withstand severe mechanical conditions longer, and to repair worn parts, e.g. the plating of bearing surfaces and the repair of worn shafts and journals by nickel or chromium deposition.
(4) Electroforming: the production of articles by electrodeposition instead of by conventional metal forming operations, e.g. the manufacture of sieves, screens, dry shaver heads, record stampers, moulds and dies.

II. CONTINUOUS COATING OF STRIP

One of the largest uses of electrodeposition is for the application of tin onto steel strip in the manufacture of tinplate. Virtually the entire world production is made in this way on high speed continuous lines operating at speeds

FIG. 1. Continuous electrotinning line. (Courtesy of British Steel Corporation.)

of up to 100 m/s. The advantages of electrodeposition here lie in the fact that the tin coating thickness on the steel strip can be much more readily controlled than in the earlier hot dipping process, whilst coatings of differing thickness can be applied to each side of the strip. The deposit is matt as plated, and is brightened by heating the strip momentarily to just above the melting point of the tin. A typical plant is shown in Fig. 1. Electrogalvanized steel is also produced by a continuous process, chiefly as material which is to be subsequently painted.

Electrodeposition is now an inseparable part of the electronics industry where it is used for the production of printed circuits by the deposition of copper, gold and other metals onto photographically reproduced conducting patterns. Solder plating is also carried out from a lead-tin solution on printed circuits and electronic components to facilitate soldering. Precious metals, such as gold, platinum and rhodium are applied to prevent tarnishing.

III. Chromium plating

One of the most important uses of electroplating is on automobiles, where chromium plating, with an undercoat, usually of nickel, provides the bright finish on both internal and external die-cast zinc alloy, and steel components. The conditions of service of such articles are often very severe, and a good deal of research has been carried out in recent years on methods of improving the corrosion resistance of the finish. The main causes of failure are corrosion of the underlying nickel by chlorides from salt put down on roads to melt snow, and by sulphur oxides in the atmosphere. The attack takes place through cracks and pores in the thin chromium deposit applied over the nickel to protect it from tarnishing. The presence of the chromium, which is more noble than the nickel beneath, results in an increased local rate of electrochemical attack at the pores, so that corrosion rapidly penetrates through to the base metal, and the entire deposit breaks down.

One approach to the problem of improving the corrosion resistance, which is now used extensively, is to apply the chromium in a highly microporous or microcracked form over the bright nickel underlayer; this tends to equalize the anodic and cathodic areas, so that the rate of attack is slowed down. Microporous chromium is produced by applying a thin deposit of a special nickel containing a large number of small non-conducting particles over the principal nickel layer. The chromium deposited over this nickel thus becomes highly microporous. Microcracked chromium performs similarly, and is generally applied from a special bath which produces a highly stressed deposit; this cracks spontaneously during the plating process. Figure 2 shows two automatic plating plants for automobile components.

FIG. 2. General view of automatic plating plants for automobile components. (Courtesy of British Leyland Motor Corporation and Oxy Metal Finishing (GB) Ltd.)

IV. ZINC, CADMIUM, LEAD AND COPPER

Zinc and cadmium plating are extensively used for the protection of steel articles against corrosion. The deposits are anodic to the underlayer, and thus act sacrificially in protecting the underlying metal. Cadmium is much more expensive than zinc, and is more protective than the latter in marine atmospheres, whilst zinc performs better in industrial atmospheres. Cadmium is also more readily solderable, and has a better appearance. However, the introduction of methods of plating zinc in a bright form has resulted in a tendency for cadmium plating to be replaced by bright zinc.

Both zinc and cadmium deposits are usually given a chromate treatment after plating to delay the formation of "white rust" as a result of attack on the zinc by the atmosphere.

Lead deposits are applied onto brass and steel, usually from fluoborate solutions, to provide protection against acids.

Copper is plated onto steel and zinc base alloys as an undercoat for nickel, and also for decorative purposes. By treatment of the deposit with a sulphide solution, the well known "oxidized copper" or bronze finish is obtained on nameplates, domestic fittings and the like. Copper deposition is also used for making electrotypes in printing, and for electroforming such articles as wave guides in the electronics industry.

V. Electrophoretic painting

Another new use of electrodeposition is for the application of paints by electrophoresis. Here the paint is in the form of an aqueous suspension, the paint particles being deposited by means of a potential applied between the work and a counter-electrode. The advantage of the process is that the paint is applied uniformly over the entire surface of the metal, including recessed areas and the interiors of box members. After removal from the processing tank, the paint is rinsed free from electrolytes, dried and cured by stoving. Automobile bodies are now being primed in this manner on a substantial scale. Also in its early stages, is the use of electrophoretic methods for the application of porcelain enamel frits prior to stoving.

VI. Anodizing

Anodizing is, in practice, almost entirely used on aluminium and its alloys; although processes have been developed for other metals (e.g. zinc) they are little used. The process is carried out by making the articles to be treated anodic in a suitable electrolyte, when an insulating, tough oxide film is produced. Anodized aluminium is highly corrosion-resistant, whilst the coating can also be coloured by dyeing if required. Much longer lasting colours can be produced by using special aluminium alloys containing elements which form intrinsically coloured compounds on anodizing. Such alloys are being increasingly used for architectural purposes, although the colours obtained tend to be muted tones. Aluminium strip is anodized and coloured continuously, the product being used for a variety of purposes, from ash trays to containers and panelling.

The basic principles of diffusion coating

Richard L. Wachtell

Chromalloy Research and Technology Division, Orangeburg, New York, U.S.A.

I. Introduction

In recent years there has been a sharp rise in the "Popularity Rating" of diffusion coatings among the technical groups concerned with such matters. This increase is in some instances merely the result of a fashion change; diffusion coatings are an "in" system at the moment. But there is as well, a cluster of technical considerations which characterize diffusion coatings and make them especially appropriate for certain groups of applications.

This paper will undertake to define those essential characteristics of diffusion coatings which separate them from coatings of other sorts, consider the major elements involved in their design and development, and provide some examples of their use.

Let us for the moment consider a "coating" as a variation in chemistry or morphology of the outer layers of an object from its substrate as a whole. Using these terms, a diffusion coating is one in which this variation is developed utilizing the influence of a diffusion process. This implies an interaction between coating and substrate on a physical/chemical level, subject to the usual laws of diffusion, phase diagram restraints, or other thermodynamic considerations which may exist in the system under review.

As some poet before me has already said, "It ain't paint". The unique aspect of a diffusion coating is the fact that the substrate is incorporated as a component part of the coating, and is reacted with or dissolved into it.

In consequence, diffusion coating systems are only partially defined by describing what is deposited (or withdrawn!). The properties of the coating are equally dependent on the substrate composition. This concept, by the way, is surprisingly little understood even in the technical community which designs for specific diffusion coating systems. It is quite common commercial practice for coating specifications to be transferred verbatim from one substrate to another, when, for design, economy or castability reasons, the substrate alloy is changed.

This is a kind of technical Russian Roulette—often one gets away with it.

Let us consider now what factors are involved in the formation of diffusion coatings, especially those which make them unusual among the general families of coatings—the things which make them "not paint"!

A Russian (Gorbunov, 1958) translated by three Israelis (Friedman, Artman, Halprin) into English, gives a definition of diffusion too delightful to refrain from quoting:

"Translocation of atoms, caused by their thermal mobility, makes possible a diffusion process in metals and alloys. Those atoms, the periodical displacements of which, under the influence of thermal oscillations, attain significant amplitudes, may leave their places in the crystal lattice, and lose their energy excess. The vacant places thus formed in the crystal lattice may be occupied by the atoms of the basic metal, or by the atoms of the diffusing substance."

Diffusion coatings, then, become diffusion coatings by virtue of alloy formation between the coating material and the substrate under the influence of heat. It is necessary only to bring together a substrate and a coating material with which it can (in the metallurgical sense) form a system, and heat. The rest is detail! Sometimes the operation is indeed just that simple. Mercury rubbed on silver will produce an amalgam which is a diffusion coating—even room temperature providing sufficient "thermal mobility" for diffusion in this case.

There are more useful coating systems, with which many of you may be familiar, employing diffusion either in whole or in part. Hot dip galvanizing and aluminizing, nitriding, carburizing (both pack and gas techniques) all utilize diffusion principles. A number of these are described elsewhere in these Proceedings.

II. PACK CEMENTATION COATINGS

This discussion will centre about pack cementation coatings, one of the oldest known systems, and yet one which, in recent years, has seen amazing new growth and technology.

In these systems, the work to be coated (the substrate) is immersed in a powdered compound containing the coating substance(s). Usually this is done in a retort, but fluidized bed versions of the process are also known. The entire charge is heated in the presence of a gas which is capable of reacting with and transporting the coating substance in the vapour stage. This gas (the halides are a versatile family, widely used for this purpose) may be introduced either by piping into the retort as a pure gas, or by mixing as a vaporizable halogen compound with the pack itself (ammonium halides are favourite choices).

In the semi-closed retort system which are probably most popular in today's practice, the sequence which follows on heating the charge is roughly as follows:

The ammonium halide decomposes, and as a vapour the nitrogen and hydrogen act to purge the retort of air and produce reducing conditions within it. The halogen reacts with other constituents in the compound to form metal halides. As heating progresses, these halides vaporize and a continuous exchange reaction among work, atmosphere and compound occurs. Consider the chromizing of iron using ammonium iodide mixed with the compound as a halogen source. With heat, the chromium powder of the compound (Cr) reacts with the vaporized ammonium iodide:

$$Cr(solid) + 2NH_4I(gas) \rightarrow CrI_2(gas) + 2NH_3(gas) + H_2(gas)$$

The chromous iodide then reacts with the iron workpiece in one or more ways as indicated below and diffuses into it.

(1) Exchange reaction:

$$Fe + CrI_2(gas) \rightarrow FeI_2(gas) + Cr(into\ work)$$

(2) Reduction reaction:

$$H_2(gas) + CrI_2(gas) \rightarrow 2HI(gas) + Cr(into\ work)$$

(3) Decomposition reaction:

$$CrI_2 \rightarrow 2I(gas) + Cr(into\ work)$$

Hydrogen (reaction 2) becomes available in large volumes when the NH_3 radical "cracks" at higher temperatures. It has also been suggested that the metallic surface of the work catalyses reaction 3.

Depending upon whether iodides, bromides, chlorides or fluorides are the chosen halide, and depending upon the temperature chosen for the reaction, either reaction 1 or 2 may predominate. Samuels (1956) contends that exchange reactions predominate when iodine is used and reduction in the case of fluorine, with bromine and chlorine intermediate.

It will be observed in the case of reaction 1, that iron(ous) iodide is formed as a gas at the surface of the work. In the retort, this reaction reverses at the surface of the chromium compound to replace FeI_2 with fresh CrI_2 and the reaction proceeds. Diffusion brings fresh Fe to the surface of the work continually, and fresh Cr to the surface of the compound. Under these conditions, the "work" develops a chromium rich coating, and the "compound" an iron rich coating—indeed, which is work and which compound depends largely on one's point of view!

Many metals can be transferred by this technique, as can some non-metallics and metalloids (carbon, silicon, boron).

III. METALLURGICAL CONSIDERATIONS

A. Binary systems

The criteria for developing coating characteristics are similar to those which would be employed in alloy development. The phase relationship of the substrate and coating substances therefore assume paramount importance in predicting the nature of a given coating/substrate system. In a classic paper, Dr. Rhines (1940) illustrated the interpretation of such coatings in the light of diffusion couples. Ranging from pure "a" on one side to pure "b" on the other, such couples display the complete phase diagram of "a"/"b" through the diffusion zone. Proper allowances must, of course, be made for the temperature at which diffusion is accomplished, and the cooling rate from the diffusion temperature when the observed structures are interpreted. In coating practice, the teachings of Rhines may be followed by viewing one arm of the diffusion couple as substrate, the other as coating, and recognizing that under specific diffusion conditions, the "coating" side of the couple frequently never reaches the 100% composition level. This latter is especially likely where diffusion rates are high and supply rates of the inward diffusing element(s) are low.

For those systems where complete mutual solid solubility is the rule, e.g. chromium/molybdenum or nickel/copper, a smooth concentration gradient curve is established through the coating, the precise slope of which is determined by the diffusion coefficients of the material/temperature parameters under consideration. Where, as is perhaps more usual, intermetallic compounds may be formed in the coating/substrate complex, sharp breaks will be found in the concentration gradient curves coincident with the appearance of the various intermetallic zones (hopefully as predicted by the phase diagram and viewed in the microscope!).

An instructive example for consideration is the iron/chromium system

(Fig. 1). At temperatures between 820 °C and 850 °C, alpha iron and chromium exist in complete mutual solid solubility. Diffusion of Cr into Fe at 830 °C may be expected to exhibit a steep but smooth concentration curve, with multiphase zones in the 25% to 60–65% Cr region developing on cooling.

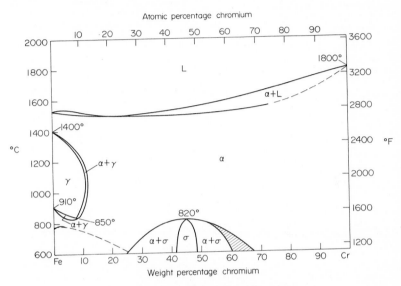

FIG. 1. The phase diagram iron/chromium.

The curve is expected to have a steep slope because diffusion rates at these temperatues are low, the diffusion coating will be thin (for normal coating times), and a fairly high Cr content should exist in the outer zone of the coating. In practice, sigma phases are not found, despite the prediction of the phase diagram, because they form so sluggishly in pure alloys (in stainless steels or when working with impure materials, they may be observed). Figure 2 is a microprobe trace of the chromium concentration resulting from diffusion for 20 h at 830 °C.

Let us now raise our diffusion temperature to 1000 °C and consider the results. At this diffusion temperature, our substrate has transformed to gamma iron. It will accept up to 12–13% of alpha chromium in solution, at which point the chromium, a ferrite stabilizer, causes the crystal habit to revert back to the alpha condition, in which iron and chromium remain mutually soluble up to 100% Cr. The concentration gradient accompanying such a diffusion temperature will show a smooth curve up to the 12–13% chromium level, where a sharp break occurs (Fig. 3).

RICHARD L. WACHTELL

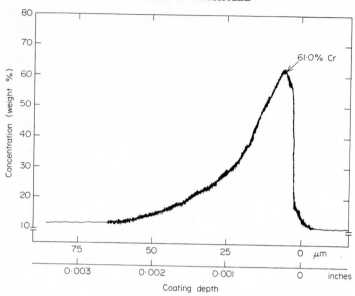

FIG. 2. Microprobe trace of chromium concentration resulting from a diffusion at 830 °C.

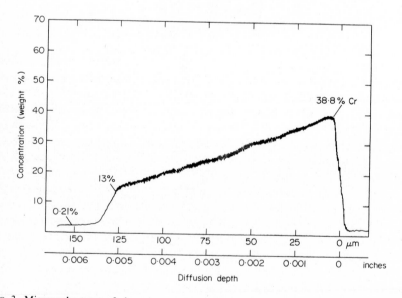

FIG. 3. Microprobe trace of chromium concentration resulting from a diffusion at 1000 °C.

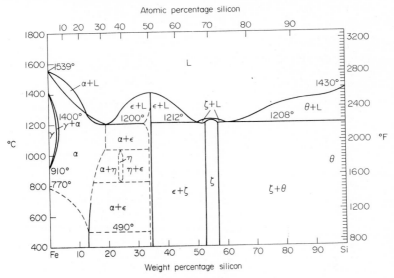

FIG. 4. The phase diagram iron/silicon.

Here a sudden drop in chromium level occurs because the diffusion rate of chromium in gamma iron is much lower than its rate in alpha iron. In this situation, the rate controlling mechanism is the diffusion rate of chromium into gamma iron. At this point the metallurgist raises his technical eyebrows. It would appear possible to increase the rate of case depth formation by insuring a ferritic (alpha) structure in the substrate. The obvious approach, to work at 1400–1500 °C (above the gamma loop), is technically sound, but not practical. Excessive grain growth damages the substrate, and the temperature is too high for normal retort and furnace materials. However, another approach is possible. Let us consider for a moment the Fe/Si phase diagram (Fig. 4).

Silicon, as is well known, is also a strong ferrite stabilizer. As little as 3% will stabilize alpha iron at 1000 °C. Its diffusion rate into iron is very high. It would appear that small amounts of silicon, made to precede chromium into iron during the process, may accelerate case depth formation. This turns out indeed to be the case. Hence, additions of silicon to chromium are common practice during chromizing of iron.

B. Multi-component systems

Unfortunately, in commercial practice, we seldom deal with simple binary systems. The substrate will, usually, have more than one component. Except for the occasional chromium coating diffusion of magnetic components made of pure iron, low to medium carbon steel is the usual substrate chromized (i.e. chromium diffusion treated). The presence of even small amounts of carbon in the substrate alter the nature of the coating radically. Because

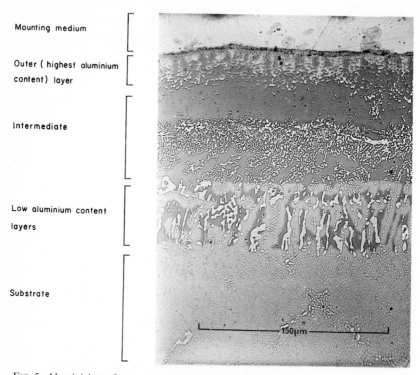

Mounting medium

Outer (highest aluminium content) layer

Intermediate

Low aluminium content layers

Substrate

150μm

FIG. 5. Aluminizing of a complex nickel-based superalloy produces this structure, in which chromium aluminides and molybdenum aluminides appear as white inclusions in a nickel aluminide (NIAL) matrix.

the heat of formation of the compound is quite high, stable chromium carbides (containing a modest percentage of dissolved iron) become the predominant phase of the coating. The carbon necessary for formation of these carbides is extracted by outward diffusion to the surface, from the core of the substrate itself. As can be appreciated, the carbide coating developed on steel will have entirely different characteristics from the alloy coating developed on iron. It will have, for example, a Vickers hardness number of about 2600, as compared with less than 200 for pure iron. It will be totally brittle as compared with elongation capabilities of 40–50% for the coating on pure iron. It will be non-magnetic compared with the ferro-magnetic surface developed on pure iron. This despite the fact that a steel and an iron part may have been coated simultaneously, the two packed side by side in the same retort, reactor chamber, or diffusion furnace when they were coated.

The iron/chromium/carbon coating system discussed has been chosen here for its simplicity. Where the substrate is a truly complex one, such as for instance, a superalloy containing cobalt, chromium, nickel, carbon, columbium (niobium), molybdenum and tungsten and we diffuse into it aluminium coupled, possibly, with silicon, the results will be very complicated indeed. Figure 5 illustrates such a typically complex structure. Aluminide intermetallics invariably form with the substrate components, precise partition being dependent upon both the energy relationships of the aluminides forming, and the amount available for combination in the substrate. The effect on the substrate is not dissimilar to that described earlier for carbon in the chromizing of mild steel, and unless steps are taken to prevent it, "dealloying" of a surface layer of the substrate may accompany coating.

IV. RATE CONTROL OF CEMENTATION SYSTEMS

Important in the case of aluminium pack cementation of superalloys, another parameter, that of rate of supply of the coating material, must be taken into account. We have loosely termed this characteristic "activity". A "high activity pack" provides ready and rapid transfer of unlimited amounts of aluminium for diffusion. Low activity packs deliver aluminium at sharply reduced rates, "metering" the supply as it were. Since the rate of diffusion of aluminium inward, and of substrate atoms outward, is affected by the temperature at which the operation is conducted, the concentration of aluminium (or other inward diffusing element) at the surface is controllable. Figure 6 shows schematically the variation which may result from compounds having differing "activity" or potential. Figure 7 shows actual microprobe traces of the coatings developed from two such packs.

Although a number of methods may be utilized, a most convenient method for controlling pack activity is through the device of combining the diffusing element (in this case aluminium) with some other substance which can act as a buffering agent. Convenient for the purpose in this instance are metals which form aluminides. Chromium and nickel, both of which form extremely stable aluminide intermetallics, will act in aluminium-containing compounds to buffer very strongly, and generate very low activity packs. Unbuffered pure aluminium packs represent the other extreme of maximum activity.

What are the practical consequences of this phenomenon; what can we accomplish through this available finesse in control of the surface chemistry of the coatings?

Firstly, coatings of higher aluminium content generally speaking have higher oxidation resistance than the more dilute systems. Where the intended use of the system involves high temperature oxidation exposure, it will be understood that degradation of the system occurs in two ways. One is by

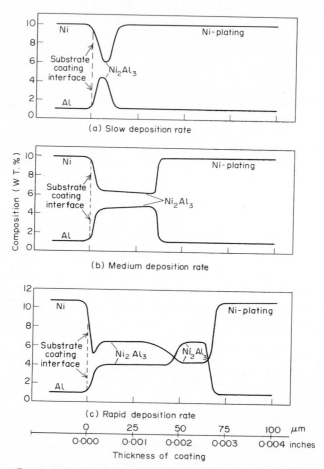

Fig. 6. Effect of pack aluminizing potential on coating chemistry.

direct oxidation of surface aluminium. The other is by gradual solution—dilution if you will—of the coating elements into the substrate. Obviously, the higher the aluminium content of the coating, the greater the reservoir of oxidation-conferring material, and hence the longer the life which may be anticipated.

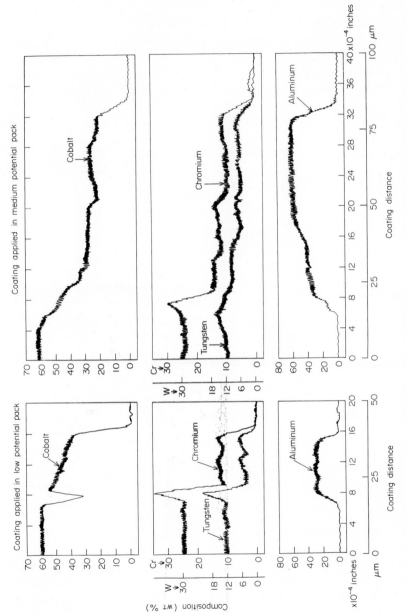

FIG. 7. Composition profiles of aluminium coatings applied to WI-52 (a cobalt-based superalloy) at 1800 °F (982 °C).

It is the engineers' particular fate, however, to be haunted by the "Daemon of the Perverse", and therefore the obvious choice of the highest possible aluminium content carries concomitant disadvantages. For one, increasing the aluminium content above a certain level sharply reduces the melting point of the system.

Secondly, where the coating formed is an aluminide or other intermetallic, its ductility is inherently low. Spontaneous cracking can occur from stresses

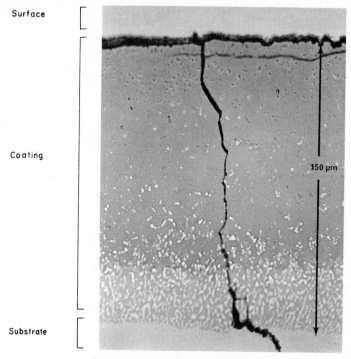

FIG. 8. A surface crack induced by thermal stress.

resulting from difference in thermal expansion coefficient and which develop during cooling from the deposition temperature. Figure 8 shows such a crack, exaggerated in this instance by the abnormal thickness (about 150 μm) of the case.

However, a measure of control of case ductility is achievable by selection of the specific surface intermetallic phase (Fig. 6) and by even further adjustment of the stoichiometry within a given intermetallic. Figures 9 and 10 are impact tests on a wedge impact bar showing the order of deformation which these optimized coatings can tolerate. Such modifications are consciously undertaken at sacrifice of *some* oxidation resistance and as a deliberate trade-off of properties. For unusual specific performance requirements, this compromise may not be indicated.

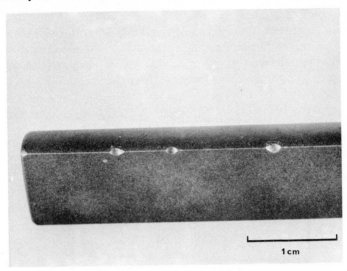

FIG. 9. Cracking limits as measured by drop impact test using a carbide tup. The two outer indents show failure at 30 inch lbs (3·4 J) of energy. No failure at 20 inch lbs (2·3 J)—centre indent.

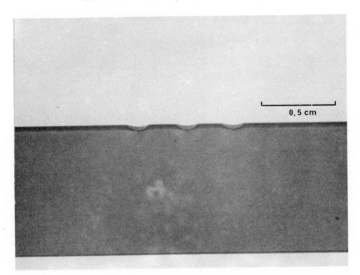

FIG. 10. Side view of indents at 20 inch lbs (2·3 J), illustrating radius of tup.

V. Conclusions

The family of coatings which may be diffusion generated is a very broad one, ranging from arsenic to zirconium and the instances treated in this review have been chosen to illustrate general approach and philosophy. Some basic properties have been defined, and a working definition offered with the admonition that diffusion coatings, because of "translocation of atoms caused by their thermal mobility", "ain't paint". If you have this much firmly fixed in your mind, you will have more than many!

REFERENCES

Gorbunov, N. S. (1958). "Diffuse Coatings on Iron and Steel" Acad. of Sci. of U.S.S.R., Moscow. (Translation by Friedman, Artman, and Helpren (1960) for the Israel Program for Scientific Translations, National Science Foundation and Dept. of Commerce, U.S.A.)

Rhines, F. N. (1940). *Am. Soc. for Metals*, October.

Samuels, R. (1956). Private communication.

Diffusion coating metals in molten fluorides

N. C. COOK

*Organic Chemistry Branch, General Electric Company, Research and
Development Centre, Schenectady, New York, U.S.A.*

ABSTRACT

The scope and technique of using molten alkali and alkaline earth fluorides
as solvents for electrolytically forming diffusion coatings on metal surfaces
will be described. The process, called "metalliding", encompasses the interaction of nearly fifty metals of the periodic table, ranging from lithium to
uranium. Several hundred combinations, not including alloy variations,
have been made using approximately twenty five different metals as the alloying or "iding" agent (anodes) and thirty different metals as the substrates
(cathodes) on which the alloy surfaces are formed.

The process is operated between 500–1200 °C in metal vessels under inert
atmospheres, principally argon. The most useful fluoride solvents are those
of lithium, sodium, potassium, calcium and barium. These fluorides are not
corrosive to most nickel-based and iron-based alloys (vessel materials) under
inert atmospheres. The fluxing action, high boiling points and thermodynamic stability of the fluoride melts, make them uniquely suited as solvents
for forming diffusion coatings. The iding ion, corresponding to the anode
metal, is usually present in the salt, at 0·1 to 0·5 mole % concentration.

In general, the purity of the fluoride solvents must be very high for forming
good and reproducible diffusion coatings. A few processes can tolerate
considerable (1–5%) oxide concentration but, for many others, oxide concentration cannot be above a few parts per million. Impurity cations that
will be codeposited with the desired iding cation must not be present if they
are harmful. They can often be beneficial. Particles in the salt, especially
carbon, should always be avoided. Means of achieving these conditions will
be discussed.

Most of the metalliding reactions function as batteries and will proceed
by themselves if the anode and cathode are connected. Generally an external

e.m.f. is supplied to secure more uniform, and higher, current densities than those provided by the battery action.

The coatings can usually be formed with quantitative coulombic efficiency and a high degree of thickness control, from fractions of a micrometre to many micrometres in thickness. The coatings are comprised of solid solutions, intermetallic compounds, and mixtures of solutions and compounds. Multiple phases of solid solutions or different compounds are frequently formed in successive layers, depending on the system. Often, however, a particular compound has so much more stability than other compounds that the coating will be essentially one compound through several micrometres of thickness.

Most of the coatings are formed in 25–75 μm thickness in two to three hours. Some, however, develop very rapidly forming coatings that are several micrometres in thickness in only a few minutes, while others are very slow and take two to three days to develop only a few micrometres. With one known exception, increasing the temperature has always speeded up the process. The compositions that are formed at the higher temperatures often have different, and sometimes less desirable, properties than those formed at a lower temperature. Increasing the salt temperature 100 °C will usually give a two- to ten-fold increase in diffusion rate. As the temperature approaches the melting point of the substrate metal, or of the alloy surface being formed, the diffusion rate usually increases very rapidly.

The property changes that result, are many and often very beneficial. They include improved hardness and erosion resistance, lower coefficient of friction, less galling and seizing tendency, improved resistance to chemical and high temperature corrosion, greater modulus of elasticity, better bonding properties to glasses and ceramics, higher strength, changes in reflectivity and emission properties, and improved appearance.

Since metalliding reagents must usually be more reactive electrochemically than the metals being treated, each reagent has a specific group of metals and alloys on which diffusion coatings can be formed when solubility or compound formation exists. The more active metals have larger scopes while the less active metals have correspondingly smaller scopes. The most useful metalliding reactions appear to be berylliding, boriding, aluminiding, siliciding, titaniding, zirconiding, chrominiding, tantaliding, and yttriding. Specific details for operating these processes and the properties of the coatings will be given.

Surface treatments and possibilities for their industrial use

J. J. Caubet and C. Amsallem

Hydromecanique et Frottement, Rue Benoit Fourneyron, 42 Andrezieux-Boutheon, France

I. Introduction

There has been considerable research concerning the various phenomena which occur in the contact zone between two components in frictional contact, and the authors have been working in this field for several years (Centre Stéphanois, 1968). Consequently it has been possible to establish rules for the design of surfaces and surface treatments.

The outer layer should prevent the possibility of the surface welding to any other surface it may contact, and in order to do this it is often necessary actually to prevent metal-to-metal contact. Fortunately there is a wide range of surface layers which perform these functions, for example an absorbed film, oxides, lubricants, or a hexagonal compound of the parent metal, such as, in the case of ferrous components, iron sulphide. Once the type of outer layer has been chosen it is important to ensure that the optimum concentration of the constituents is deposited to develop the best operational life. It is also necessary to ensure that the outer layer is homogeneous to avoid stress concentration. Tensile stresses, such as can occur when one component is sliding over another with appreciable friction between them, can be very damaging; weak phase boundaries can be particularly detrimental. In an application where a material is likely to undergo any sort of deformation during operation, such as creep or running-in, the layer should be ductile to enable it to accomodate deformation of the substrate without breaking.

The outer layer need only be thin, for example a few micrometres or, in some circumstances, only a fraction of a micrometre in thickness.

II. APPLICATION OF MULTI-LAYER COATINGS—THE DELSUN PROCESS

Our work has shown that for components that are to be subjected to high stresses, the best results are achieved by applying a coating in which there are three distinct zones—an outer, very thin, weld-inhibiting zone; an intermediate zone that is both hard and ductile; finally, the innermost zone where the mechanical properties progressively decrease until similar to those of the substrate.

This article describes the "Delsun" process, one of the surface treatments based on this principle which has been developed by the authors (Esteveny *et al.*, 1970). This process provides protection against seizure and wear for copper alloys and in addition enhances resistance to corrosion.

The first stage in this treatment is to deposit the appropriate alloy on the surface and it is normally most convenient to do this by means of electroplating. The composition of the deposited alloy can be varied to some extent depending on the exact operating conditions. In the Delsun treatment, the deposited alloy contains antimony, tin, and cadmium. Where, as is often the case, a component is at least partially protected from environmental oxidation, such as when operating in oil or water, then the alloy deposited can contain a higher proportion of tin. However, where there is a danger of oxidation the proportion of cadmium should be increased.

After the alloy has been deposited the component is soaked at a temperature of 420 °C in a suitable salt bath or an inert atmosphere furnace. This causes diffusion at the boundary between the component and the coating, this diffusion giving the junction considerable strength and providing a smooth transition.

The total process time, including depositing and diffusing the coating, varies between 8 and 15 h. The total thickness of the deposited alloy is normally about 30 μm. When a component has been treated in this way the three characteristic zones described previously can be clearly distinguished at and near to the surface (Fig. 1). The composition and properties of these zones depend on the composition of the deposited coating.

The hard ductile intermediate zone is intended to resist wear caused by abrasion and ploughing by asperities at the surface of the mating component, and also to counter the risk of failure if excessive stress were to be accidentally applied. It is also a good conductor of heat. Should the outermost zone be worn away in any part, this intermediate zone will function reasonably satisfactorily if the component has to undergo further wear. The three zones are composed of metals and intermetallic compounds, for instance when a high proportion of tin is deposited during the Delsun treatment, the main constituents are Cu_2Sn and Cu_4Sn.

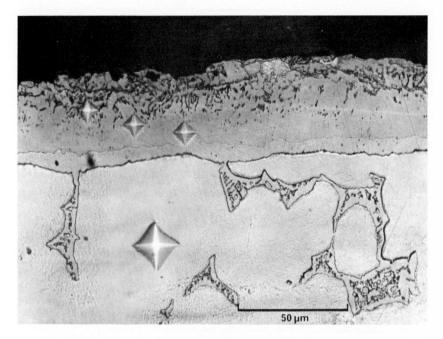

Fig. 1. Delsun layer on a bronze to specification UE 12 P. Etched electrolytically.

III. Performance of coated components

Figure 1 shows a microsection from a Delsun treated bronze part, the impressions visible being microhardness indentations. In this case the hardness of the surface layers is HV 480 (50 g load), whilst the surface layers of treated brass parts would have a hardness of about HV 600 (50 g load). The outer surface zone is too thin to be detected under an optical microscope, but electron microprobe analysis shows it to consist of tin or tin and cadmium depending on the type of alloy initially deposited.

Figure 1 also shows that there is no break between the innermost zone and the eutectoid constituent (delta).

The majority of copper alloys can be treated by the standard Delsun process. However in alloys containing more than 3% aluminium, 2% manganese or 4% nickel, special precautions are necessary when preparing the surfaces

TABLE I   Wear tests on untreated and

Type of test machine	Faville falex (Centre Stéphanois, 1968, pp. 15–22)	H.E.F. tribometer (Michalon and Gibert, 1972)
	Testpiece in Copper Alloy Jaws in case hardened EN 353 steel Speed 0·1 m/s	Plate in Copper Alloy Ring in case hardened EN 353 2 m/s
Tested material Bronze 88/12 untreated	Load: 100 kg Operating in water Scuffing after 28 s	Load: 12 kg Lightly greased Scuffing after 26 min

Bronze 88/12
  Delsun treated
	Conditions as above
	Scuffing after
	  1 h 50 min
		Conditions as above
		Surface still good
		  after 12 h 30 min

Bronze 88/12
  untreated
	Load: increasing
	Operating in oil
	Oil film rupture and
	  scuffing at 520 kg
		Load: 30 kg
		Operating in oil
		Scuffing after 36 min

Bronze 88/12
  Delsun treated
	Conditions as above
	Operating well at
	  machine limit of
	  2000 kg
		Conditions as above
		Surface still good
		  after 30 h

Brass (39% Zinc,
  2% Lead)
  untreated
	Load: 100 kg
	Operating in water,
	  scuffing after 13 s
		Load: 12 kg
		Lightly greased,
		  scuffing after 21 min

Brass (39% Zinc,
  2% Lead)
  Delsun treated
	Conditions as above
	Scuffing after
	  5 h 30 min
		Conditions as above
		Surface still good
		  after 15 h

Brass (39% Zinc,
  2% Lead)
  untreated
	Load: increasing
	Operating in oil
	Rupture of film
	  and scuffing at
	  500 kg
		Load: 30 kg
		Operating in oil
		Scuffing after
		  26 min

Brass (39% Zinc,
  2% Lead)
  Delsun treated
	Conditions as above
	Operating well at
	  machine limit of
	  2000 kg
		Conditions as above
		Stopped after 30 h
		  operation

Delsun treated copper alloys

Oscillation machine (Centre Stéphanois, 1968, pp. 247–299)	H.E.F. gear testing machine (Michalon, 1969)
Sleeve in Copper Alloy Shaft in case hardened EN 353	Driven Gear Copper Alloy Driving Gear hardened EN 353
0·04 m/s Frequency 1 Hz Projected Pressure 100 bars (10 MPa)	5·25 m/s Hertz Pressure 50 h bar (5 MPa)
Operating in water Lightly greased Scuffing after 150 h	Degreased Operating in air Scuffing after 1 min
Conditions as above Surface still good after 625 h	Conditions as above Scuffing after 26 min
Lightly greased Operating in oil Scuffing after 68 h	Operating in oil Scuffing after 10 min
Conditions as above Light scratching after 386 h	Conditions as above Surface still good after 12 h
Lightly greased Operating in water, scuffing after 136 h	
Conditions as above Slight scratching after 1650 h	
Lightly greased Operating in air Scuffing after 53 h	
Conditions as above Slight scratching after 350 h operation	

of parts to be treated and the environment where the diffusion heat treatment is carried out, must be carefully controlled. The degree of roughness of a surface before treatment should ideally be 1–3 μm c.l.a. (centre line average). Treated parts grow slightly, the exact amount of growth varying and depending upon the alloy being treated, but it is usually about 7·5 μm per surface, i.e. 15 μm on a diameter. The corrosion resistance and friction characteristics of treated parts are considerably better than those of untreated parts as shown in Table I which summarizes the results from a number of tests.

FIG. 2. Lubricated steel shaft after rotating in an untreated bronze bearing; the oil film broke down at a load of 250 kg.

The load required to destroy the oil film on a treated part is considerably higher than the load for an untreated part. For example, in the case of a lubricated steel shaft rotating in a bronze bearing at 900 r.p.m. the oil film broke down at a load of 250 kg (Fig. 2), but when the test was repeated with a Delsun treated bearing the load required to destroy the oil film was 650 kg (Fig. 3).

Significant improvements in performance are also obtained in components that are "lubricated for life". The performance of treated copper alloys rubbing together unlubricated in water is improved by a factor of up to one hundred. For example a certain type of water ball valve, made of brass with

FIG. 3. Lubricated steel shaft after rotating in a Delsun treated bearing; the oil film broke down at a load of 650 kg.

a nylon gasket, usually leaked between the brass and gasket after 5000 operations. However when the brass valve had been treated by the Delsun process there was no sign of leakage after 50,000 operations.

Treated parts are compatible with most materials. The following are listed in order of decreasing performance: hard chromium, hardened steel, struc-

TABLE II
Industrial applications of the Delsun process

Gear Boxes—synchronizers —bearing sleeves —forks	Plumbing—valves —flap valves —pump rods
Pumps—gears	Various—electrical components —die plates for stainless steel
General engineering—sleeves —rings —nuts —screws —segment gear —worm wheel	

tural steel and stainless steel. Table II shows some of the industrial applications of the Delsun process.

IV. OTHER COATING PROCESSES

Other processes which work on a similar principle have been developed by the authors (Centre Stéphanois, 1972). The "Stanal" treatment provides resistance to scuffing, wear, and corrosion of ferrous parts. The "Forez" treatment provides resistance to seizure and wear of ferrous metals and because the coating, consisting of tin and copper, is easily deformed plastically, treated components can replace bronze in certain cases.

The "Zinal" treatment has been developed to provide seizure and wear resistance for aluminium.

REFERENCES

Centre Stéphanois de Recherches Mécaniques (1968). "Traitements de surface contre l'usure". Dunod, Paris.

Centre Stéphanois de Recherches Mécaniques (1972). *Mémoires Techniques du CETIM* 11.

Esteveny, S., Caubet, J. J., Polti, J. L. and Selva, J. (1970). *Mémoires Techniques du CETIM* 4, 51–70.

Michalon, D. (1969). *Revue Générale des Transmissions* 14, 7–17 and 16, 7–18.

Michalon, D. et Gilbert (1972). *Mémoires Techniques du CETIM* 11, 41–50.

Spark-hardening—a technique for producing wear-resistant surfaces

H.-U. STEIN*

Technical University, Rheinisch–Westfälische Technische Hochschule Aachen, Aachen, West Germany

I. INTRODUCTION

The production of wear-resistant surfaces by "spark-hardening" (also named "carbide impregnation") is a Russian development. Because of the various possibilities of application, this technique has been increasingly used in Germany recently. Hardening is partly due to the transfer of metal carbide in an arc discharge. To give a first impression of spark-hardening, Fig. 1 shows a hardened layer on a test piece of 20 × 20 mm. Figure 2 illustrates a microsection of the test piece, where the hardened layer appears to be white. Because of this it is called the "white layer" in the international literature. The "white layer" of the hardened surface in Fig. 2, for example, has a thickness of about 25 μm. Below the white layer, a dark-appearing heat-affected zone can be recognized and beneath this zone is the unaffected base material.

Present address:
* 5074 Odenthal-Voiswinkel, Hoher Wald 16, W. Germany.

H.-U. STEIN

FIG. 1. Spark-hardened layer on a test piece (20 × 20 mm).

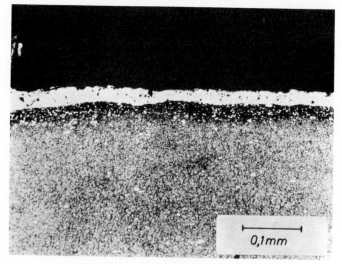

0,1mm

FIG. 2. Microsection of the hardened layer. Base metal: high speed steel
S18-1-2-5; electrode: tungsten carbide K10.

II. THE PRINCIPLE OF SPARK-HARDENING

The basic circuit diagram for spark-hardening is illustrated in Fig. 3. Generally d.c. sources are used for spark-hardening. The electrode (anode) is connected to the positive pole and the workpiece (cathode) to the negative pole. Usually atmosphere is the surrounding medium.

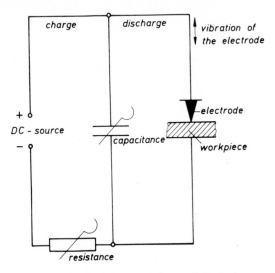

FIG. 3. Basic circuit diagram for spark-hardening.

The treatment is carried out with a vibrating electrode. Between the electrode and the surface to be hardened there occurs a continuous spark discharge of the capacitor energy produced by a short-circuit. The time of discharge, which is of the order of milliseconds, is determined by the magnitude of the electrical and mechanical parameters and by the condition of the atmosphere. It is assumed that at the time of discharge the workpiece is heated up to about some 1000 °C in the zone of contact and is then quenched in an extremely short time. At the same time, material of the electrode melts and is transferred to the workpiece. The processes cause a considerable physico-chemical change in the structure of the surface layer and effect the hardening at the treated spot. The hardened layer is well bonded to the base material. It is possible to build up all electrically conducting materials. A special preliminary treatment of the surface is not necessary; it must only be clean.

The decisive parameters of spark-hardening are the electrical power of the equipment, the capacity of the capacitor bank, the amplitude and the frequency of the vibrating electrode, and the time of treatment, which is also called "specific welding time". The "specific welding time" is a standard term in the international literature. It is defined as welding time per unit surface area.

III. Effect of the Variable Welding Parameters on the Layer's Quality

A. Specific welding time

It is sometimes questioned in the literature whether material of the electrode is transferred to the workpiece by spark-hardening.

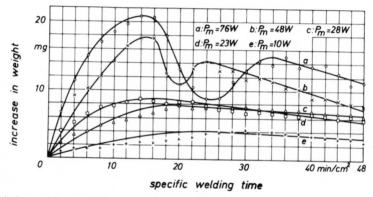

FIG. 4. Increase in weight of the workpiece depending on the specific welding time, for different electrical powers P_m.

Because of this, tests were made to measure the change of weight of the electrode and the workpiece as a function of the specific welding time and the electrical power. The results of these tests are shown in Fig. 4. The increase of weight of the workpiece is an obvious proof of the transfer of material from the electrode to the workpiece but it has also been recognized that only some of the material of the electrode is transferred to the workpiece —part of the material splatters away. A highspeed film of spark-hardening shows the same results. Although there is a linear relationship between the decrease of the electrode weight and the specific welding time, there is no corresponding linear build-up of material on the workpiece. Therefore it is not possible to reach a particular thickness by an increase of welding time. The increase of weight of the workpiece is proportional to the thickness of the layer.

Only in the first phase of the welding process is the thickness of the layer proportional to the welding time. It can be seen that curve (a) in Fig. 4 illustrates in the first part a steep, almost linear, slope. Beyond the maximum the curve shows a decrease. The hardened layer diminishes despite a longer welding time and despite a further decrease of the electrode weight. Beyond the minimum, the weight of the workpiece increases again and reaches a second maximum. A further increase of the welding time produces a slight, constant drop in weight of the workpiece.

B. Current density and voltage in the welding process

As well as the specific welding time, the current density and the voltage, i.e. the electrical power in the welding process, strongly influence the quality of the surface and the thickness of the hardened layer. The more the electrical power increases, the greater becomes the thickness of the layer, but at the same time the roughness of the layer increases. A layer of thickness more than about 1 mm is so rough that it cannot be used in practice. Thick layers also have a deep penetration into the base material, the surface layer thickness remaining the same as on a thin layer. Because of this it is not possible to polish the layer by a mechanical treatment after the hardening process. Therefore it is advisable not to work with a large electrical power, so that thin, but regular layers can be produced. Figure 4 shows that up to a power of about 100 W, a definite dependence of the increase of weight (and that means a dependence of the thickness of the layer) on the electrical power can be realized. Obviously to each electrical power there corresponds a maximum thickness of the layer. The more the electrical power is increased, then the more the specific welding time for maximum deposition is reduced. The loss of material after the maximum increases with increasing electrical power. Welsh (1963) explains this effect as a thermal fatigue. He assumes that the rapidly fluctuating temperatures tend to disintegrate the surface and effect, therefore, an erosive process which counteracts and eventually dominates the transfer process.

C. Capacity of the capacitor bank

At the moment of discharge, a greater capacity effects a greater flow of current. At the same time the greater flow of current produces a greater electrical power.

Therefore it can be deduced that the capacity has the same influence on the hardened layer as the electrical power. The greater is the capacity of the capacitor bank, the thicker (but also rougher) is the surface. Equipment for practical application must have a capacity between 0 and 1000 μF.

D. *Amplitude of the vibrating electrode*

The manner and the time of contact between the electrode and the workpiece depends on the amplitude of the vibrating electrode. The discharge process also depends on the amplitude, and therefore the choice of the amplitude directly influences the weld quality. To achieve a good weld quality, it is best to choose an amplitude between 0·6 mm and 2·0 mm.

IV. SURFACE ROUGHNESS

Evaluation of surface topography shows that the roughness increases with increasing electrical power during the discharge process. The transferred particles of material grow with the increase of discharge energy. A layer thickness of about 30 μm has a maximum roughness of 20 μm (c.l.a.). To obtain a layer of sufficient thickness and good surface quality, it is advisable to produce the layer in two stages. The first stage is operated at a high electrical power to achieve the required thickness. The second stage is then operated at a lower power to polish the surface. So it is possible to diminish the maximum roughness of a layer of 30 μm to 13 μm (c.l.a.).

V. STRUCTURE TESTS

Layers hardened with tungsten carbide have, on average, maximum hardness values of about VHN (50 g load) 15,000 N/mm^2 (1530 kgf/mm^2); single values greater than 20,000 N/mm^2 (2040 kgf/mm^2) are often produced. The layer is composed of nearly 80% electrode material and 20% base

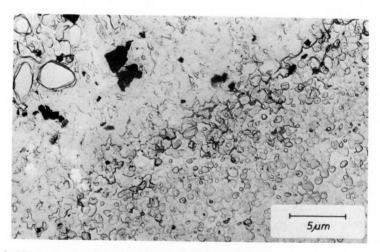

FIG. 5. Microstructure of the hardened layer, the heat-affected zone and the base metal. Base metal: high speed steel S18-1-2-5; electrode: tungsten carbide K10.

material. Structure tests show that the tungsten carbide in the layer and in the electrode differs in structure. In Fig. 5, a microsection can be seen in the electron microscope. At the upper left you can see some great carbides of the base material. At the bottom of the right side of the figure, the hard layer can be seen. Between the hard layer and base material, there is the heat zone. The layer does not appear "white"; a clear structure can be recognized. Because of the high temperatures during the transfer process and because the material is quenched in an extremely short time, the structure of the tungsten carbide changes. The tungsten carbides are disintegrated in the welding process and form a complex combination with the base material. Only some single carbides, which correspond to the structure of cemented carbide product, can be found in the hardened layer. The structure tests have also shown that the hardened layer has pores and cracks. These faults are unavoidable and originate in the special welding process of spark-hardening.

VI. Service life tests

In a practical application of spark-hardening, highly heat-resistant materials are splintered off with spark-hardened high speed steel tools. In milling tests, spark-hardening leads to an increase of service life over the whole range of cutting rates. A threefold service life can be reached.

Rotation tests show an increase of service life only in the range of lower cutting rates ($\leqslant 24$ cm/s); this is caused in the high thermal claim at the zone of contact of the tools, which occurs beside the mechanical strain. At a higher cutting rate, the zone of thermal effect in the base material weakens because of the high thermal claim. The hardened layer will be destroyed.

VII. Discussion

An evaluation of the feasibility of using spark-hardening in practice, suggests that this welding process is applicable to those surfaces which need to be wear-resistant, but are not subjected to very high temperatures. Therefore tools for punching, deep drawing, bending cold press forming, extrusion, cold- and warmforming, can be treated by spark-hardening. Some industrial test reports show that spark-hardening may be successfully used in the near future to increase wear-resistance.

REFERENCE

Welsh, N. C. (1963). "Spark-Hardening of Tools", pp. 257–263. International Research in Production Engineering, New York.

Conversion and conversion/diffusion coating

J. C. GREGORY

Imperial Chemical Industries Ltd., Cassel Heat Treatment Service, Kynoch Works, Witton, Birmingham, England

I. INTRODUCTION

Many raw materials used in manufacturing processes do not have all the properties likely to be required of them. It may be that the core properties of a certain component, such as strength and ductility, are satisfactory, whilst the surface properties, such as visual appearance, resistance to wear or resistance to corrosion, are unsatisfactory.

A logical way of tackling this problem is first to select a basic material having the necessary core properties and then to modify the properties of the surface in the desired direction by processes which will not adversely affect those of the core.

It is found possible, convenient and often a marked advantage to convert some metallic surfaces, by reaction with suitable chemical reagents, to compounds having quite different properties. Processes working on this principle are available for iron and steel, brass, bronze, aluminium bronze, light aluminium alloys, titanium, copper and its alloys, cadmium, zinc, aluminium and magnesium, and others are constantly being sought.

Two important surface properties of metals which can be modified by conversion and conversion/diffusion processes are resistance to wear and resistance to corrosion. The need for improvement in these properties is emphasized by two recent reports which show that in the U.K. alone more than £500 million per annum is lost through wear (H.M.S.O., 1966) and more than £1000 million through corrosion (H.M.S.O., 1971).

Our own particular interest in surface treatments has been concentrated on processes designed to improve the wear resistance of iron and steel, particularly those using molten salt baths as the treatment media, and we propose to illustrate the principle of conversion and conversion/diffusion processes by reference to iron and steel.

II. WEAR OF IRON AND STEEL

A. Mechanism of wear

Wear can be caused by various mechanisms such as adhesion, abrasion, erosion and fatigue, of which the first two are the most common. Abrasion is due to a cutting action occurring for example when hard particles are in contact with a softer material; the remedy is to use harder materials or to convert the surface of a soft material into a harder material. This conversion can often be done in the case of steel by a case hardening operation (I.C.I., 1968) which involves diffusion of carbon into the surface, or by a nitriding operation (Firth Brown, 1963) involving diffusion of nitrogen into the surface. Adhesion on the other hand is due to welding together of the asperities of two surfaces in frictional contact. Often these microwelds are broken by the momentum of the moving parts but in some cases welding may be so serious as to cause total stoppage (seizure) of the mechanism. There are many remedies, some more effective than others, but where iron and steel are concerned conversion and conversion/diffusion coatings are most useful for this purpose since they produce surfaces which resist welding. The damage to surfaces due to adhesion or welding is often called scuffing or galling and is illustrated in Fig. 1. This is taken from a photomicrograph of a transverse section through the surface of a case hardened steel disc which has been rubbing on a similar disc without adequate protection by lubrication or other means. It will be seen that a fragment of one surface has been welded to the other by the pressure applied and the frictional heat produced during operation.

B. Combating wear of iron and steel

Conversion and conversion/diffusion processes for the production of scuffing resistant and wear resistant coatings on iron and steel can be carried out by solid, liquid or gaseous methods but we shall limit our consideration here to liquid methods, i.e. those where the operating medium is a molten salt mixture or an aqueous solution of salts. The term "conversion coating" has for many years been accepted as referring only to the well known phosphating processes (I.C.I., 1967) whereby coatings of iron, manganese and/or

FIG. 1. Photomicrograph taken from a transverse section of an Amsler wear testpiece. This is of case hardened mild steel and had been rubbing on another similar testpiece. A small fragment of the second testpiece has welded to the first. The surface has become tempered and hence appears darker after etching. Etched in 2% Nital.

zinc phosphates are applied to steel, this sometimes being extended to the rather less well known oxide coating processes (Tomlinson). For purposes of resistance to scuffing and wear, however, conversion of iron and steel surfaces to carbides and nitrides (Degussa, 1962); carbides, nitrides and sulphides (Gregory, 1968); or even sulphides alone (Amsallem, 1970; Porte, 1971) is much more effective.

1. *Application of nitride coatings, commonly called nitriding*

(*a*) *Principles.* All the different salt bath nitriding processes convert iron and steel surfaces to iron carbide as well as to iron nitride. The principle is the same in all cases, that where a supply of nascent nitrogen is in contact with hot iron at temperatures between, for example, 500 °C and 750 °C, iron nitride will be formed. Some nitrogen will also diffuse through the metal beneath this conversion coating to give a diffusion layer which contains iron nitride but is not entirely composed of this compound. This iron nitride may be visible in the microstructure as a separate constituent or may be retained in solid solution in the iron, depending upon whether the metal is slowly or quickly cooled after treatment. Iron carbide found to be present along with the nitride in the conversion coating results from reaction between the iron and carbon monoxide which, along with nitrogen, is liberated from the salt mixture. Figure 2 shows the microstructure of mild steel treated for 1·5 h at 570 °C in a nitriding bath and quenched into water, whilst Fig. 3 shows a similar steel treated in the same way but afterwards tempered at 300 °C to

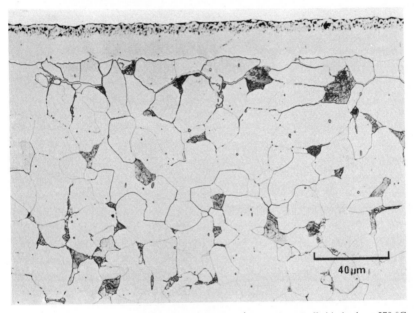

FIG. 2. Photomicrograph of mild steel treated for $1\frac{1}{2}$ hours in a Tufftride bath at 570 °C and water quenched. Note the white compound layer at the surface showing some porosity at its outer edge. No nitride needles are visible. Etched in 2% Nital. Transverse section. (Courtesy Degussa, Frankfurt am Main.)

show the extent to which nitrogen is present in the diffusion zone. The needles visible are thought to consist of Fe_4N.

Sodium or potassium cyanate is the principal source of the carbon and nitrogen which produces the conversion coating, this often being produced in the bath itself by atmospheric oxidation of molten cyanide, thus:

$$4NaCN + 2O_2 \rightarrow 4NaCNO$$

This cyanate decomposes at the operating temperature of, for example, 500–600 °C to give nitrogen in the nascent state, thus:

$$4NaCNO \rightarrow 2NaCN + Na_2CO_3 + 2N + CO$$

The carbon monoxide also liberated breaks down at the surface of the parts being treated thus:

$$2CO \rightarrow CO_2 + C$$

The carbon thus formed is in the nascent, and therefore active, condition and reacts with the iron to form iron carbide Fe_3C. The nitriding potential

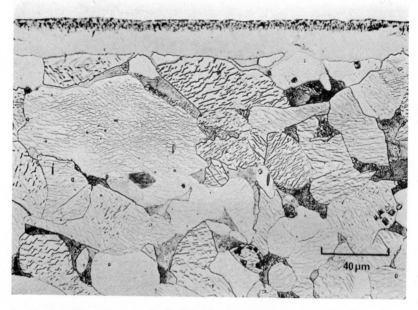

FIG. 3. Photomicrograph of mild steel treated for $1\frac{1}{2}$ hours in a Tufftride bath at 570 °C and water quenched. Tempered 1 hour at 300 °C and water quenched. Shows the white compound layer and a profusion of nitride needles precipated by the tempering treatment. Etched in 2% Nital. Transverse section. (Courtesy Degussa, Frankfurt am Main.)

of these baths can be increased by adding oxidants such as manganese dioxide, by blowing air through the molten salt as in the "Tufftride" ("Tenifer") process (Degussa, 1962), or by adding a sulphur compound to the melt as in the "Sulfinuz" process (Gregory, 1968). In the latter case the sulphur compound also impregnates the nitride/carbide coating with iron sulphide, which itself has anti-scuffing properties.

(b) Composition and properties of the coating. The chemical composition of the compound layer, i.e. the portion of the coating which has been totally converted, has been shown to consist of FeN and Fe_3C; thus the Tufftride coating is said to contain 80% epsilon iron nitride (FeN) with 20% iron carbide (Fe_3C), this iron carbide also containing nitrogen. This layer is completely integrated with the substrate and normally 10–15 μm thick, and the diffusion zone beneath may extend to 0·5 mm.

There is a degree of porosity in the coating, which can be minimized by careful control of the process, although when conventional lubricants are used porosity can be an advantage in providing oil retaining properties.

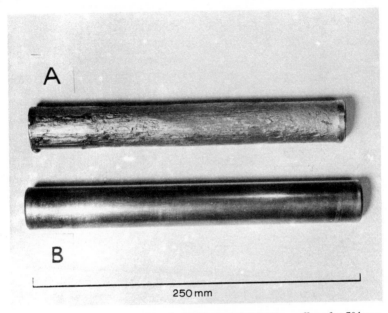

250 mm

FIG. 4. Marquette rods for press tools. (A) Untreated, failed by scuffing after 70 h use.
(B) Noskuff treated, not scuffed after 1400 h use.

There is some roughening of the surface during treatment which again can be minimized by careful control of the nitriding bath analysis. For example a steel component having an original surface finish of 12 microinches (0·3 μm) centre line average, may be roughened to 25–30 microinches (0·6–0·8 μm) c.l.a.

The efficiency of the conversion coating produced in nitriding baths is well illustrated in Fig. 4 which shows treated and untreated marquette rods after use as press tool guides. These were subjected to a specially modified nitriding treatment called "Noskuff" (I.C.I., 1970) which in contrast to some other nitriding processes allows the substrate to remain hard. The untreated rod failed by scuffing after 70 h use whilst the treated rod showed no scuffing after 1400 h use. Figure 5 again demonstrates the efficiency of the nitriding treatment using Amsler wear testpieces treated by the Sulfinuz process. Treated and untreated discs were tested under a Hertzian stress of 20 kgf/mm² (196 MPa) at a speed of 400 r.p.m. using SAE 30 (32–40 mm²/s) oil as a lubricant. The extent to which the conversion coating, in this case rich in carbides, nitrides and sulphides, prevented scuffing and wear is clearly visible. The untreated samples lost more than 1 g in 20,000 revolutions whilst the treated samples lost only 0·1 g in 70×10^6 revolutions.

FIG. 5. Amsler wear testpieces after testing. The pair on the left were of untreated mild steel, the pair on the right were of Sulfinuz treated mild steel. The untreated samples lost 1 g in 20,000 revs. whilst the treated samples lost only 0·1 g in 70×16^6 revs.

(c) *Pretreatment of the substrate*. The surface finish before treatment is not important except in so far as the removal of material from the surface after treatment to improve the finish constitutes a removal of the conversion coating itself and so must be kept to a minimum. Pretreatment consists in ensuring comparative freedom of parts to be treated from oxide and grease, and preheating to 300 °C to minimize the danger of salt splashing and thermal shock.

(d) *Different processes and limitations*. As already indicated, there are several proprietary salt mixtures available for nitriding, the two most common being Sulfinuz and Tufftride (Tenifer). Noskuff has been designed especially to give a similar conversion coating to these processes whilst retaining a high hardness in the substrate.

All ferrous materials can be treated by these processes although it must be borne in mind that the comparatively high operating temperature in some cases may cause softening of the substrate and there may be some loss of corrosion resistance in stainless materials. The normal operating temperature of Sulfinuz and Tufftride is 570 ± 5 °C.

(*e*) *Parts treated commercially*. Examples of parts treated commercially by these processes are high speed steel cutting tools, 5% chromium steel extrusion dies, cylinder liners, gear type couplings, drop hammer slides, tappets, valve guides and sintered iron components.

As an indication of the popularity of these salt bath nitriding treatments, it has been estimated that about 100,000 tons (10^8 kg) of engineering components are treated by them each year, and the quantity is continually increasing.

2. *Application of sulphide coatings*

(*a*) *Principles*. The process of applying a sulphide coating to iron and steel is variously known as sulphurizing, sulphiding or sulphidizing. Sulphur and sulphides, by virtue of their suitable crystal structures and in the case of sulphur because of its low melting point, have long been recognized as having good lubricating properties.

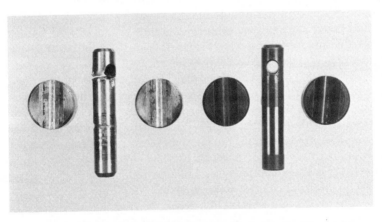

FIG. 6. Falex (Faville–Levally) mild steel testpieces after testing. The testpiece on the left seized and sheared quickly when rotated dry between the jaws, whilst that on the right, tested whilst surrounded by powdered sulphur, showed no evidence of seizure.

Figure 6 shows Falex testpieces made from mild steel when tested dry and when tested whilst immersed in powdered sulphur. When tested dry, scuffing and seizure occurred with an applied jaw load of 700 lbf (3·1 kN), whereas when sulphur was used, although the testpieces became hot and melted the sulphur in immediate contact, no surface damage occurred at a load of 3000 lbf (13·3 kN).

(*b*) *Comparison with nitriding and other processes.* Whilst the nitride and carbide conversion coatings described above have excellent properties, the high operating temperatures (570 °C) required to produce them in the case of most steels and cast irons, precludes the retention of a high hardness in the substrate. The Noskuff process already mentioned goes a long way towards solving this problem, but may cause unacceptable distortion due to the need for a final quenching operation. There is therefore a definite need for a process which will apply similar anti-weld properties to ferrous metals as conferred by the nitriding treatments but using operating temperatures below 200 °C. Sulphur and sulphides suggest themselves for this purpose and many processes have been developed to apply such coatings. One process, known as the "Neely" process (Caterpillar Tractor Co.), pits the metal surface by the use of boiling concentrated caustic soda solution and fills in the pits with elemental sulphur. Many others are quoted in the references at the end of this paper (Luceveanu and Pavlencu, 1962; Michev and Ivanova, 1965; Vinogradov, 1964; Zot, 1957). There is also a profusion of methods of bonding molybdenum disulphide to metal surfaces, but none seems to give the same degree of protection as the nitriding treatments described above. It would be anticipated that sulphide coatings produced by conversion of the metal *in situ* rather than by relying on the adhesive powers of a particular bonding material would give the best results.

(*c*) *The Sulf BT process.* (i) *Principle.* The most recent coating process of the conversion type, and probably the most efficient of the sulphide coating techniques, is the "Sulf BT" process (Amsallem, 1970; Porte, 1971), also known in the U.S.A. as the "Caubet" process. This produces a coating of iron sulphide integral with the ferrous metal substrate. It may be argued that molybdenum disulphide is preferable to iron sulphide for frictional applications because of the comparatively small force required to displace individual lamellae of MoS_2 crystals, a virtue that has long been recognized for this material, but Bartz and Müller (1972) have recently suggested that the good lubricating properties of MoS_2 when used with ferrous materials are due to the formation of a surface layer of iron sulphide and not to its lamellar structure.

Sulf BT is based on the principle that a tenaciously adherent coating of iron sulphide can be applied to many ferrous materials electrolytically using a mixture of sodium and potassium thiocyanates as electrolyte. The components to be treated are made the anode and the metal container for the molten thiocyanates is made the cathode. A low voltage current of density about 320 A/m^2 is passed through the molten bath, which is maintained at about 190 °C.

At the operating temperature the thiocyanates are completely ionized and the negative CNS ions containing sulphur migrate towards the anode, i.e. towards the parts being treated. Here the sulphur forms iron sulphide FeS at the surfaces of parts being treated. Amsallem (1970) has given an interesting explanation of the reason why the sulphide adheres so tenaciously to the substrate. This is based on the premise that the iron sulphide coating acts as

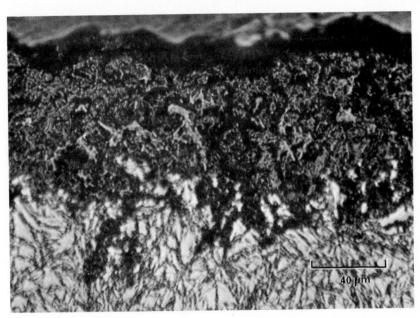

FIG. 7. Photomicrograph from an oblique section of a case hardened and ground mild steel sample subsequently treated by the Sulf BT process. The dark etching surface layer is iron sulphide (FeS). Note its integration with the martensitic (light etching) substrate. Etched in 2% Nital.

a dielectric between the substrate, i.e. the iron or steel, and the molten electrolyte. The high temperatures produced at positions of local electric discharges result in superficial fusion of both sulphide and metal with consequent integration of coating and substrate.

(ii) *Structure and physical effects of coating.* The structure of the sulphide coating is shown in Fig. 7 which is a photomicrograph of a field on a transverse section through a Sulf BT treated case hardened and ground mild steel sample. The coating is clearly visible as a dark etching zone and its complete integration with the hard martensitic substrate is also shown. Electron diffraction analysis confirms that it consists mainly of ferrous sulphide FeS. There is some degree of porosity which can be an advantage in that it assists the surface to carry oil when used in lubricated mechanisms. The effect on the topography of the surface is to reduce the roughness of rough machined parts and to increase the roughness of finely finished parts. For example a surface having an 80 microinch (2·0 μm) c.l.a. finish might be improved to 60 microinches (1·5 μm), whilst a 20 microinch (0·5 μm) finish might be roughened to 50 microinches (1·25 μm). However, this sulphide conversion coating allows of rapid running-in by plastic deformation of the asperities without scuffing, thus resulting in considerable improvement in the surface finish. For example, after running-in for only 60 minutes at 300 r.p.m., a case hardened and Sulf BT treated bush running against a case hardened flat plate had its finish improved from 42 to 15 microinches (1·1 μm to 0·4 μm) c.l.a.

Parts are normally treated for 10 minutes at 190 °C, which gives a conversion coating thickness of 7·5 μm. The special properties of this coating are its resistance to scuffing and wear such as might occur when ferrous metal parts rub together, and the patentees also claim an increase in corrosion resistance.

(iii) *Parts treated commercially.* The process is suitable for cast iron, mild steel, carbon steels and all alloy steels containing up to 13% chromium. It is normally operated at 190 ± 5 °C, and it is this low operating temperature which makes it so useful in the engineering field. Since it is a conversion process, the surface finish of the original metal as far as efficiency of coating is concerned is relatively unimportant. It should be noted however that where the bore surfaces of tubular parts require treatment this may involve provision of auxiliary cathodes to ensure uniformity of treatment, and that where the bore is 1 mm or less in diameter treatment is very difficult.

Where the substrate is required to be hard the necessary heat treatment should be applied before this conversion treatment. The parts should be cleaned free from grease and rust by first degreasing, pickling, neutralizing and rinsing in water, before immersion in the treatment bath of molten thiocyanates. They should be allowed to attain the operating temperature of 190 °C before passing the electric current, and the molten salt is agitated during the treatment period by means of compressed air.

Examples of parts treated are gears, bearings, heavy duty rear axle spiders, slitting machine saws and textile machine parts, indeed any ferrous materials containing up to 13% chromium and requiring to have an anti-scuffing surface whilst retaining a hard substrate. Cast iron cylinder liners are also currently being treated on a large scale to improve their wear resistance. In this case this method is being used not to conserve a high hardness, but to minimize distortion which might occur when using processes such as nitriding, which are operated at an appreciably higher temperature.

3. *Other types of conversion coating*

We had intended in this paper to deal also with the more common conversion processes such as phosphating and oxide coating processes but with the limited space available have concentrated on the less well known nitriding and sulphiding processes. Readers interested in such processes for improving resistance to wear and to corrosion are however urged to consult the available literature, some of which is mentioned in the references below (A.S.M., 1964; Drysdale, 1969; I.C.I., 1967; James, 1971; Mock, 1970.

ACKNOWLEDGEMENTS

The author is indebted to Imperial Chemical Industries Limited for permission to publish this paper and to the following companies for illustrations and/or technical information: Degussa Wolfgang (W. Germany), Hydromécanique et Frottement (France), and Tool Treatments (Chemicals) Limited (U.K.).

REFERENCES

American Society for Metals (1964). "Chemical Conversion Coating", Metals Handbook 8th edn. Vol. 2, pp. 627–628.
Amsallem, C. (1970). *Revue de métallurgie* **11**, 945–957.
Bartz, W. J. and Müller, K. (1972). *Wear* **20**, 3, 371–379.
Caterpillar Tractor Co. U.S.A. "The Neely Process".
Degussa Durferrit Division, W. Germany (1962). "Tufftride".
Drysdale, R. F. (1969). *Trans. Inst. Metal Finishing* **47**, 149–155.
Firth Brown Ltd., Sheffield 4 (1963). "Nitriding Steels".
Gregory, J. C. (1968). *Metal Forming*, August/September/October.
H.M.S.O., London (1966). "Lubrication (Tribology) Education and Research".
H.M.S.O., London (1971). "A Survey of Corrosion Protection in the U.K.".
Imperial Chemical Industries Ltd., London (1967). "Granodine Phosphating Processes".
Imperial Chemical Industries Ltd., London (1968). "Cassel Salts for Heat Treatment and Casehardening".
Imperial Chemical Industries Ltd., London (1970). "The Cassel Noskuff Process".
James, D. (1971). *Wire Industry*, December.
Luceveanu, G. and Pavlencu, L. (1962). *Met Constructia Masini* **14**, 874–880.
Michev, V. and Ivanova, M. (1965). *Mashinostroene (Sofia)*, pp. 159–161.
Mock, J. A. (1970). *Finishing* **31**, 4, 77–85.

Porte, M. (1971). *Materiaux et Techniques*, February.

Tomlinson, A. "Black Oxide Coatings". Reprinted from *Product Finishing* for Tool Treatments (Chemicals) Ltd., West Bromwich.

Vinogradov, Y. M. (1964). *Protsessy Uprochneniya Detalai Mashin, Akad. Nauk. SSSR*, pp. 83–89.

Zot, A. I. (1957). *Materialy Nauch Tech. Konf. Rabotnikov Zaved Lab.*, *Rostov*, pp. 23–40.

The basic principles of chemical vapour deposition (CVD)

J. C. Viguié

*Laboratoire d'Étude des Matériaux Minces, Département de Métallurgie,
Centre d'Études Nucléaires de Grenoble CEA, Grenoble, France*

Chemical vapour deposition uses a chemical process occurring between gaseous compounds when in contact with a heated material. The deposition takes place as long as the reaction produces a solid. The heating attachment must provide two temperature plateaux (Fig. 1): the first is held not far below the evaporation temperature T_E in order to have a constant vapour pressure of the source material above the crucible. The second is higher and allows the reaction to take place. A carrier gas brings the vapour from the evaporation zone to the reaction zone. In many cases, a reactant gas such as hydrogen or oxygen is added to achieve either metal or oxide deposition. Metallic carbides, borides and nitrides, are deposited by using CH_4, BCl_3 and NH_3 respectively.

I. THE SOURCE MATERIAL

For a given layer, the choice of a source material will very much depend on the following practical considerations:

(i) The use of a volatile compound will greatly simplify the technology of the gas transport. Otherwise, the pressure must be held below atmospheric pressure.

(ii) It must not decompose when heated, otherwise an attempt must be made to evaporate at low temperature and low pressure (in the 0·1–10 torr range (0·01–1·3 kPa)).

(iii) It must react (or decompose) at a temperature consistent with usual technology, namely not higher than 1200 °C. The lower the temperature the more versatile is the method.

From the data available in the thermodynamic literature, a choice can be made at once of the metallic compound for the source material, the reactant gas and the reaction temperature.

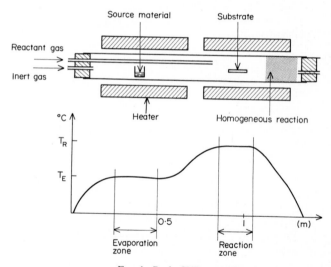

FIG. 1. Basic CVD operation.

A. The Ellingham diagrams

The Ellingham diagrams give a convenient presentation of the whole of these data by evaluating the standard free energy change ΔG_T° in the formation of metallic compounds versus temperature. The calculation uses the standard relationship (Kubaschewski et al., 1967) between changes in free energy, enthalpy H and entropy S:

$$\Delta G_T = \Delta H - T\,\Delta S$$

For a standard partial pressure of 1 atmosphere ($1 \cdot 01 \times 10^5$ Pa),

$$\Delta G_T^\circ = \Delta H^\circ - T \, \Delta S^\circ$$

ΔG_T and ΔG_T° are related by

$$\Delta G_T = \Delta G_T^\circ + RT \ln Kp$$

Kp, the equilibrium constant is a function of pressure. Introducing heat capacities:

$$\Delta G_T^\circ = \Delta H_{298}^\circ + \int_{298}^T \Delta C_p \, dT - T \, \Delta S_{298}^\circ - T \int_{298}^T \frac{\Delta C_p}{T} \, dT$$

It may be permissible, as a first approximation, to neglect the heat capacity terms. Thus:

$$\Delta G_T^\circ = \Delta H_{298}^\circ - T \, \Delta S_{298}^\circ$$

The standard values ΔH_{298}° and ΔS_{298}° are calculated by elementary operations from thermodynamic tables. The Ellingham diagrams represent ΔG_T° as a function of temperature.

B. The choice of a suitable source

Figure 2 gives data for some metallic chlorides. Their stability increases when descending the scale of ΔG_T°. As a result, the curves for $MoCl_5$, $ReCl_3$, $AsCl_3$ being above the HCl curve, these chlorides can be reduced with production of hydrochloric acid. For example:

$$MoCl_5 + \tfrac{5}{2} H_2 \rightarrow Mo + 5HCl$$

The curves for chlorides of the transition elements cross over. So $NiCl_2$, $FeCl_2$ and $CoCl_2$ are reducible only above the intersection temperature. $SiCl_4$ is reducible at high temperature only.

ΔG_T° for chromium chloride does not intersect, but it will, as we adjust the partial pressures p:

$$\Delta G_T = \Delta G_T^\circ + RT \ln \frac{p_{CrCl_2}}{p_{Cl_2}}$$

For $p_{CrCl_2}/p_{Cl_2} = 10^3$ at $1400\ ^\circ K$, the free energy change for one atom of chlorine is:

$$\Delta G_T = \Delta G_T^\circ + 10 \text{ kcal (42 kJ)}$$

Owing to the partial pressure effect, attempts can reasonably be made to deposit metals from chlorides whose ΔG_T° approaches $(\Delta G_T^\circ)_{HCl}$ to within 10 kcal (42 kJ).

Descending towards the electropositive elements, gallium trichloride is hardly reducible. Further on, titanium, aluminium, and the rare earth chlorides are too stable to be used as a source material for metal deposition.

On the other hand, fluorides are very stable compounds. None thermally

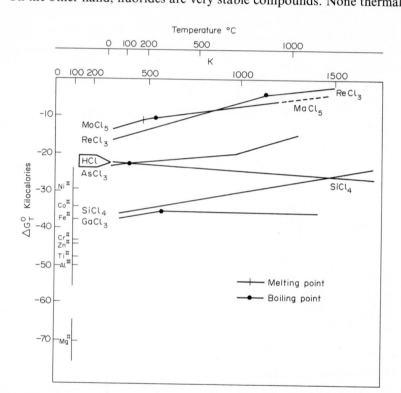

Fig. 2. Ellingham diagram for chlorides. For clarity, not all curves are drawn. Some are merely situated along the scale of ΔG_T°. For example, Co^{II} is for CoCl$_2$, Al^{III} for AlCl$_3$ etc.

decompose. Only fluorides of refractory metals such as tungsten, molybdenum and rhenium, which are close in stability to hydrofluoric acid, will react with hydrogen to give the metals. For example:

$$MoF_6 + 3H_2 \rightarrow Mo + 6HF$$

Bromides and iodides are less stable than chlorides; their representative lines intersecting $\Delta G_T^\circ = 0$, they thermally decompose above the temperature of the intersection. So, few compounds are stable in the vapour phase. These ones have been used for years in the metal refining industry. They include iodides of silicon, chromium, zirconium, titanium and hafnium. From such sources, metal deposition is possible even when hydrogen is missing. For example:

$$SiI_4 \rightarrow Si + 2I_2$$

Still less stable are organometallic compounds. This feature allows very low temperatures for metal deposition. Electropositive elements can be deposited by this method alone. At 250 °C the tri-isobutyl aluminium yields the metal:

$$Al\left[CH_2{-}CH\begin{array}{l} \diagup CH_3 \\ \diagdown CH_3 \end{array}\right]_3 \rightarrow Al + 3CH_2 = C\begin{array}{l} \diagup CH_3 \\ \diagdown CH_3 \end{array} + \tfrac{3}{2}H_2$$

Similar considerations apply to oxide, carbide and nitride formation. For example, consider the reaction:

$$SiCl_4 + CH_4 \rightarrow SiC + 4HCl$$

The behaviour at all temperatures of the $(SiCl_4 + CH_4)$ mixture will be evaluated by comparing $(\Delta G°_{SiCl_4} - \Delta G°_{HCl})$ read on the chloride diagram, with $(\Delta G°_{SiC} - \Delta G°_{CH_4})$ read on the carbide diagram.

II. THE PROCESS

A. Reactor geometry

Having chosen the source material, the geometry of reactors must:

(i) maintain a constant transport of the vapours,
(ii) avoid condensation losses on the cooler parts of the tube,
(iii) take into account the opposite effects of gravity and natural convection.

As all these parameters are difficult to handle together, empiricism will be the golden rule. Undoubtedly this explains the variety of reactor shapes, which is rather perplexing for a non-specialist in this field.

As a guideline, we can say that a volatile source material gives greater flexibility in managing the apparatus (Fig. 3). When a solid source material is to be used, the furnace must heat both the source and the substrate. This obliges us to use a more compact installation with a simpler reactor shape. Owing to the short transport path, the mixing of components could be a problem, but fortunately the diffusion of gases multiplies tenfold when the temperature is increased from room temperature to 1000 °C.

For a few years, attempts have been made to describe the complex situation prevailing in the vicinity (a few millimetres) of a substrate. Hydrodynamics and heat transfer are taken into account in calculating the mass transfer rate towards the substrate through the boundary layer. This is the subject of a young science, the so-called aerothermochemistry.

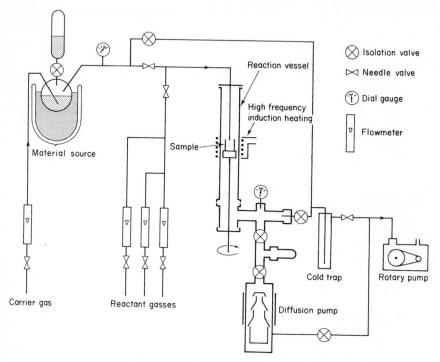

FIG. 3. CVD from liquid source material.

B. Kinetics of deposition

Let us now give some attention to what happens on the substrate surface. It is important to emphasize that the process of interest is the one which happens only on contact with the substrate surface. We are not concerned with the homogeneous reactions which might take place in the bulk of the vapour phase; as they yield powdery and non-adherent deposits, everything is done to get rid of them. Fortunately the heterogeneous reaction is more favourable owing to the "seed effect" of the solid.

Thermodynamic considerations show what is possible but not what actually happens; this is given by the knowledge of the reaction rate. Let us take for example the reduction of a metallic chloride:

$$MCl_2 + H_2 \rightarrow M + 2HCl$$

Often, the reaction rate Y can be written in the form:

$$Y = \frac{d(M)}{dt} = K\left(\exp{-\frac{E}{RT}}\right)p_{MCl_2}^{\alpha}p_{H_2}^{\beta}$$

K is independent of temperature and pressure. E is the activation energy, R the gas constant, p_{MCl_2} and p_{H_2} are the partial pressures of gaseous components, and α and β are in the simplest cases small integers (0, 1, 2) or fractions $(\frac{1}{2}, \ldots)$.

It is clear that the higher the temperature, the higher is the reaction rate. From a very practical point of view, the temperature can be set at once at 1200 °C and one examines what turns up! Then, as the deposition temperature is lowered, one observes how the coating modifies.

A more thorough analysis of the deposition process will help us in seeking the conditions which yield a high growth rate with a good quality layer. More precisely, we have to find out the temperature, the flow rate and the component partial pressures. Table I presents a generally accepted scheme for heterogeneous reactions.

TABLE I

Steps of the heterogeneous reaction

Steps	Whether activated
(i) diffusion to the substrate	non-activated
(ii) adsorption of one or more gaseous components	activated
(iii) surface reaction	activated
(iv) desorption of gaseous products	activated
(v) diffusion away from the substrate	non-activated

Plotting growth rate versus flow rate, gives the conditions where saturation occurs. In this case (high concentration and/or high flow rate, and/or low temperature, and/or high pressure), the limiting process is an activated step. In that case the reaction rate must satisfy the Arrhenius relation

$$\log Y = A - \frac{B}{T}$$

The constants A and B are independent of temperature. According to the initial definition of Y one obtains:

$$B = \frac{E}{2 \cdot 3R}$$

Then the experimental determination of B from the Arrhenius plot gives us the activation energy of the limiting step of the process. Finally, plotting growth rate versus the compound partial pressures, will indicate the mechanism.

III. THE COATING

A. Structure

CVD coatings are very varied. Metals as well as metallic compounds can be deposited. The limitations are only practical: a suitable source material is not always available.

The structure of deposits is closely related to the rate determining step of the reaction. So information taken from kinetic investigations is very valuable for choosing the conditions which will give the desired structure with a good reproducibility.

In saturation conditions the growth rate is high. Nuclei have no time for rearrangement. The deposit is fine grained and the crystallites are oriented at random. Below saturation, the whole process is diffusion controlled. This means that gaseous molecules impinge on the substrate from time to time, the growth rate is low and rearrangement on the substrate surface is possible. These conditions yield large, oriented crystallites. As the mass transfer rate is lowered and the temperature increased, surface diffusion allows a rearrangement of the nuclei. Epitaxy is then possible provided that the substrate is a single crystal.

It is worth noting the influence of foreign gases on the structure of deposits. As the heterogeneous reaction includes an adsorption step, it is felt that a strongly adsorbed gas will hinder the reaction. As the magnitude of this phenomenon varies according to the crystallite face, there is a tendency towards growth in a particular direction. This has a determining effect on the crystallization of the layer.

As adsorption can produce a concentration effect, even p.p.m. (parts per million) levels of gaseous impurities are of importance. They are sometimes reactive gases such as oxygen and chlorine, and sometimes gaseous intermediates of the reaction. It is clear that it will be difficult to foresee what will happen as long as we cannot detect them. This is not the least important factor of irreproducibility.

Gas adsorption effects on crystal growth should be an attractive research field which could add a great deal to our knowledge of CVD.

B. *Thickness*

Thickness in some particular cases might be as much as one millimetre. But generally such a large value is not necessary to achieve the required mechanical properties. A few hundred micrometres are sufficient.

More often a limitation occurs when the crystallites grow larger and larger. The coarseness and the lack of cohesion oblige us to stop. From this point of view, corner effects have a drastic influence on the quality of the coating.

C. *Substrate requirements*

The nature of the substrate will determine the reactor design. Insulators are generally heated in electric furnaces whereas conductors are heated with radio frequency power.

Given the material to be coated, the chemist has to find a reaction temperature which does not change the properties of the substrate. Refractory metals, graphite and insulators can be heated up to 1200 °C. But, for steels, keeping the temperature below the transformation temperature is a rather limiting condition.

The shape and size of samples are limited by the heating possibilities. A sophisticated geometry yields non-uniform growth rates. Practically, the method is limited to samples having a rotation axis and planes of a few square centimetres in size.

Substrate treatment before deposition, is rather a perplexing subject. When crystals of an electronic quality are to be grown by epitaxy from a monocrystalline (single crystal) substrate, attention must be paid to surface preparation. The substrate must be mechanically and chemically polished. Annealing is necessary to get rid of long range damage of the crystal lattice induced by the previous abrasive treatment. With reference to gas adsorption, the atmosphere prevailing before deposition has a definite influence on the quality of the layer. This is the reason for baking the substrate *in situ* for one or two hours. For example, one gets rid of the oxygen adsorbed layer on the substrate surface by baking it in hydrogen flow.

In every case attention should be paid to the gas adsorbed in the bulk of the substrate, particularly when it is of porous material such as graphite. Otherwise, the evacuation of the gas during deposition has a deleterious effect on the quality of the layer. It is better to add a vacuum attachment onto the apparatus in order to have the facility for degassing substrates. Typical figures are 1200 °C for the temperature, and 10^{-5} torr (1·3 mPa) for the residual pressure.

IV. CONCLUSIONS

Until now, the main industrial applications of CVD have been the preparation of refractory coatings (e.g. Mo, W, Re and SiC) and CVD epitaxy of silicon films. The field of investigation is still expanding at laboratory level.

REFERENCE

Kubaschewski, O., Evans, E. V. and Alcock, C. B. (1967). "Metallurgical Thermo-chemistry", pp. 23–24. Pergamon Press, Oxford.

Deposition of metal, carbide and oxide films by thermal decomposition of metal acetylacetonates

A. POLITYCKI AND K. HIEBER

Forschungslaboratorien der Siemens AG, Munchen, West Germany

I. INTRODUCTION

One of the various methods of Chemical Vapour Deposition (CVD) is pyrolysis, which is understood as the thermal decomposition of a substance to either smaller molecules or its elements. Examples are: the metal powder production by pyrolysis of carbonyls and the manufacture of carbon resistors by high temperature decomposition of hydrocarbons.

II. DECOMPOSITION OF METAL ACETYLACETONATES

Recently, metal acetylacetonates have been used as a starting material for pyrolytic decomposition as demonstrated by Schmeckenbauer and Schneider (1960), Papke and Stevenson (1967) and Ryabova and Savitskaya (1968a, b). These crystalline compounds are not poisonous and they can be evaporated and decomposed at low temperatures. Their chemical constitution is indicated in Fig. 1. Acetylacetone is derived from acetone by replacing a hydrogen atom by the group $CO-CH_3$. It exists in the tautomeric forms keto and enol. The hydrogen of the OH-group in the enol can be replaced by many metals. The acetylacetonates (Acac) are not considered to be organometallic compounds because the metal bonding is not direct to the carbon atom.

The thermal decomposition of acetylacetonates can result in different products, depending on the metal and conditions of pyrolysis. The metal may be deposited as an element, carbide or oxide and sometimes mixtures will appear. Examples of the first group are copper, iron and cobalt. Nickel can be deposited as a metal or a carbide. Chromium belongs to the second group, which forms carbides or oxides. Decomposition of compounds which contain metals with a strong bonding force to oxygen, such as aluminium

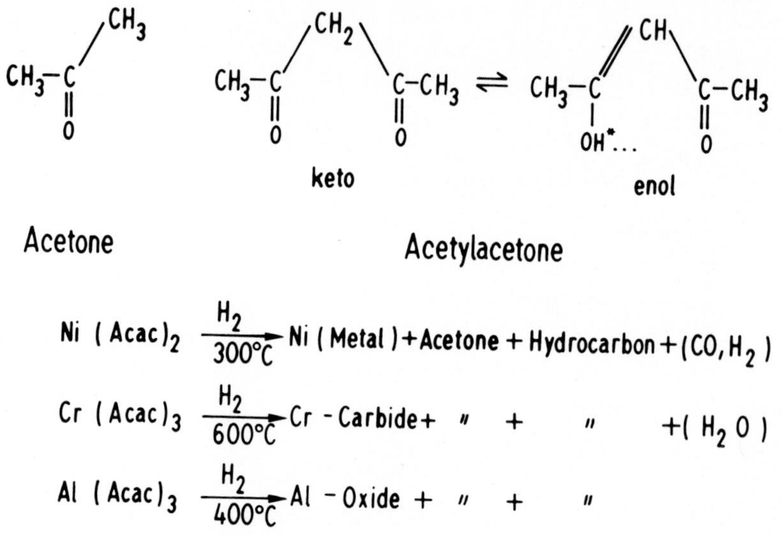

FIG. 1. Decomposition of metal acetylacetonates.

and vanadium, results in oxides, even if hydrogen is used as a carrier gas. In all cases, volatile hydrocarbons are to be observed as coproducts of the decomposition. They are partly trapped by the film growing at the surface of the hot substrate. By subsequent annealing, these hydrocarbons can be evaporated or decomposed. In the latter case carbon will be incorporated in the deposit. The carbon content of the films generally increases with the temperature of pyrolysis.

III. EXPERIMENTAL

A. Apparatus

Our investigations were restricted to thin films deposited on glass or ceramic plates of 40×40 mm. The apparatus for the pyrolysis is shown in Fig. 2.

The reaction chamber consists of a 10 cm diameter quartz tube, which can be evacuated. Furnaces, heated by direct current and containing crucibles with the starting material, are introduced from the right-hand side. The substrate is mounted on a rotating metal disc which is placed opposite the evaporation source. The disc supports the wires for the current supply, the

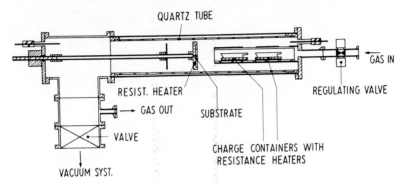

FIG. 2. Apparatus for thermal decomposition of acetylacetonates.

thermocouple and two contacts. Before carrying out resistivity measurements, two strips of Cr/Ni and gold were evaporated onto the substrate and connected to spring wires. Hydrogen or ammonia was used as a carrier gas, thus favouring reducing conditions during the reaction. The flow rate was maintained at one litre per minute by a needle valve at the gas entrance on the right.

B. Procedure

After having mounted the cleaned substrate onto the furnace disc and having filled the evaporator with acetylacetonate crystals, the tube was evacuated to remove moisture and oxygen. Then, the gas flow was started and the rotating substrate disc was heated to a constant temperature. The evaporator was heated and pyrolysis began. Film growth at the surface of the substrate was controlled by recording the resistance between the two contact strips with a chart recorder. Resistance curves obtained during the growth of nickel films, showed that after a certain incubation time, depending on the amount of evaporated material and carrier gas, films of good conductivity were built up.

IV. RESULTS

A. Deposition of metal

The film growth depends largely on the nucleation rate at the surface, which can be influenced by the choice of substrate and the pretreatment. Figure 3

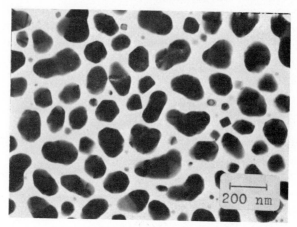

FIG. 3. Copper deposit on glass substrate.

shows electron micrographs of copper deposited onto ultrasonically cleaned glass. The crystallites are agglomerated into islands. According to Brenner (1954), prenucleation of the glass by dipping it into diluted tin chloride and palladium chloride, however, resulted in a continuous metal film (Fig. 4) even after a shorter deposition time.

Thickness measurements, performed by applying interferometric and mechanical precision methods ("Talystep" Taylor Hobson Ltd.), confirmed that there were no local differences in film thickness. In spite of the fact that the electron diffraction patterns sometimes indicate the presence of a small

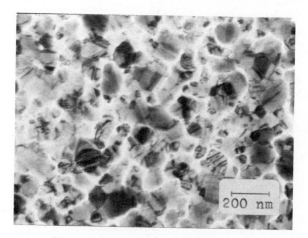

FIG. 4. Copper deposit on prenucleated glass.

amount of carbon within the film, the conductivities of nickel and copper films with a thickness of about 0·1 µm were very good, at least after annealing for a short time at 450 °C. Table I shows that the resistivities of copper and nickel films are not very different from those of evaporated films. There is, however, an important influence from the temperature of pyrolysis.

TABLE I
Resistivities of metal films

Metal	Resistivity (µΩcm)*		Decomposition Temperature (°C)
	Evaporated	Pyrolyt. deposited	
Cu	4	4	300
Fe	42	87	390
Co	38	79	400
Ni	12	17	300

*After annealing 15 min. at 450 °C

According to chemical analyses, the carbon content of nickel films increases linearly with the temperature of decomposition. Electron micrographs and electron diffraction patterns revealed the structural difference between a low and a high temperature film. The low temperature (350 °C) deposit consists of metal with only small amounts of carbon; the high temperature (550 °C) product, however, is an amorphous film of carbon, containing isolated metal particles. These observations are consistent with resistivity measurements which showed a change for nickel from 17 µΩ cm, to about 10^5 µΩ cm as demonstrated by Schaefer et al. (1972). In the same range of temperature, the temperature coefficient of resistivity changes from + 4000 p.p.m. per °C to about − 1000 p.p.m. per °C the latter corresponding to the value for carbon resistors as found by Morgan (1971).

B. Deposition of carbide

As previously mentioned, the decomposition sometimes results in the formation of carbides. We observed that pyrolysis of nickel acetylacetonates between 380 °C and 400 °C resulted in films consisting of only nickel carbide. Figures 5 and 6 show an electron micrograph and the electron diffraction pattern of such a deposit, corresponding to the Ni_3C structure. This compound is not very stable. After annealing for a short time at 450 °C it decomposed into nickel and carbon, as seen from the diffraction pattern which showed characteristic lines of the face centred cubic metal and amorphous carbon.

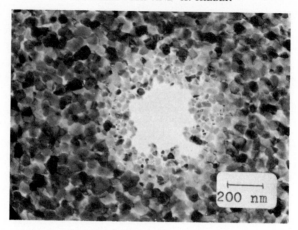

FIG. 5. Electron micrograph of a Ni₃C film.

Another film of carbide, which is stable up to very high temperatures, can be produced by pyrolysis of chromium acetylacetonate. Growth of films with poor conductivity starts at 350 °C. Above 600 °C the decomposition results in nearly amorphous films (Fig. 7). From the diffuse electron diffraction pattern (Fig. 8) it was concluded that the deposit consists of the carbides Cr_7C_3 or $Cr_{23}C_6$. Heating the specimen to very high temperatures by electron bombardment did not change the structure. As known from the literature, chromium carbides are extremely hard and very resistant to chemical agents. A pyrolytic deposition of chromium carbide films seems, therefore,

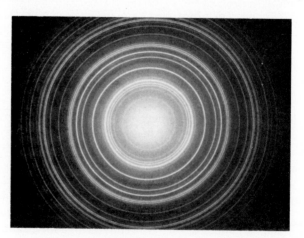

FIG. 6. Electron diffraction pattern of a Ni₃C film.

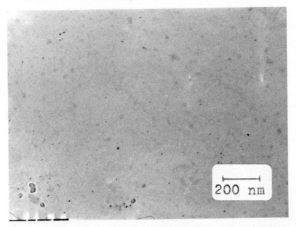

FIG. 7. Electron micrograph of an amorphous chromium carbide film.

FIG. 8. Electron diffraction pattern of an amorphous chromium carbide film.

advantageous for the protection of metal surfaces against wear or corrosion. The production temperature of 600 °C is very low in comparison with other methods of preparing chromium carbides. In addition to this, the pyrolytic surface coating technique is not too expensive.

C. Deposition of oxide

Mixtures of carbides and oxides result when vanadium acetylacetonates are decomposed at 500 °C. The films are nearly amorphous and very dense.

Only oxide was detectible, however, when ammonia was used as a carrier gas instead of hydrogen. The diameter of the grains corresponded to 5–10 nm. According to the diffraction pattern, the oxide corresponds to the structure V_2O_3. After annealing by strong electron bombardment, a coarsened grain structure was obtained as shown in the micrograph in Fig. 9.

Much more importance is to be attached to the deposition of alumina. On account of their good insulation properties and stability, alumina films have several applications, e.g. for the passivation of diodes in the semi-conductor industry. Krongelb (1969) and Tung and Caffrey (1967) showed

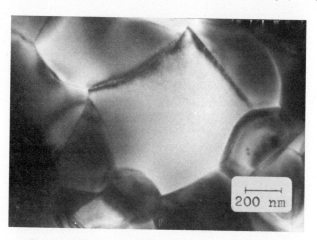

FIG. 9. V_2O_3 film coarsened by electron bombardment.

that alumina films are more resistant to diffusion of alkaline ions and moisture than silica deposits. It is, therefore, of importance to have a suitable coating technique. Besides sputtering and some other CVD reactions, there is a good chance that pyrolysis may be used also. Thermal decomposition of aluminium acetylacetonate in hydrogen results in film growth above 350 °C. Electron micrographs did not reveal any pores or cracks and voltage tests demonstrated good insulation. A change in structure was observed only after annealing above 1000 °C. Up to this temperature the insulation properties of these films should, therefore, be preserved. In addition, this coating technique has an excellent throwing power as shown by Glaski (1967). Figure 10 shows an alumina deposit on etched copper which follows exactly the profile of the rough substrate. Profiled specimens may, therefore, be coated with continuous films.

It should be mentioned that pyrolysis above 500 °C resulted in a codeposition of carbon. The deposits appear grey or even black in colour and the electrical properties are changed. These films may be of interest for the production of special resistors, similar to the nickel–carbon films. For the preparation of highly insulating deposits, the low temperature pyrolysis is adequate. The presence of hydrogen prevents thermal oxidation of substrate metals such as iron or copper. The reducing conditions are, however, not essential for the decomposition. On the contrary, high temperature pyrolysis in oxygen or mixtures of oxygen with argon or nitrogen results in alumina deposits without any carbon content.

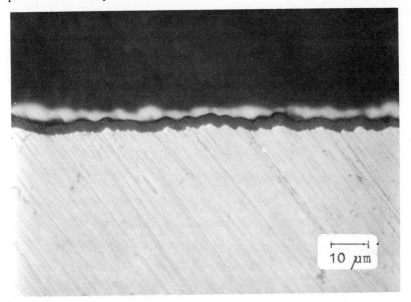

FIG. 10. Pyrolytically deposited alumina on Cu.

V. CONCLUSION

It has been shown that pyrolysis may be applied to produce surface films, consisting of a metal, carbide or oxide. The thickness of the deposit is not restricted and may be increased in a wide range. Furthermore, it is possible to deposit pyrolytically, certain films (e.g. alumina, chromium carbides) at low temperatures which, by other methods, require expensive equipment.

REFERENCES

Brenner, A. (1954). *Metal Finishing*, Dec., 61–68.
Glaski, F. A. (1967). Proc. Conf. on CVD Gatlinburg, Tennessee, pp. 275–289. Am. Nuclear Soc.

Krongelb, S. (1969). *J. Electrochem. Soc.* **116**, 1583.

Morgan, M. (1971). *Thin Solid Films* **7**, 313–323.

Papke, J. A. and Stevenson, R. D. (1967). Proc. Conf. on CVD Gatlinburg, Tennessee, pp. 197–198. Am. Nuclear Soc.

Ryabova, L. A. and Savitskaya, Ya. S. (1968a). *Thin Solid Films* **2**, 141–188.

Ryabova, L. A. and Savitskaya, Ya. S. (1968b). *J. Vac. Sci. Techn.* **6**, 934–937.

Schaefer, A., Hieber, K. and Politycki, A. (1974). To be published in *J. Materials Techn.*

Schmeckenbauer, A. and Schneider, E. F. (1960). Sci. Report No. 1 Contract AF 19(604)–4978.

Tung, S. K. and Caffrey, R. E. (1967). *J. Electrochem. Soc.* **114**, 2750.

Chemical vapour deposition from an aerosol

J. C. Viguié

*Laboratoire d'Étude des Materiaux Minces, Département de Métallurgie,
Centre d'Études Nucléaires de Grenoble CEA, Grenoble, France*

I. Introduction

The idea of a solution spray process is not new. Transparent, electrically conducting films of tin dioxide, prepared by spraying a solution of tin chloride onto preheated glass substrates, have been known and used for years. The interesting feature is that the source material is not transported

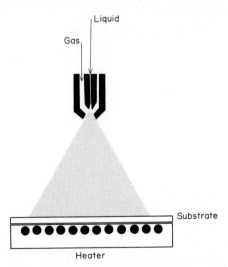

Fig. 1. Simple experimental set-up, using pneumatic atomization.

as a vapour but as a solution in droplets. This allows transport at room temperature. When the droplets reach the heated substrate, the solvent evaporates and a pyrolytic chemical reaction occurs.

II. EXPERIMENTAL DETAILS

A simple experimental set-up (Fig. 1) (Vorob'eva and Bessonova, 1964; Chamberlin and Skarman, 1966) can give a large variety of coatings such as metallic sulphides and oxygen-containing inorganic salts, e.g. silicates, phosphates and sulphates.

A. The deposition process

Two possible processes can occur with this arrangement. In the first of these (process A), the solvent evaporates as it approaches the substrate, then the solid salt melts and vaporizes in its turn. The gaseous species diffuse towards the substrate and the heterogeneous reaction occurs on the substrate surface (Fig. 2). This is genuine chemical vapour deposition. For example:

$$2FeCl_3 + 3H_2O \rightarrow Fe_2O_3 + 6HCl \tag{1}$$

$$SnCl_2 + \tfrac{1}{2}O_2 + H_2O \rightarrow SnO_2 + 2HCl \tag{2}$$

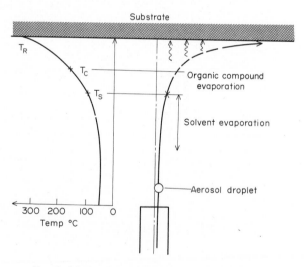

FIG. 2. Mechanism of deposition from an aerosol.

The second possibility (process B) is that the droplets splash onto the substrate and decomposition occurs in the solid. This happens when the temperature is too low or when the concentration of solution is too high.

Process A is believed to yield smooth, adherent films whereas process B yields coarser films. We imagine therefore that generally both processes occur. The larger droplets reach the substrate, the smaller ones evaporate before reaching it. In only a few instances can experimental conditions be adjusted to give process A alone.

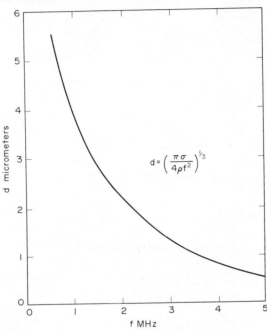

$$d = \left(\frac{\pi \sigma}{4\rho f^2}\right)^{1/3}$$

FIG. 3. Ultrasonic atomization—variation of mean diameter d with frequency f.

B. Atomization

In our attempts to produce droplets similar in size, we found some interesting properties when using ultrasonic atomization rather than the more conventional pneumatic atomization:

(i) The mean diameter d of droplets is related to the surface tension σ, the solution density ρ and the ultrasonic frequency f by:

$$d^3 = \frac{\pi \sigma}{4\rho f^2} \quad \text{(Bisa et al., 1954)} \tag{3}$$

This is shown in Fig. 3.

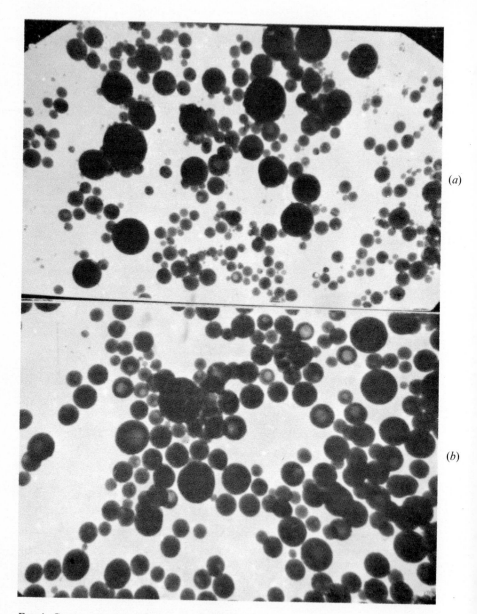

(a)

(b)

FIG. 4. Comparative size spectra of atomized droplets: (*a*) pneumatic generation; (*b*) ultrasonic generation.

(ii) The size spectrum of droplets is narrower (Fig. 4).

(iii) The production rate of the aerosol is independent of the gas flow. Since this property allows very much lower flow rates of the carrier gas, the substrate is not cooled. This in its turn helps process A.

C. Practical conditions

In the case of the preparation of ferric oxide from a solution of ferric chloride, the practical conditions of atomization are summarized in Table I.

TABLE I

Comparison of pneumatic and ultrasonic atomization techniques

	Pneumatic	Ultrasonic
Solution	$FeCl_3$ 0·2 mol/l	$FeCl_3$ 0·2 mol/l
Amount of solution consumed	1·4 ml	0·1 ml
N_2 flow rate	6·2 l/min	6·2 l/min
Duration of deposition	5 min	25 min
Temperature of the substrate	115 °C	225 °C
Surface area	5 cm^2	3 cm^2
Thickness	1500 Å (150 nm)	1500 Å (150 nm)
Distance between the nozzle and the substrate	4 cm	1·5 cm

The deposited layers are either amorphous or finely crystallized, depending on the temperature. Metals are often in the amorphous state.

The deposition rate decreases with thickness. 2000 Å (200 nm) layers are prepared quite rapidly. But very often the layers become coarse as the thickness approaches one micrometre.

CONCLUSIONS

The range of compounds which can be deposited is presumably smaller than for classical CVD, owing to the lower reaction temperature. On the other hand, the aerosol method produces several advantages of practical interest:

(i) The source material does not need to be heated.

(ii) No early decomposition will occur until the aerosol is in the neighbourhood of the substrate.

(iii) Deposition of mixed compounds in a well defined composition is achieved by adjusting the solution composition. No separation during the transport occurs.

(iv) The use of organometallic compounds allows moderate reaction temperatures (from 300 to 500 °C). Thus a larger variety of substrates can be used.

(v) Owing to its simplicity, the method is less limited in sample size. Large surfaces can be covered. Continuous coating operations can be set up. Figure 5 shows an installation for continuous coating operations. In this example, glass substrates of 2×2 in. (5×5 cm) are covered with a 2000 Å (200 nm) layer of transparent ferric oxide within 10 minutes.

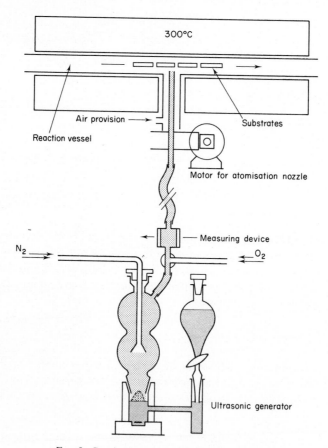

FIG. 5. Continuous coating from an aerosol spray.

REFERENCES

Bisa, K., Dirnagl, K. and Esche, R. (1954). Siemens-z, 341.
Chamberlin, R. R. and Skarman, J. S. (1966). *J. Electrochem. Soc.* **113**, 86–91.
Vorob'eva, O. V. and Bessonova, E. S. (1964). *Steklo i Keramika* **21**, 9, 9–13.

The basic principles of electroless deposition

M. SCHLESINGER

Department of Physics, University of Windsor, Windsor, Ontario, Canada

I. INTRODUCTION

Electroless deposition of thin metal films has recently attracted the attention of many scientists and technologists interested in surface coating. This is largely so because of the considerable technical potential that this film has in the area of printed circuits, magnetic tapes, etc. Electroless cobalt, for instance, may be prepared with magnetic characteristics applicable to digital recording media. From a basic scientific point of view, the electroless system represents a unique case of a metal or alloy midway between the solid and liquid states.

One of the rather attractive characteristics that these films have is the relative ease and simplicity with which they can be deposited.

Although quite a large number of chemical systems for electroless deposition have been developed (e.g., Ni, Co, Cu, Au, Fe) little is understood about the actual mechanisms governing the physical chemistry of these systems. Some progress has been made, however, in recent years and this gives good reason to believe that in the not-too-distant future many of the processes involved will be recognized and understood. Such understanding, no doubt, will broaden even more the many uses of electrolessly deposited systems.

II. GENERAL

The electroless deposition of some metals and their alloys on catalytic surfaces is well documented (see, for example, Marton and Schlesinger, 1968).

Deposition on dielectric (non-catalytic) surfaces has also been demonstrated. A dielectric surface may be sensitized and activated by immersion in solutions of $SnCl_2$ and $PdCl_2$ for electroless deposition.

Clean substrates of, say, glass, quartz, mica or Formvar, can be sensitized and activated at room temperature by immersion in the proper solutions. Possible variations in activation of substrates are shown in Table I. As an actual example the composition of a solution to be used in electroless nickel (Ni–P) deposition is given in Table II. Formulae for other systems are available in the scientific literature.

TABLE I

Variations in Activation of Substrates

Method (a)	Dip clean substrates in activating solutions listed in Table II.
Method (b)	Dip clean substrates as in (a) in activating solutions modified by the addition of 10 ml/l isopropyl alcohol.
Method (c)	Dip clean substrates as in (a) in activating solutions modified by the addition of 5 ml/l Kodak "Photo-Flo" (Kodak Ltd.).

TABLE II

Compositions of solutions used in the experiments

Activating solutions			Electroless solution	
$SnCl_2$ solution	$SnCl_2$	0·1 g/l	Nickel sulphate	29 g/l
	HCl	0·1 ml/l	Sodium hypophosphite	17 g/l
Rinse	H_2O at pH = 7		Sodium succinate	15 g/l
$PdCl_2$ solution	$PdCl_2$	0·1 g/l	Succinic acid	1·3 g/l
	HCl	0·1 ml/l	Solution pH	5·3
Rinse	H_2O at pH = 7		Solution temperature	25 °C

In Fig. 1 (after Marton and Schlesinger, 1968) we present the plot of what is essentially equivalent to the film thickness (at early stages) versus time of immersion in the metallizing solution for Ni–P (or electrolessly deposited nickel). A somewhat faster rate of deposition is observed for Cu and a slower one for Co.

III. NUCLEATION AND GROWTH

Metal films deposited chemically by the electroless method have a different growth pattern than those deposited in vacuum. This difference is two-fold: first, the substrates have to be catalytic with respect to the specific metal deposited and second, the metal deposition occurs with the substrate immersed in a solution rather than in vacuum.

Cleaned dielectric substrates may be catalysed as described above. This treatment produces nucleation sites for electroless Ni, Cu, Co and Au growth with density of sites between $10^{10}/cm^2$ and $10^{12}/cm^2$ depending on the

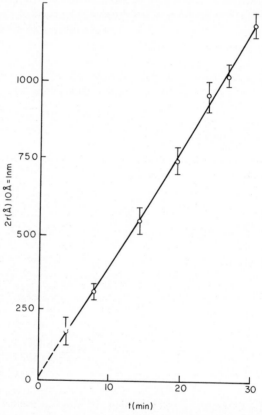

FIG. 1. Plot of the mean diameter $2r$ of the growing Ni–P islands against time of deposition. (Note that the rate of growth dr/dt is a constant.)

hydrophilic or hydrophobic nature of the substrate. This range of activation site densities is of the order of that of nucleation densities found for vacuum deposited films. The sites are of the size of about 10 Å (1 nm) (Marton and Schlesinger, 1968). The exact mechanism by which active sites develop on the substrate is uncertain. The most likely process involves the adsorption of tin oxides on the substrate and then an exchange of tin and palladium ions in the $PdCl_2$ solution. The tin ions precipitate during the water rinse that follows the $SnCl_2$ immersion.

The mechanism of the actual metal deposition on the catalytic sites is also not accurately known. Chemically the ionic reaction has been described (Goldenstein *et al.*, 1957) but physically no accurate model has yet been proposed. It is possible that the processes involved are similar to those in electrodeposits. It has been shown (Marton and Schlesinger, 1968), however, that deposition occurs on the active sites only. The growth of the metal on

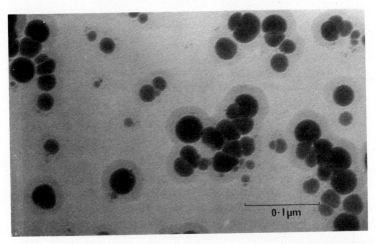

0·1 μm

FIG. 2. Transmission electron micrograph of an electrolessly deposited thin Ni–P film. Deposition time = 5 min. Note the regular shape of islands.

the sites is isotropic for Ni, this is shown in Fig. 2. Such does not, however, seem to be the case for, say, Cu or Co grown electrolessly (see Figs 3 and 4). In these latter cases it is still likely that the initial stages of growth are similar to the ones shown for Ni except that at relatively early stages islands may wander and diffuse together on the surface rendering bigger islands with irregular shapes.

Naturally at early stages of growth the films will be discontinuous and only at later stages will the individual islands touch and form a truly continuous film. This manifests itself in "anomalous" electrical properties for very thin [< 200 Å (20 nm)] i.e. discontinuous films (see below).

Generally, the structure of metal films is sensitive to impurities. In the case of electrolessly deposited metal films, first there is palladium in atomic quantities at the core of each island, second there is phosphorous present in various quantities. The latter is a result of the chemical reduction of the metal

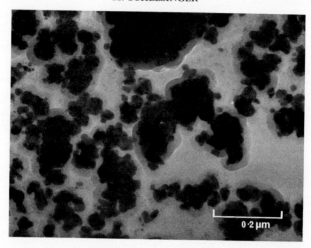

FIG. 3. Transmission electron micrograph of an electrolessly deposited Cu–H film. Deposition time = 0·5 min. Note the irregular
shape of islands and their partial crystallinity.

ions from solution (see Table II above). Finally there is hydrogen present as
the by-product of the chemical deposition process. The composition, for
instance, of freshly deposited electroless nickel is (Pai *et al.*, 1972) a Ni–P–H
system with a liquid-like structure. If it is annealed at an elevated temperature
(Schlesinger *et al.*, 1972) hydrogen is desorbed, the atomic mixture of Ni and

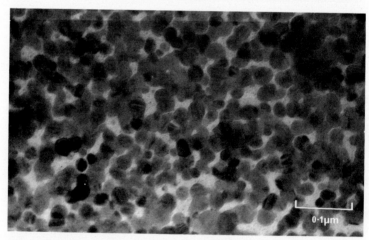

FIG. 4. Transmission electron micrograph of an electrolessly deposited Co–P film.
Deposition time = 2 min. Note the irregular shape and sizes of islands and their
degree of crystallinity.

P forms the compound Ni_3P and crystalization will take place. The mixed state consists of tetragonal Ni_3P and f.c.c. Ni. Other chemical effects, such as oxidation in air, further complicate the situation, and will not be discussed here.

IV. Selective deposition

An important property of electroless systems is their ability to deposit metals selectively. This has been demonstrated by D'Amico et al. (1971), who de-sensitized portions of a surface with ultra-violet (u.v.) radiation. The same effect has been observed by other workers. In all this, u.v. was applied after sensitization (the $SnCl_2$ bath). We have observed (Chow et al., 1972) that for cobalt and nickel, u.v. radiation is effective in inhibiting electroless de-position even if it is applied after activation (the $PdCl_2$ bath). The u.v. source in our experiments was a 50 watt Hg lamp held at about 2 in. (5 cm) from the substrate for 10 minutes. Utilizing this difference between, say, Cu and Ni, we could deposit their alloys in different proportions selectively, depending on the time and stage of u.v. application. A detailed study (Chow et al., 1972) of electron diffraction patterns exhibited by sensitized and activated Formvar, before and after u.v. irradiation, was conducted by us. Our conclusion is that microstructural changes in the sensitizing and activating agents induced by the u.v. light are connected with the inhibition of metal deposition. In con-cluding this section, it should be noted that the utilization of "u.v. inhibition" in the production of printed circuits is already a fact and "resolutions" of about 1 mil (25 μm) are easily achieved.

V. Electrical properties

The electrical characteristics of thin electrolessly deposited films are of interest from two points of view. Firstly the practical applications of electroless systems involve electric charge transfer (current) or charge storage in one way or other. Consequently a better knowledge of the electronic processes might be expected to lead to a more efficient use of these films. Secondly, thin discontinuous electroless films can be considered as one of the very few reproducible systems approaching amorphous semiconductors. These are presently considered as one of the most interesting problems in basic solid state physics.

Essentially only one aspect of electrical conductivity in electroless films will be discussed here. Some information as well as a partial list of references relating to Ni is available (Schlesinger and Marton, 1969). An example of I (current) versus V (voltage) on a log–log scale is given in Fig. 5 for some electrolessly deposited Cu–H films. The slope of each of the curves is representative of the film's resistivity. In the thinner films it remains constant while in the thicker ones it is field (voltage) dependent. This is an indication that the electric transport properties are quite different in films of varying thicknesses These differences, while only partially understood at present, are an important key to the further development of electroless systems as useful active and passive electronic elements.

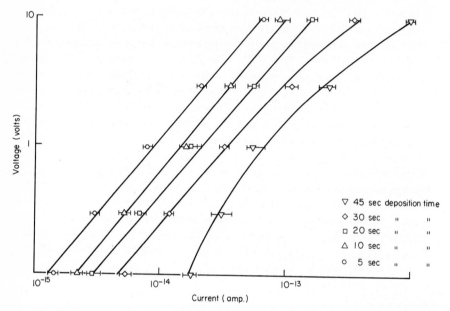

Fig. 5. Current versus electric field curves for Cu–H films of different thicknesses.

REFERENCES

Chow, S. L., Hedgecock, N. E., Schlesinger, M. and Rezek, J. (1972). *J. Electrochem. Soc.* **119**, 1013.

D'Amico, J. F., De Angelo, M. A., Henrickson, J. F., Kenney, J. T. and Sharp, D. J. (1971). *J. Electrochem. Soc.* **118**, 1695.

Goldenstein, A. W., Rostouer, W., Schossberger, F. and Gutzeit, G. (1957). *J. Electrochem. Soc.* **104**, 104.

Marton, J. P. and Schlesinger, M. (1968). *J. Electrochem. Soc.* **115**, 16.

Pai, S. T., Marton, J. P. and Brown, J. D. (1972). *J. Appl. Phys.* (in Press).

Schlesinger, M., Hedgecock, N. E. and Chow, S. L. (1972). Unpublished results.

Schlesinger, M. and Marton, J. P. (1969). *J. Appl. Phys.* **40**, 507.

Application of electroless and electrolytic metal deposition in semiconductor connection technology

A. POLITYCKI AND W. STOEGER*

Siemens Aktiengesellschaft, Forschungslaboratorien, München, West Germany

I. ONE-LAYER INTERCONNECTION STRUCTURES

The manufacture of semiconductor devices is based on elaborate surface coating techniques. One of them, the photolithographic method (photoresist technique), which enables smooth surfaces to be subdivided into very small regions, was a prerequisite for the development of integrated circuits. Today the manufacture of semiconductor chips of a few square millimetres size supporting more than a thousand transistors and the necessary conducting lines is possible. Sometimes, however, it is a problem to find a suitable technology for the connection of the extremely small conducting paths (drive lines) on the surface of the semiconductor with an appropriate system of more spacious and broader metal lines on a substrate, which serves for the mounting and may be handled with less difficulty. One solution to this problem is the "flip-chip technique" published by Sideris (1965). Others are the "beam-lead technique" developed by Lepselter (1966) and the "inverted beam-lead technique" demonstrated by Cohen et al. (1971).

Recently another multibond technique for integrated circuits was described by Clark (1971); a polyimide foil is punched to generate apertures for chip insertion and then clad on one side with a copper foil. Following the photolithographic technique, most of the copper is dissolved, leaving a pattern of metal lines with free supporting ends (inverted beam-leads) cantilevering into the apertures of the insulating foil. This interconnection

Present address:
*Max Planck Institut für Festkörperforschung, Heilbronner Str. 69, Stuttgart, Germany.

structure is aligned by a micropositioner and lowered into contact with the surface of a semiconductor chip. Then, in a single operation, the beam-leads are bonded to the metal pads of the integrated circuit. The diagram in Fig. 1 shows a connection structure of this type. The chip is welded with its back surface to a ceramic plate, which acts as a support and a heat sink. The copper lines are attached to the insulating foil while their free supporting ends, the beam-leads, are bonded to the chip in a way that yields flexible

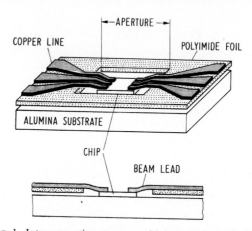

FIG. 1. Interconnection structure with inverted beam-leads.

connections. At the outer terminals of the interconnection lines, spacious exit tabs can be bonded to conducting wires or to a thick film metal pattern, deposited onto the ceramic substrate.

II. TWO-LAYER STRUCTURES WITH BEAM-LEADS

A. Preparation by plating-up technique

In memory devices a great number of connection lines are required and they should be as short as possible. Two-layer interconnection systems with many crossovers and through-holes (feed-throughs) are, therefore, indispensable.

A suitable preparation technique has been published by Marley and Trolsen (1969): the base material is a polyimide foil of 25 μm thickness which is pre-shrunk by a treatment at an elevated temperature. Thin films of evaporated chromium (as adhesive) and copper are reinforced by electrolytic deposition of copper in a total thickness of about 12 μm. The next processing steps are indicated in Table I and Fig. 2. By application of the conventional photoresist and etching technique, a pattern of small holes in the metal deposits of both sides is produced (step 1). The next step is the perforation of the polyimide by hot concentrated sodium hydroxide solutions. During this procedure the copper serves as a mask. Afterwards the metal is

TABLE I

Processing steps for preparation of two-layer structures by plating-up technique

1. Etch holes into the metal deposit
2. Hence perforate polyimide, then remove metal
3. Re-metallize the foil
4. Coat with resist and develop a reversed pattern of conductor lines
5. Deposit nickel by electrolysis
6. Etch rectangular openings (windows) in the copper
7. Generate apertures by dissolving polyimide
8. Remove all metal between the nickel-lines

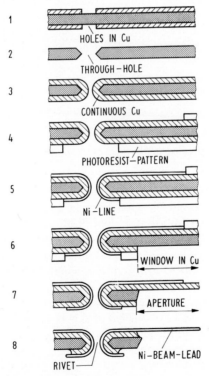

FIG. 2. Processing steps for preparation of two-layer structures by plating-up technique.

removed (step 2) and the foil is coated again with chromium and copper. This time, care has to be taken to ensure that the metal deposits onto the inside walls of the holes. The copper film is reinforced by electrolysis to about 15 μm (step 3). Then the two sides of the surface are covered with photoresist of at least 4 μm thickness. A reversed pattern of the desired conducting line (width, e.g. 50 μm) is generated by using an appropriate photomask for the illumination (step 4). The unprotected parts of the surface are now electrolytically filled with nickel, thus delineating the system of conductor paths (of about 50 μm width and 4 μm thickness) on a continuous copper deposit of 15 μm thickness (step 5). The resist is then illuminated a second time by using another photomask, and rectangular openings (windows) in the copper deposit are produced by etching (step 6). Then, the apertures with the cantilevering beam-leads are generated by dissolving those parts of the polyimide which are not protected by copper (step 7). Finally, the copper deposit and the thin chromium film between the nickel lines are removed by etching (step 8). With this step the structure of the bi-metal lines is accomplished and the system is ready for application.

The metal lines on both sides of the insulating foil are connected at certain points by through-holes (feed-throughs). Near the edges of the apertures they are attached to the support by rivets. They differ from feed-throughs by the fact that the anchored lines do not continue to the other side of the foil. Metallographic cross sections revealed the undercutting of the copper and showed that the beam-leads consist of only nickel with a thickness of 4 μm.

B. Preparation by etch-back technique

There is no doubt that the flexibility of extremely thin beam-leads will be advantageous for the bonding technique. It seemed, however, doubtful if their conductivity would be good enough for application in high speed memory devices. In addition, the deposition of adhesive films by vacuum evaporation is time-consuming and expensive. We therefore developed a different technique with electroless metallization.

The polyimide foil, held by a frame, is roughened and pre-shrunk by the following treatment for half an hour at 240 °C. After activation by dilute tin chloride and palladium chloride, the foil is electrolessly metallized with silver. The thickness amounts to about 0·2 μm. It is increased by electrolytic deposition of copper to 5 μm. The following processing steps are indicated in Table II and Fig. 3. The through-holes in the organic foil are produced in the same way as before (steps 1 and 2). After removal of the metal, the perforated polyimide is re-metallized with silver, which is also deposited within the walls of the holes. The electrolytic copper deposit reinforces both the metal film on the surface and the thin metal tubes within the holes. This electrolysis is performed until a desired copper thickness, e.g. 12 μm, is achieved (step 3). We used a commercial copper sulphate bath.

TABLE II

Processing steps for preparation of two-layer structures by etch-back technique

1. Etch holes into the metal deposit
2. Perforate polyimide, then remove metal
3. Re-metallize the foil
4. Produce the line structure by photolithographic technique
5. Coat with negative resist and develop an image of apertures
6. Generate apertures by dissolving polyimide
7. Coat beam-leads electrolytically with gold
8. Remove the resist

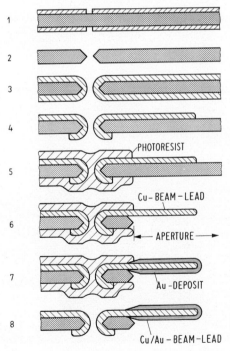

FIG. 3. Processing steps for preparation of two-layer structures by etch-back technique.

The surface is now covered with photoresist on both sides and, after having generated a positive pattern of the desired line structure, the metal is etched using a warm solution of iron chloride (step 4). In order to remove the silver,

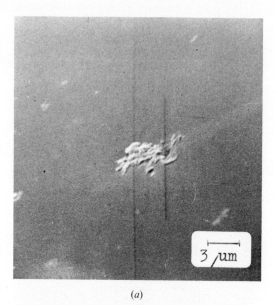

(a)

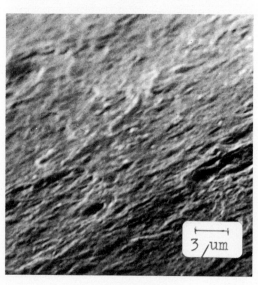

(b)

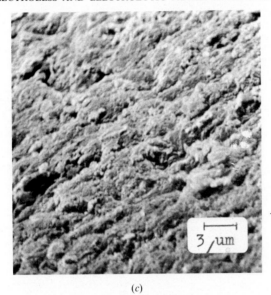

(c)

FIG. 4. Scanning electron micrographs of polyimide surfaces (a) untreated, (b) chemical
roughened, (c) sandblasted.

the sample is dipped several times into aqueous solutions of bromine and
potassium bromide, followed by diluted sodium thiosulphate. By this procedure the line structure is accomplished. The next steps are to cover the
whole sample with negative photoresist (withstanding hot alkalines) and to
develop an image of the desired apertures for the chips (step 5). Afterwards
the apertures can be produced by dissolving the unprotected parts of the
polyimide in hot concentrated alkalines (step 6). The ends of the copper lines,
which are now relieved from their support (beam-leads), are electrolytically
coated with gold or tin (step 7). Finally, the resist is removed (step 8) and the
samples are treated for some hours at an elevated temperature to improve
adhesion of the metal.

This reported technique seems to be advantageous. It is of importance,
however, to know the conditions for achieving sufficient adherence of the
metal to the foil. Figure 4 shows scanning electron micrographs of the polyimide surface before and after treatment. The chemical roughening was performed by dipping the foil into a mixture of sulphuric and hydrochloric acids,
then into a hot solution of sodium hydroxide. Sandblasting yielded the most
fissured surface. The dependence of the adherence of the deposited metal
coatings on the pretreatment of the foil, the metallization and the ageing
temperature, was tested by measuring the peel strength of etched lines of

TABLE III

Peel strength of copper lines on polyimide

Pretreatment of the polyimide	Shrinkage	Metallization	Ageing temp. (1 hour, vac.)	Peel strength [N/cm]
HCl/H_2SO_4; NaOH	30 min 240 °C	0·2 μm Ag + 12 μm Cu	80 °C	0·20
HCl/H_2SO_4; NaOH		0·2 μm Ag + 12 μm Cu	160 °C	0·23
HCl/H_2SO_4; NaOH		0·2 μm Ag + 12 μm Cu	240 °C	0·07
Sandblasting	30 min 240 °C	0·2 μm Ag + 12 μm Cu	80 °C	1·90
Sandblasting		0·2 μm Ag + 12 μm Cu	160 °C	2·21
Sandblasting		0·2 μm Ag + 12 μm Cu	240 °C	1·75
NaOH	30 min 240 °C	0·5 μm Cr + 12 μm Cu	80 °C	3·60
NaOH		0·5 μm Cr + 12 μm Cu	160 °C	3·75
NaOH		0·5 μm Cr + 12 μm Cu	240 °C	4·10

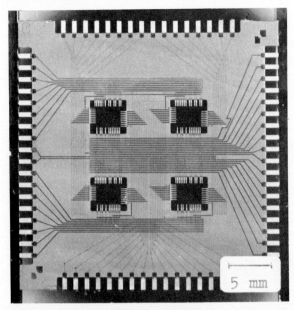

FIG. 5. Two-Layer multichip interconnection structure.

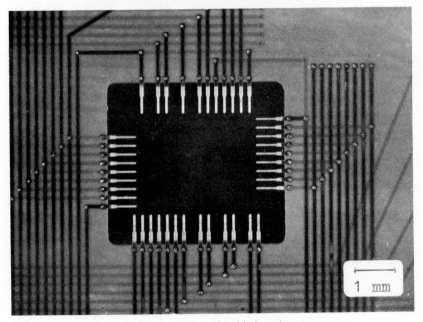

FIG. 6. Aperture for chip insertion.

1 mm width. They were attached to a spring balance and peeled at a constant speed in a direction 180 degrees against the lines. The results are presented in Table III. The sandblasted foils with their very rough surfaces proved to be much better than the chemically treated ones. An increase of the peel strength was noticed when the copper covered samples were heated for two hours at 160 °C. Although the peel strength is lower than that of lines

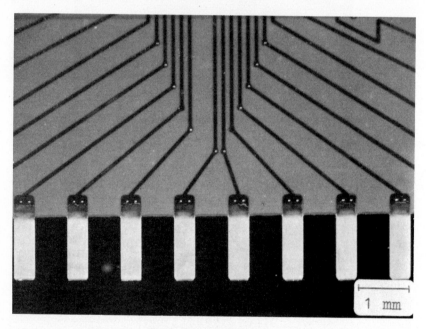

FIG. 7. Protruding exit tabs.

deposited on evaporated chromium–copper films (and lines etched out of commercial copper-clad polyimide laminates), the adherence of the lines on sandblasted foils proved sufficient.

Figure 5 shows a multichip interconnection structure, which has been fabricated by the preparation method described above. It has apertures and the complete drive-line system for four integrated circuits. This structure also comprises many major exit tabs protruding from the outer edges, which are to be attached to metal lines on the ceramic substrate. The enlarged sections in Figs 6 and 7 show details of the structure. The bright parts have an overlayer of gold. Metallographic cross sections showed that the thickness of the copper was 12 μm; the thickness of the gold covered beams was 20 μm. The line resistance of the beams is, therefore, very small. In the structure of

Fig. 8, an integrated circuit has been introduced and the beams are bonded to the aluminium pads of the chip by ultrasonic welding.

It is almost certain that there are still other ways to produce connection structures. The problem will always be how to meet the demands of high operation performance, reliability and low manufacture costs. This report has been intended to indicate that surface coating techniques play a very important role in this field.

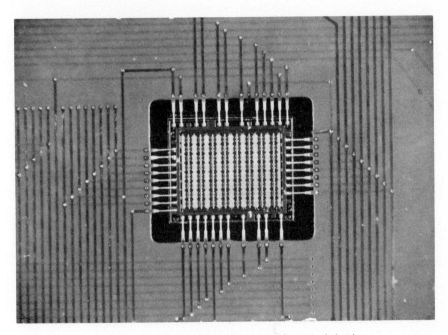

FIG. 8. Connection structure with bonded integrated circuit.

REFERENCES

Clark, M. E. (1971). Colloque International sur les Applications des Techniques du Vide à l'Industrie des Semiconducteurs, Versailles, 13–17 September, 1971.

Cohen, R. A., Bachner, F. J. and McMahon, R. E. (1971). *Metallurgical Transactions* **2**, 723–727.

Lepselter, M. P. (1966). *Bell Lab. Rec.* **44**, 9, 298–303.

Marley, J. and Trolsen, G. (1969). *Electronics*, December 22, 105–110.

Sideris, G. (1965). *Electronics*, June 28, 68–73.

The basic principles of wetting processes

ALBERTO PASSERONE

Centro Studi di Chimica e Chimica Fisica Applicata alle caratteristiche di Impiego dei Materiali, Genova, Italy

I. INTRODUCTION

Under the name of "wetting processes" we can include all the very large number of processes which aim to coat one or more solid phases with materials, such as varnishes, polymers, glasses or metals, which are in the liquid state when touching the solid. This process is widely used in the manufacture of composite materials and cermets, in the coating of metallic wires and in the sizing of glass fibres etc.

The liquid can be brought into contact with the solid by methods ranging from the hand-dipping of glass or graphite fibres into polymeric resins, to the hot-pressing process (sintering in the presence of a liquid phase), and to the covering of wires by continuous dipping in the coating bath. The conditions which govern these processes are not always the same. Even though it is possible to assert the hypothesis of thermodynamic equilibrium in some instances, in other cases (for example with continuous coating) kinetic factors are usually dominant.

II. Solid–Liquid Systems: Principles Governing their Equilibrium

A. The free surface energy and the surface tension of liquids

We will deal, first, with the conditions for the wetting of solid substrates by liquids. We will examine solid–liquid systems and especially interfacial interactions (indeed, the characteristics of the final product are due not so much to the bulk properties as to the surface properties of the phases).

The molecules on the surface of a liquid are in a very different condition from those contained within it. Because of the asymmetry of their potential field, the surface molecules are attracted more towards the bulk liquid and we must consequently expend some additional energy to move a molecule from the interior of the liquid to its surface.

1. Mechanical approach

Let us suppose, following Defay *et al.* (1966, p. 2), that we draw a line across the liquid surface (i.e. the liquid–vapour interface) so that it divides the liquid surface into two regions (Fig. 1). Region 2 produces a tensile action on region 1; this force can be expressed as $\gamma \, \delta l$, where δl is an infinitesimal element of the dividing line at the point P, and γ is a vector, named the "surface tension" at the point P whose modulus has the dimension of a force per unit length. This vector usually varies from point to point, but if it is

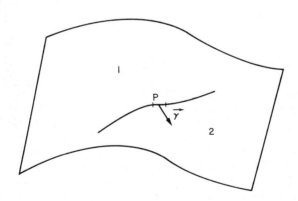

Fig. 1. Ideal interface divided by a line on which acts the surface tension.

always perpendicular to the dividing line, no matter how the line has been chosen, and if it has the same magnitude over all the surface (or interface), then the latter is said to be in a condition of "uniform tension".

It is possible to show that any liquid surface, under the action of surface tension, reaches a condition of mechanical equilibrium when the relation

$$p'' - p' = \gamma \left(\frac{1}{R_1} + \frac{1}{R_2} \right) \tag{1}$$

is satisfied. This expression, due to Laplace (1806), describes, in general, the profile of the interface between two fluid phases, whose interfacial tension is γ and in which the pressures are p' and p''. R_1 and R_2 are the principal radii of curvature at the point where eqn. (1) is applied.

From eqn. (1) we can immediately deduce that the interface between two fluid phases can be a plane (i.e. $R_1 = R_2 = \infty$) only if the pressures inside the two phases are identical. On the other hand, if $R_1 = R_2$ at every point, then the interface takes on a spherical shape.

Equation (1) does not take into account the pressure gradient due to the density of fluids; indeed in most cases the consequences of the gravitational field cannot be neglected and eqn. (1) must be rewritten as:

$$(p'' - p') + (\rho'' - \rho')gh = \gamma \left(\frac{1}{R_1} + \frac{1}{R_2} \right) \tag{2}$$

where h is the vertical distance (measured inside the denser phase) of a point of the interface from a plane of reference, and ρ', ρ'' are the densities of the phases.

Bashforth and Adams (1892), in their classical work, re-expressed eqn. (2) in the non-dimensional form

$$2 + \frac{\rho g}{\gamma} bZ = \frac{1}{R/b} + \frac{\sin \phi}{X/b} \tag{3}$$

which they solved numerically for the case of a surface formed by a liquid drop resting on a solid plane (Fig. 2). In eqn. (3), b is the radius of curvature at the apex of the drop and ϕ is the slope of the tangent at a point on the

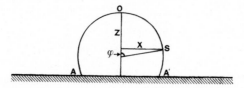

Fig. 2. Profile of a sessile-drop with the parameters used in eqn. (3).

profile of the drop. Z is the vertical distance of point S below the apex O, and X is the radius of curvature in the horizontal plane at point S.

It is possible, using eqn. (3) and, for example, the sessile drop method, to calculate the liquid surface tension when the density and the coordinates of some points of the drop profile are available. (For the various methods, see Ellefson and Taylor, 1938; Banneister and Hamill, 1968; Padday, 1969, 1972; Maze and Burnet, 1971.)

2. Thermodynamic approach

Suppose now we have a flat interface between two fluid phases ($p' = p''$) and extend it by an infinitesimal quantity dA; since this extension is made against the surface tension force, we must consider it according to the first law of thermodynamics:

$$dQ = dU + p\, dV - \gamma\, dA \qquad (4)$$

where dQ is the heat exchange, dU is the energy change, dV and dA the variations of volume and surface area respectively.

If the process is reversible, we can replace the exchange quantity dQ with $T\, dS$ (second law of thermodynamics) so that eqn. (4) becomes:

$$\gamma\, dA = dU - T\, dS - p\, dV \qquad (5)$$

As $dU - T\, dS$ is the free energy change dF in our system at constant temperature, it is easy to verify that, at constant volume,

$$\left(\frac{\partial F}{\partial A}\right)_{T,V} = \gamma \qquad (6)$$

In this case, and only in this case, the surface tension of a liquid and the free energy per unit area coincide.

The fundamental thermodynamic quantities on the surface of separation between the phases can be defined in a way completely analogous to that of bulk phases.

If f is the free energy per unit surface area, as a function of the variables: temperature, concentration in the bulk phases, and surface excess concentration (Γ_i), it is possible to show (Defay et al., 1966, p. 58) that between the free surface energy and the surface tension, there exists the relationship:

$$\gamma = f - \sum_i \Gamma_i \mu_i \qquad (7)$$

where the sum covers all the i components of chemical potential μ_i.

From eqn. (7) it is again evident that the surface tension and the free surface energy are two non-coincident quantities unless the term due to surface adsorption is zero (or negligible). This approximation is valid, for example, when dealing with pure liquids.

B. The solid surface tension

The extension to solids of the concept of surface tension involves some difficulties.

Following Gibbs (1948) treatment more closely, we can interpret the above defined f as the work done in forming a new surface of unit area. Gibbs points out that actually this quantity does not necessarily measure the mechanical tension existing in the surface. The latter arises in fact from a stretching of the surface whilst the quantity f, as defined in eqn. (7), depends on the work required to create a new surface of unit area. For liquids, owing to the great mobility of atoms (or molecules), the two quantities can be (and usually are) indistinguishable but for a solid this does not happen.

Let us consider a crystal: its state of surface stress is linked to movements and configurational changes of surface atoms which aim to minimize the total energy of the system. Moreover, if we fix a certain surface of the crystal, these stresses depend on the orientation, while f is a scalar quantity.

Indeed it is necessary to bear in mind that, as pointed out by Hondros (1970), every time we deal with the surface energy of a solid, its "surface energy" does not usually coincide with its "surface stress".

III. WETTING

A. Derivation of Young's equation

Following the analysis of Johnson (1959), let us consider a system formed by a homogeneous, continuous and isotropic solid in contact with a liquid in which it is not soluble; moreover both solid and liquid are in the presence of their vapours.

Equilibrium requires that the variation of the total free energy, at constant temperature (T), volume (V), and mass (n_i), be zero:

$$(\delta F)_{T,V,n_i} = 0 \tag{8}$$

F, the total free energy, is defined as

$$F = \int_v dF^v + \int_s dF^s + \int_v gz \, dm^v + \int_s gz \, dm^s \tag{9}$$

where v and s refer to volume or surface quantities, g is the gravity constant, z the height from a reference plane and m the mass of a volume (m^v) or surface (m^s) element.

Applying eqns. (8) and (9) and calculating all terms in their explicit form, we obtain a rather complex relationship describing in a completely general way the equilibrium of the system. Such a relationship, when dealing with a rigid solid plane surface and (if it is possible) neglecting the effect of the surface tension and the curvature of the liquid–vapour interface on the pressure inside the liquid, becomes

$$\int_{V^1} (\delta p + \rho g\, \delta z)\, dV^1 + \int_L (\gamma_{sl} + \gamma_{lv} \cos \theta - \gamma_{sv})\, \delta T\, dL = 0 \qquad (10)$$

where V^1 = liquid volume; p = pressure; δT = virtual shift of the line L where the three phases meet.

The first term in eqn. (10) takes into account the change of pressure due to the gravitational forces inside the liquid while the second term

$$\gamma_{sl} + \gamma_{lv} \cos \theta - \gamma_{sv} = 0 \qquad (11)$$

represents the Young (1805) relationship. Eqn. (11) must always be verified by the surface and interfacial tensions, but it cannot, by itself, completely represent the equilibrium conditions of the system unless, in addition to the hypothesis already made, we can neglect the effects of the gravitational field.

In eqn. (11) the contact angle between liquid and solid appears as a relevant parameter in describing the equilibrium conditions. The angle θ, relatively easy to measure, gives an empirical but widely accepted criterion to define the wettability of a solid by a liquid. If $\theta > 90°$, we assume that the liquid does not wet the solid while we assume the contrary if $\theta < 90°$ (Figs 3, 4). Only if $\theta = 0$ do we have complete spreading of the liquid, i.e. the system tends to replace the solid–vapour interface with the liquid–vapour one, evidently in a lower energetic state.

B. Effect of the surface roughness

Equation (11) requires that the measured contact angle coincide with the microscopic one. This happens, in theory, only if the solid surface is absolutely flat and homogeneous. Usually, however, the solid surface is not flat, being wrinkled and scratched, and these irregularities destroy the usefulness of eqn. (11) in predicting wettability.

Wenzel (1936), Cassie and Baxter (1944), Good (1952), and Shuttleworth and Bailey (1946), have fully treated the effects of roughness on the contact angle. The last two authors, in particular, have examined in detail the thermodynamic factors which lead to Wenzel's equation

$$\cos \theta^* = r \cos \theta_0 \qquad (12)$$

where θ_0 is the contact angle the liquid forms with the solid plane surface, while θ^* represents the apparent contact angle the same liquid forms on the surface of the same material, but of roughness r; r is defined as the ratio between the true area A and the apparent (or geometric) area A'

$$r = \frac{A}{A'} \tag{13}$$

Johnson and Dettre (1959), with the support of extensive experimental work, have used to a great extent the concept of surface roughness in order to explain the phenomenon of contact angle hysteresis.

FIG. 3. A non-wetting sessile drop.

FIG. 4. Sessile drop wetting the substrate.

Although from eqn. (11) it follows that, for a given solid–liquid–gas system, only one equilibrium contact angle exists, it is experimentally verified that a liquid drop advancing on a plane (possibly obtained by the addition of liquid inside the drop itself) often has a contact angle greater than for equilibrium, whilst a receding drop (obtained by evaporation) shows a lower contact angle.

The cited authors, using a model of rough surface formed by concentric grooves, showed that on grounds of geometric criteria the contact angles of a liquid with a rough surface are not infinite but are bounded by a maximum value $\theta_0 + \alpha_{max}$ and a minimum value $\theta_0 - \alpha_{max}$ (α_{max} is the maximum slope of the surface inside the grooves) and that, inside the above interval, only two values can exist if the volume and the spherical form of the drop have to be kept constant.

Thermodynamics show that the metastable configurations assumed by the drop within the aforesaid limitations, are divided by energy barriers which reach an absolute minimum for $\theta = \theta^*$, i.e. when the system verifies the Wenzel relationship. From all that, it is clear how additional sources of energy, such as vibrations, can reduce hysteresis so that the advancing angle tends to coincide with the receding one. It is easy to comprehend why the surface roughness helps the wetting of a surface. From eqn. (12) we can deduce that for $r = 1/\cos \theta_0$ and $\theta_0 < 90°$ the liquid wets the surface spontaneously. Moreover a liquid spreads more easily along the preferential orientation of the roughness because the barriers of potential met with in crossing the grooves are always higher than those along the axis of the groove itself.

Similar remarks can be made when the hysteresis phenomena are induced, even on plane surfaces, by surface heterogeneities. These surfaces can in fact be formed by zones in which the liquid has high contact angles and by zones in which it has low contact angles. This causes irregularities on the border of the drop which cannot reach the theoretical circular shape.

Johnson and Dettre have also derived, using a thermodynamic treatment similar to the previous one, the minimum of energy for all possible metastable systems which coincides with the value given by Cassie's (1948) equation

$$\cos \theta = Q_1 \cos \theta_1 + Q_2 \cos \theta_2 \qquad (14)$$

where θ_1 and θ_2 are the contact angles on the surfaces of area (fractional) Q_1 and Q_2 ($Q_1 + Q_2 = 1$). An interesting test of the effect the surface heterogeneities have on the border of a sessile drop was given by Patrick and Brown (1971), who observed the irregular borders of silicon oil drops by means of the scanning electron microscope.

C. Effect of temperature and of additives on γ_{lv}

In eqn. (11) there are four variables; it is possible to act on each of them to modify the wetting conditions of a given liquid–solid system.

It is self-evident that the liquid surface tension plays an important role as it is often the easiest parameter to modify—in fact when it decreases, the wettability increases since the contact angle decreases. We have, in practice, only two methods of modifying the surface tension of a liquid; we can vary the temperature or we can add to the liquid itself other substances which depress its surface tension (surfactants). For pure liquids the gradient $d\gamma/dT$ is always negative and for liquid metals can be evaluated to be about 0·5 mN/(m K). Many expressions, based on a large number of experimental results, have been proposed, describing the dependence of surface tension on temperature. Typical of these is the well-known equation of Ramsey and Shields (1893):

$$\gamma(MV)^{2/3} = P(T_c - T - 6) \tag{15}$$

where M is the molecular mass, V the specific volume and P a constant which for most liquids is equal to 2·12 erg/K (212 nJ/K). As can be seen from eqn. (15), γ decreases with increasing temperature, reaching zero near the critical temperature T_c.

When to one liquid we add another liquid completely miscible, the surface tension of the mixture is usually different from the surface tension of the two constituents. If the solution is perfect and if the surface tensions of the constituents are not too different we can write

$$\gamma = \gamma_A X_A + \gamma_B X_B \tag{16}$$

where X_A and X_B are the molar fractions of liquids A and B which have surface tensions γ_A and γ_B. However, this model is only valid under limited conditions.

Eberhart (1966) proposed the following equation for liquid binary mixtures:

$$\gamma = \frac{SX_A\gamma_A + X_B\gamma_B}{SX_A + X_B} \tag{17}$$

The parameter S, which is a function only of the temperature, was defined by Eberhart as the surface enrichment factor for the component with lower surface energy. In a later paper, Ramakrishna and Suri (1967) derived S thermodynamically, showing that it is linked to γ_A and γ_B and to the surface area A_B of B molecules by the relationship

$$S \simeq \exp\left\{(\gamma_A - \gamma_B)\frac{A_B}{RT}\right\} \tag{18}$$

Eberhart (1966) gives various values of S for different mixtures of molten salts of alkali metals, these values range between 1·5 and 3; taking data from different sources Eberhart moreover calculated $S = 1·77$ for the Cu–Ni

alloy at 1550 °C, and obtained a good agreement among the calculated values
of γ from eqn. (18) and experimental data.

Figure 5 reports the data from Sternberg and Terzi (1971) for binary
systems formed by fused chlorides, and Fig. 6 the data for gases in molten
iron taken from Kozakevitch (1968).

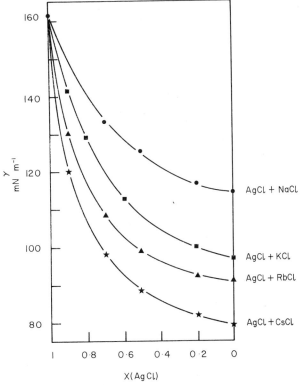

FIG. 5. Dependence of the surface tension of various fused chlorides on the content of AgCl.

D. The interfacial tension

Equation (11) takes into account solid–liquid interactions by means of
the term γ_{sl}.

The interfacial tension arises from a large number of intermolecular forces.
Among these the most important are the London dispersion forces and the
ones originating from ionic and dipolar interactions.

ALBERTO PASSERONE

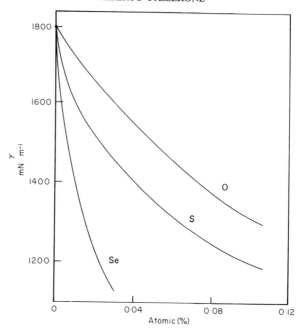

FIG. 6. Dependence of the surface tension of liquid Fe at 1550 °C on the content of O, S, and Se.

Good and Girifalco (1960) proposed that the interfacial tension for a given solid–liquid system be expressed by

$$\gamma_{sl} = \gamma_{sv} + \gamma_{lv} - 2\phi(\gamma_{sv} \cdot \gamma_{lv})^{1/2} \qquad (19)$$

where ϕ is a parameter, typical of the given system, which has a value of about 1 for liquid systems built up by similar molecules with regular interactions. In case the molecules do not have similar dimensions, ϕ can be approximated by

$$\phi = \frac{4(V_a \cdot V_b)^{1/3}}{(V_a^{1/3} + V_b^{1/3})^2} \qquad (20)$$

where V_a and V_b are the molar volumes of the two phases.

Fowkes (1967) extended this treatment, making the basic hypothesis that the solid–liquid system interacts essentially through dispersion forces and, sometimes, through forces due to hydrogen bonds. Therefore Fowkes presumes that for many liquids the surface tension is made up of two additive parts—the first, γ^d, due to dispersion forces, and the second, γ^h, due to hydrogen bonds. For mercury and possibly for other liquid metals, the surface tension may be the sum $\gamma^d + \gamma^m$, where γ^m is the contribution to the γ_{total} of the metallic bond. So the interface tension can be expressed as

$$\gamma_{sl} = \gamma_{sv} + \gamma_{lv} - 2(\gamma_{sv}^d \cdot \gamma_{lv}^d)^{1/2} \tag{21}$$

if the geometric mean can be used to calculate the molecular interactions.

If we take together eqn. (21) and eqn. (11) (which is possible if vapours are not adsorbed onto the solid), we get the following relationship for the contact angle:

$$\cos \theta = 2(\gamma_{sv}^d)^{1/2} \cdot \frac{(\gamma_{lv}^d)^{1/2}}{\gamma_{lv}} - 1 \tag{22}$$

E. Adsorption on the solid surface and critical surface tension

A solid surface, which has a surface tension γ_{sv}, adsorbs vapours if the newly formed surface is in a lower energetic state. The adsorbed layer causes a decrease in the solid γ_{sv} and acts as if, in eqn. (11), there were a new term π_s which resists the spreading of the drop

$$\gamma_{sv} - \pi_s = \gamma_{sl} + \gamma_{lv} \cos \theta \tag{23}$$

Actually this happens every time the liquid in contact with the solid has a lower surface tension than the solid and a correspondingly high volatility.

Zisman and coworkers (1964) thoroughly studied systems made of solids with low and high surface energy in the presence of liquids both pure and in solution. While these liquids do not spread over low energy solids they should spread over high energy solids. Since this is not always true Zisman and coworkers supposed that such behaviour could be due to a pre-adsorption of some component of the liquid onto the solid. A layer, and often, a monolayer of the new component is formed on the solid surface and affects the solid surface properties depressing, in general, the solid surface tension.

Research made on the system n-alkanes–polyethylene and then extended to various liquids, such as esters, hydrocarbons, fluorinated hydrocarbons etc., on different substrates led them to establish, for a given substrate, a correlation between the liquid surface tension and the cosine of the contact angle. This relationship, linear in most cases, enabled the authors to define as "critical surface tension γ_c", the surface tension value obtained by extrapolation to $\cos \theta = 1$. Therefore γ_c represents the maximum γ_{lv} value the liquid can still have if it is to spread completely on the solid considered.

Recently Rhee (1971a, b; 1972a, b), has tried to utilize the γ_c even in solid–liquid systems with high surface energy and at high temperatures, like molten metals–ceramics, with the aim of getting the solid surface tension by the sessile drop method, directly.

IV. KINETICS OF SPREADING

In the systems so far examined we have always made the hypothesis of equilibrium. Nevertheless, in real processes, this hypothesis is often no longer consistent and consequently we need to introduce some new parameters when studying wettability. Let us suppose that we melt a piece of substance B onto a solid surface A. As soon as the liquid melts, it assumes, owing to the effect of surface tension, a spherical shape more or less flattened by the gravitational force. The liquid drop does not show, however, a contact angle identical to that for equilibrium (θ_∞), but a somewhat higher value θ_t. So the edge of the drop undergoes a drawing force, per unit length:

$$F = \gamma_{lv} \cos \theta_\infty - \gamma_{lv} \cos \theta_t \tag{24}$$

Owing to the force F the liquid spreads onto the solid surface until $\cos \theta_t = \cos \theta_\infty$. Obviously the rate of this process depends strictly upon the viscosity of the liquid, for liquid metals the process is nearly instantaneous.

Naidich and Perervetailo (1971), Eremenko $et\ al.$ (1971), Ischimov $et\ al.$ (1971), and Naidich $et\ al.$ (1971), recently studied the spreading of liquid metals on solid metals and oxides at high temperatures. They pointed out that the spreading process exhausts itself in extremely short times, of the order of some hundredths of a second. Eremenko and coworkers established that, for the Al–Fe system, the spreading rate decreases with time following the parabolic relationship:

$$d^2 = kt \tag{25}$$

where d is the diameter of the drop at the time t. With increasing temperature the spreading rate increased. From the dependence of $\ln k$ on $1/T$ it was possible to calculate the (apparent) activation energy of the process, which is, for the system already cited, of about 18 kcal/g atom (75 kJ/mol).

The same authors have remarked that around the aluminium drop, after a few seconds of contact, a thin halo shaped film is formed, onto which the spreading process is continued. This is due to the surface diffusion of Al atoms, which is greatly helped by the chance to form new intermetallic compounds Fe–Al.

Schonhorn $et\ al.$ (1966), Kwei $et\ al.$ (1968), Cherry and Holmes (1969), and Passerone $et\ al.$ (1972), have studied (among other researchers) the spreading of molten polymers on surfaces of oxides and metals. The liquids used are highly viscous so that the time the drops need to reach equilibrium is very long, sometimes taking many hours.

In this case the spreading can be described by the relationship:

$$\cos \theta_\infty - \cos \theta_t = \cos \theta_\infty . \exp(-Ct) \qquad (26)$$

or

$$d_\infty - d_t = d_0 B \exp(-Ct) \qquad (27)$$

where C is the kinetic constant of the process and d_∞, d_t and d_0 represent the diameter of the base of the drop at equilibrium, at time t, and when $\theta = 90°$. B is a factor depending on the particular system studied.

We find rather different conditions in the case of the continuous coating of wires or ribbons by plastic materials, enamels, molten metals and so on.

Peters and coworkers (1968) have made some original research into wetting under dynamic conditions using an equipment in which a glass fibre could be drawn into the coating bath with a speed ranging from 0·05 cm/min (8 μm/s) to 60,000 cm/min (10 m/s), measuring the contact angle between fibre and liquid which forms when the fibre enters the liquid.

The results they obtained show that with low viscosity liquids, the contact angle does not vary considerably up to speeds of 50–100 cm/min (0·83–1·67 cm/s). On the contrary, with highly viscous liquids there is a nearly uniform decrease of the contact angle from the low range side of the speeds: the authors already cited moreover, have shown that the results given by liquids of different viscosities (η) can be superimposed in the plane $\cos \theta - \log \eta V$.

This means that in wetting under dynamic conditions, the influence of the surface tension may become negligible while other parameters, like viscosity, surface diffusion, velocity of the coating process, pretreatment of the surface, and so on, gain more and more importance.

REFERENCES

Banneister, K. J. and Hamill, T. D. (1968). NASA TN D-4779.

Bashfort, F. and Adams, J. C. (1892). *In* "An attempt to test the Theory of Capillary Action". Cambridge University Press, Cambridge.

Cassie, A. B. D. (1948). *Discussions Faraday Soc.* **3**, 11.

Cassie, A. B. D. and Baxter, S. (1944). *Trans. Faraday Soc.* **40**, 546.

Cherry, B. W. and Holmes, C. M. (1969). *J. Colloid Interf. Sci.* **29**, 1, 174.

Defay, R., Prigogine, I., Bellemans, A. and Everett, D. H. (1966). *In* "Surface tension and adsorption". Longmans, London.

Eberhart, J. G. (1966). *J. Phys. Chem.* **70**, 4, 1183.

Ellefson, B. S. and Taylor, N. W. (1938). *Am. Ceramic Soc.* **21**, 193–205.

Eremenko, V. N., Lesnik, N. D., Pestun, T. S. and Rjabov, V. R. (1971). *In* "Physical Chemistry of Surface Phenomena in Molten Materials", p. 203. Naukova Dumka Pub., Kiev. (In Russian.)

Fowkes, F. M. (1967). *In* "Surfaces and Interfaces". Proc. 13th Sagamore Army Mat. Research Conf. (J. J. Burke, N. L. Read and V. Weiss, eds.), Vol. 1, p. 197. Syracuse Press, New York.

Gibbs, J. W. (1948). "The collected works of J. W. Gibbs", Vol. 1, p. 315. Yale University Press, London.

Good, R. J. (1952). *J. Am. Chem. Soc.* **74**, 5041.

Good, R. J. and Girifalco, L. A. (1960). *J. Phys. Chem.* **64**, 561.

Hondros, E. D. (1970). *In* "Surface Energy Measurements". Techniques of Metals Research (R. A. Rapp, ed.), Vol. IV, part 2, Chap. 8A.

Ischimov, V. I., Cluinov, V. V. and Esin, Q. A. (1971). *In* "Physical Chemistry of Surface Phenomena in Molten Materials", p. 213. Naukova Dumka Pub., Kiev. (In Russian.)

Johnson, R. E. Jr. (1959). *J. Phys. Chem.* **63**, 1655.

Johnson, R. E. and Dettre, R. H. (1969). *In* "Surface and Colloid Science" (E. Matijeviĉ, ed.), Vol. 2, p. 85. Wiley-Interscience, New York.

Kozakevitch, P. (1968). *In* "Surface Activity in Liquid Metal Solutions" (S. C. J. Monograph "Surface Phenomena of Metals"). S. C. J., London.

Kwei, T. K., Schonhorn, H. and Frish, H. L. (1968). *J. Colloid Interf. Sci.* **28** (3/4), 543.

Laplace, P. S. (1806) "Mechanique céleste", suppl. 10th vol. Coureier, Paris.

Maze, C. and Burnet, G. (1971). *Surface Sci.* **24**, 335.

Naidich, Yu. V. and Nevodnik, G. M. (1971). *In* "Physical Chemistry of Surface Phenomena in Molten Materials", p. 238. Naukova Dumka Pub., Kiev. (In Russian.)

Naidich, Yu. V. and Perervetailo, V. M. (1971). *Russian J. Phys. Chem.* **45**, 7, 1025.

Padday, J. F. (1969). *In* "Surface and Colloid Science" (E. Matijeviĉ, ed.), Vol. 1, p. 106. Wiley-Interscience, New York.

Padday, J. F. (1972). *Proc. Roy. Soc. Lond.* **A330**, 561.

Passerone, A., Lorenzelli, V. and Biagini, E. (1972). *Ann. Chim.* **62**, 276.

Patrick, R. L. and Brown, J. A. (1971). *J. Colloid and Interface Sci.* **35**, 2, 362.

Peters, R. H., White, E. F. T. and Inveraty, G. (1968). p. 23/1. "Sixth Int. Conf. Reinf. Plastics". The British Plastics Fed. Pub., London.

Ramakrishna, V. and Suri, S. K. (1967). *Indian J. Chem.* **5**, 7, 310.

Ramsey, W. and Shields, J. (1893). *J. Chem. Soc.* 1089.

Rhee, S. K. (1971a). *J. Am. Cer. Soc.* **54**, 7, 332.

Rhee, S. K. (1971b). *J. Am. Cer. Soc.* **54**, 8, 376.

Rhee, S. K. (1972a). *J. Am. Cer. Soc.* **55**, 3, 157.

Rhee, S. K. (1972b). *J. Am. Cer. Soc.* **55**, 6, 300.

Shonhorn, H., Frisch, H. L. and Kwei, T. K. (1966). *J. Appl. Phys.* **37**, 13, 4967.

Shuttleworth, R. and Bailey, G. L. J. *Discussions Faraday Soc.* **3**, 16.

Sternberg, S. and Terzi, M. (1971). *J. Chem. Thermodynamics*, **3**, 259–65.

Wenzel, R. N. (1936). *Ind. Eng. Chem.* **28**, 988.

Young, T. (1805). *Phil. Trans. Roy. Soc.* **95**, 65. London.

Zisman, W. A. (1964). *In* "Contact Angle, Wettability and Adhesion". Adv. Chem. Ser. 43, Am. Chem. Soc. Pub.

Use of brush applied coatings in black and white television tubes

D. DE GRAAF

Acheson Colloiden B.V., Scheemda, Holland

I. INTRODUCTION

Because graphite is a fairly good semiconductor, it has been used for a long time in the electronic and electrochemical industry to serve a variety of purposes. Particularly, the television industry is using large amounts of this material, applying it in the form of coatings on the exterior and interior walls of television picture tubes. The coating material usually consists of four elements:

A. Graphite

Some properties of graphite are:

1. Electrical resistivity

Graphite is a semiconductive material. The resistivity of pure graphite is approximately 5×10^{-6} Ωm along the planes of the crystal structure. However the resistance depends on such factors as particle size distribution, particle shape, type of graphite and the impurities content. Pure flake graphite with a wide particle size distribution will mostly give low resistivities.

2. *Chemical inertness*

Graphite is not attacked by acids, alkalis, or oxygen under normal conditions. It starts to oxidize at about 500 °C depending on particle size, particle shape, type of graphite and oxygen available. It is also resistant against radiation and ultra violet light.

3. *Heat conductivity*

Graphite shows a good heat conductivity. This can be important when heat must be conducted from large areas.

B. Binders

Binders are needed to produce good adherent and coherent coatings. Resins, and even inorganic materials, are used. In conductive coatings for television picture tubes, mostly sodium and potassium silicate are used to obtain a good adhesion to the glass and to withstand the curing cycle at 420–450 °C. The silicates show a different behaviour with regard to adhesion, cohesion, blistering and cracking of the coating during heating.

C. Carrier liquids

The carrier liquid serves as a medium for transporting the conductive material to the substrate. The type of carrier liquid plays an important part in many applications. A fast evaporating solvent for instance, will give coatings in which the graphite particles are standing on their edge resulting in a fluffy coating with a high resistance. In too-slow-drying coatings the graphite will settle, resulting in a low graphite/binder ratio at the surface. This will cause a high resistance.

In other words the graphite particles must have the time and possibility to arrange themselves in horizontal planes. In internal television picture tube coatings, water is used as the carrier liquid for the graphite and the silicate. This cheap material has attractive properties such as non-toxicity and no air pollution.

D. Dispersants

Dispersants are needed for dispersing the graphite in the water, to obtain a product with a good workability, and to form a homogeneous, attractive coating. These dispersants are mostly of an organic nature.

During curing this organic material must disintegrate completely. Organic substances left in the tube will decompose slowly under high vacuum and electron bombardment. The gaseous products formed will cause a drop in emission when the tube is in use for several years.

II. The internal coating

A. Introduction

A television picture tube consists of a glass screen on which a fluorescent coating is applied, a conical section and a cylindrical part, usually called the neck. The electron gun is inserted in the neck of the cone (Fig. 1). After assembling, the tube is exhausted to a high vacuum. The internal coating acts as an electrically conductive layer between the anode contact and the accelerating lens of the electron gun (Fig. 2). Because of the high voltage on the anode of the gun, the connection cannot be made through the tube base holder since discharge sparks to other leads could occur. Therefore a chrome-nickel-steel anode contact is melted into the conical section of the tube. This contact is connected on the outside with the high voltage circuit and on the inside via the graphite coating, with the electron gun.

Screen   Cone   Electron Gun

FIG. 1. The three parts of a black and white television picture tube.

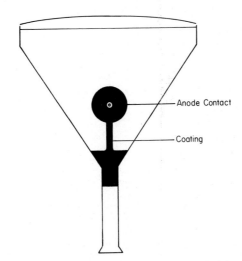

FIG. 2. The internal coating in a black and white television tube.

B. The application

The internal coating is applied after the melting together of screen and cone and the application of the screen coatings. The glass cone and screen are cleaned thoroughly with solutions of hydrofluoric acid and sodium carbonate before applying any coating.

To apply the coating, the tube is fitted into a device which holds and rotates the tube with the neck downwards. A specially designed, bendable brush is used. A small brush is fitted on a piece of spring steel which is fitted on a rod. By means of a lever and a wire the spring can be bent, so that the brush will press on the inside of the cone. The brush is dipped into a flat basin containing the aqueous graphite dispersion. Then the brush is inserted through the neck and the anode contact is coated. Next a line is drawn from the anode contact to the neck. The tube is rotated and at the same time the brush is drawn into the neck giving this area a ring shaped coating. Finally the brush is removed and carefully withdrawn so as not to dirty the lower part of the neck. After drying by infra red through the glass in a constant air stream, the coating is degassed in an oven at 420–450 °C. Then the normal processing of the picture tube can proceed.

C. Required properties

1. Good workability

The dispersion must wet the glass and anode contact and must flow to form an even coating. The coating should not show brush marks, drop formation or running on the tube. The quantity of product taken up by the brush on dipping must be sufficient to coat one tube.

As a standard practise a very slightly thixotropic product with a yield value of 20–80 dynes/cm^2 (2–8 N/m^2) is used to get a good brushability (Fig. 3). Moreover the product should be rather thin to avoid thick coatings around the anode contact.

2. No sedimentation

The graphite should not settle too fast on standing in the flat basin. Before the application the product should be well homogenized to restore the original viscosity and to redisperse the sediment. Usually the dispersion is sieved through a nylon cloth to avoid dried-out material on the tube.

3. Good covering power

The covering power must be such that the coating will cover in one treatment. The graphite used should have a small particle size.

4. *Long shelf life*

The dispersion should not increase in viscosity during transportation and storing. It has a high pH due to the presence of silicates. In such an alkaline environment, graphite tends to flocculate, which causes a non-reversible increase in viscosity.

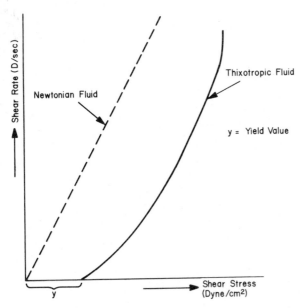

FIG. 3. Typical viscosity curve of an internal coating product.

5. *Good adhesion*

The coating should adhere well to glass and not flake off. The binder should still be present after the curing cycle. Therefore inorganic binders, such as sodium and potassium silicate, must be used.

6. *Good cohesion*

No loose particles should be present as these can affect the electrical functioning of the tube by causing discharge. Such particles also can fall on the fluorescent screen and either damage the fluorescent coating or make black spots by forming shadow areas due to electronic excitation of the screen.

7. *High vacuum stability under electron bombardment*

As mentioned previously, to guarantee a long life for television tubes, the coating should not give off gaseous products under high vacuum and electron bombardment, since such products will cause a drop in emission. Therefore organic materials such as dispersants and wetting agents must disintegrate completely during curing. The temperature stability of the coating can be tested by a thermogravimetric analysis during which a small sample is heated at a constant rate and in a constant air stream. This analysis gives a very good impression of the degassing properties of the coating.

8. *No coating defects*

The coating should not show cracks and must therefore have a low shrinkage. The area around the anode contact is quite uneven and in the irregularities a thick layer is formed, which on drying may show cracks. The type of graphite and silicate should be selected carefully to avoid crack formation. The coating should show no blisters. These blisters can be caused by poor degassing of the coating during curing.

Porosity promoters, which decompose during curing causing gaseous substances, are sometimes used to obtain a porous and easily degassing coating. Particularly in areas where the graphite/silicate binder ratio is comparatively low, blisters may occur. A much lower ratio in the edge of a coating than in the standard coating has been found. The reason is that the ratio changes during the first stage of drying. The graphite dispersion flows evenly over the surface of the substrate. However the solid graphite will stay behind, resulting in a coating, around the edges of which less graphite is present.

9. *Good scratch hardness*

Contact between the graphite coating and the electron gun system is made by three contact springs. The coating must be hard enough so that no graphite particles are detached during insertion of the electron gun. On the other hand too hard a coating will give a poor electrical contact.

10. *Electrical resistivity*

Because the silicate is an insulating material, compared with graphite, somewhere a compromise has to be found between conductivity, scratch hardness, adhesion and cohesion.

The resistance of such coatings is measured by coating microscope slides on one side. After curing at 450 °C the coating weight and the surface electrical conductivity are measured. The multiplication of coating weight and resistivity per unit area will give a constant value.

The basic principles of welding as a surface coating process

HAROLD N. WATSON

Hard Face Welding and Machine Co., Inc., Buffalo, New York, U.S.A.

I. INTRODUCTION

Welding processes used in the application of surface coatings can be divided into seven major classifications as follows: manual electric arc welding; oxy-acetylene welding; gas tungsten arc welding; gas metal arc welding; submerged arc welding; spray welding; plasma arc welding. In all of these processes, we will limit discussion to surfacing for improved properties of heat-, wear-, or corrosion-resistance, and not to the mere rebuilding of surfaces to regain original size because of size alone. All of these processes require localized high temperatures, developing base part temperatures ranging from a low of 100 °C or 150 °C to a high of over 1100 °C. The effect of the heat input upon the metallurgical structure of the base part must be considered along with the possibilities of distortion.

Coatings applied by welding are usually thicker than those applied by other processes.

The thinnest coatings are usually applied by spray welding, with 0·75 mm being average although 0·25 mm is practical.

The heavier coatings, applied by manual electric arc or submerged arc welding, although normally in the range of 6 mm to 10 mm, may on severe applications, go to 25 mm or more.

II. MANUAL ELECTRIC ARC WELDING

Almost everyone is acquainted with the ordinary arc welding process wherein the power source is a motor-generator set, a transformer, or a rectifier. The high temperature of the developed arc between a consumable electrode and the subject part to be coated, melts the electrode and the contacted surface of the base part, producing a true welded deposit with a certain amount of base metal diluted into the deposited metal and some alloying of the deposited metal into the base.

Dilution is controlled by operator technique and control of the amperage output of the welding machine. Higher amperages produce a faster weld deposit but also more dilution.

Stick electrodes, usually covered with a flux, are available in a wide range of alloys, with varying properties of high temperature resistance, hardness, and corrosion resistance. The application is manual, consequently slow, and the deposit quality and smoothness subject to the skill of the operator.

Metals available as surfacing materials, include the heat-treatable steel alloys, and tool steels, martensitic and austenitic stainless alloys, copper and nickel base alloys and a wide range of superalloys designed for special applications. Hardness ranges go as high as 66 Rockwell on the "C" scale or 750 BHN.

The more common uses are in the rougher applications where the coating is used as applied. Coating thicknesses usually range from 3 mm to 6 mm. Deposition rates are normally in the range of 1·5–2 kilograms per hour (0·4–0·6 g/s). Typical applications are the surfacing of steam shovel teeth, coke pusher shoes, catalyst valves in refinery applications, and crusher rolls.

III. OXY-ACETYLENE WELDING

This type of welding covers, perhaps, the widest range of activity, from very small parts to very large, hundreds of alloys and combinations of base metals and surfacing alloys. The welding equipment, a torch, hoses and regulators and tanks of oxygen and acetylene are universal the world over and represent one of the smallest investments of all types of welding equipment.

To produce quality in surface coatings requires a high degree of operator skill, usually acquired in years of practice.

Most applications of surface coatings by this method require precision finishing to high standards. Flaws in the surface are usually cracks, porous or spongy areas or very small bubbles in the weld which may be exposed upon finishing, as small pits or craters. I have known welders who, working as a team, have applied as much as 100 kilograms of surfacing weld metal on one base part with no detectable flaws. This can represent approximately 250 hours of continuous work. Final inspection was by dye penetrants and X-rays wherein bubbles as small as 0·25 mm could be detected subsurface. A very steady hand, clean working habits and a good eye are imperative for this quality of work.

Deposition rates vary, depending upon the type of alloy used and size of the base parts. Most applications fall within the range of 0·3–1·5 kilograms per hour (0·08–0·4 g/s).

Surfacing alloys include the copper base, nickel base, and cobalt base alloys and composite alloys containing additions of tungsten carbide.

Hardness ranges vary from very soft to well above the machinable range where finishing requires grinding with aluminium oxide, silicon carbide and even diamond wheels. With top quality work, finishes as fine as 0·6 micro-inches (15 nm) r.m.s. have been attained.

Typical applications include seating surfaces on valves for combustion engines and compressors, rolls for precision rolling and forming of plastics and glass, power plant equipment parts such as steam valves and nozzles, extrusion screws for the manufacture of plastics and rubber, and bearing sleeves for precision equipment operating in unusual environments of temperature and corrosion.

IV. GAS TUNGSTEN ARC WELDING

This process is a hybrid of the two processes just described. The power source is a generator, transformer, or rectifier as used in electric arc welding.

The torch does not use a gas but produces its heat by means of an electric current arcing between a tungsten electrode within the torch and the base piece to be surface welded. The tungsten electrode and the molten pool of weld deposit plus adjacent areas of the base piece are protected from oxidation by the flow of an inert gas through the torch. Weld quality is normally high and is closely tied to the skill of the operator.

Deposition rate is quite slow. Highest quality work has a deposition rate of 0·25–0·5 kilograms per hour (0·07–0·14 g/s).

Choice of surfacing alloys is broader than with oxy-acetylene applications due to the protective influence of the inert gas and the higher temperature attained with the inert arc torch.

Typical applications include those for oxy-acetylene welding and the determining factor in deciding which process to use is usually that of heat input into the base part.

(i) The part may be impractical to heat to the high temperature required for oxy-acetylene welding and therefore demand the gas tungsten arc process with its higher temperature and more localized heat zone.

(ii) The weld quality requirement may call for a high level of purity of deposit and therefore rule out gas tungsten arc welding with its higher factor of base metal dilution due to its heat concentration.

V. GAS METAL ARC WELDING

This process has all of the advantages of the gas tungsten arc process plus several extras:

(i) With a wide choice of wire sizes and power sources, deposition rates are considerably higher.

(ii) Due to automatic control of arc length, the need for operator skill is not so great.

(iii) Equipment can be machine mounted and fully automated.

In this process, the surfacing metal is fed automatically through the torch and the feeding wire becomes the consumable electrode. The arc and the molten pool are protected from oxidation by a flow of inert gas as in the gas tungsten arc process.

Deposition rates vary from a low of approximately 0·4 kilogram per hour (0·1 g/s) in manual applications using wires as small as 0·5 mm diameter, to a high of 4·5–5 kilogram per hour (1·2–1·4 g/s) in automated set-ups using wires as large as 4 mm diameter. Typical applications include the overlay of seating surfaces on valve seats, the working areas on large piston rods and pistons wherein low friction and non-galling properties are required.

VI. SUBMERGED ARC WELDING

This process differs from gas metal arc welding in several ways:

(i) The equipment is much heavier, using larger wires of surfacing material and usually considerably higher amperages and is therefore beyond the range of hand-held equipment, so must be fully automated.

(ii) Instead of an inert gas, a flow of granular flux completely surrounds the arc and covers the molten weld metal, serving the same purpose as a protective cover of inert gas and adding three further advantages:

(a) A certain amount of the flux becomes molten, solidifies rapidly and in addition to preventing oxidation, insulates the deposited coating and slows cooling.

(b) With the use of flux, metallic additions can be used which will alloy with the depositing metal, changing chemical analysis, hardness, and deposition rate, thus giving greater flexibility to the process.

(c) The granular flux, being an opaque material, acts as a shield, completely covering the arc and removing the radiation hazards normal to the open arc.

Repeatable quality can be obtained without the highly skilled operator requirements of the previously mentioned processes.

Deposit materials are either in the form of wires varying from 2 mm diameter to 4 mm diameter, or in the form of metal strip with a cross section of up to 2 by 50 mm.

Deposition rates vary from approximately 2 kilograms per hour (0·6 g/s) with the smaller wires to as high as 70 kilograms per hour (20 g/s) with the large strip electrode.

Typical applications include the surface coating of large valves, rolls, the lining of large chemical vessels for corrosion resistance. Coatings applied are usually 3 mm to 6 mm thick.

VII. Spray Welding

This process differs radically from all the previous processes. It is in reality a two-part process. In the first part, the surfacing material, in the form of a fine powder, is sprayed through an oxy-acetylene torch or gun onto a surface previously prepared by a rough grit blast. At this point you have a thermal sprayed or metallized coating. The powder particles passing through the flame become semi-molten and upon impingement against the roughened grit blasted surface, deform and interlock forming a mechanical bond to the base and to each other. In the second part of the process, both the part and the surface coating are brought to fusion temperature.

Fusion temperature varies from about 1050 °C to approximately 1300 °C and is usually accomplished by means of hand-held oxy-acetylene or oxy-propane torches although some production parts, especially in the smaller sizes, are furnace fused, preferably in a dry hydrogen atmosphere furnace.

Materials available, designated as self-fluxing alloys, include nickel and cobalt base alloys usually with additions of boron and/or silicon to lower the melting point and improve wettability.

Coating thickness is normally in the range of 0·25 mm to 0·75 mm; however where need has merited the cost, coatings as thick as 10 mm have been applied. Uniformity of coating can be controlled accurately enough so that usually no more than 0·25 mm stock removal is required to produce fine finishes.

Quality is very high, hardness ranges of 230 BHN to 650 BHN (225 to 700 VPN) are attainable and surface finishes of 2–4 microinches (50–100 nm) on ground parts are regularly obtained. The actual deposition of the surfacing metal does not require a high degree of skill and one operator can deposit metal at a rate as high as 20 kilogram per hour (6 g/s). The fusing operation, which is normally manual, requires a high degree of skill.

The fusion temperature lies within a fairly narrow range—in the nickel chrome boron alloys between 1030 °C and 1080 °C, and in the cobalt type fusable alloys, between 1130 °C and 1180 °C. Below the low limit a good weld is not obtained, while above the high limit the metal will become fluid and run off the part. The operator normally judges temperature by the colour and surface appearance and the skill is acquired only through experience. At the proper fusion temperature, all of the sprayed particles will wet and bond to the base metal and to each other, forming a very dense and impervious coating weld bonded to the base material.

Of all the welded coating processes, this one supplies the smoothest finish as welded, the closest tolerance as to size and requires the least finishing to produce precision parts.

Typical applications include bearing sleeves, valve stems, piston rods, process rolls.

As spray welding is a variation of the spray processes known as flame spraying, please refer also to the paper "The Basic Principles of Flame Spraying" by C. W. Smith, in these Proceedings.

VIII. PLASMA ARC WELDING

In plasma arc welding, the power source is one or more direct current rectifiers supplying power at levels as high as 1000 amperes. The power is supplied to a torch through water cooled cables and the torch is also water cooled. An inert gas or combination of gases is supplied. An independent powder feeder meters and feeds the surfacing material to the torch, conveying it in an auxiliary stream of inert gas.

Two different methods are used to produce the high temperature arc. In the non-transfer arc method, the positive and negative power supplies provide the arc within the nozzle of the torch. In the transfer arc method, the negative power is fed to the torch and the positive power to the base work piece and the high temperature arc is formed between the two. In both cases, the surfacing material is fed through the torch onto the base piece, the highly localized heat producing a true weld.

The process is ideally suited to automation and in fact, practically demands it as distance of torch to work and traverse speed must be uniform, or extreme variations in base metal dilution, thickness of coating and quality of weld deposit will occur.

The choice of materials for surfacing is wide and includes a number of nickel, cobalt and iron base alloys either with or without additions of tungsten carbide.

Deposition rates are as high as 6 kilogram per hour ($1 \cdot 7$ g/s). Weld quality is excellent and operators are easily trained. Although the cost of equipment is high and the cost of automation usually even higher, the process holds great promise for production. I know of one case where a production part required 2 hours time for the highest skilled arc welder, using the gas tungsten arc process; after setting up for automated plasma transfer arc, the same job was produced to a higher quality standard by an unskilled operator in exactly two minutes.

Typical applications include edges on trim dies, flights on extrusion screws, process rolls, and parts subject to severe wear where composite alloys with high percentages of tungsten carbide are required.

As the plasma arc welding process is a variation of the plasma spray process, the reader should also refer to the following papers included in these Proceedings; "The Basic Principles of Plasma Spraying" by A. R. Moss, and "Production of Thick Film Circuits by Plasma Spraying" by R. T. Smyth.

BIBLIOGRAPHY

Ballard, W. E. (1963). "Metal Deposition and the Flame Deposition of Ceramics and Plastics", 4th edn. Charles Griffin, London.

"Welding and Brazing" (1971). Vol. 6 of "Metals Handbook", 8th edn. American Society for Metals, Ohio.

"Welding, Cutting and Related Processes" (1970). Section 3A of "Welding Handbook", 6th edn. American Welding Soc., New York.

The basic principles of printing as a surface coating process

SIMO KARTTUNEN AND PIRKKO OITTINEN

The Technical Research Centre of Finland, Graphic Arts Laboratory, Helsinki, Finland

I. INTRODUCTION

A. *Printing in general*

As a background to this paper the general production stages of the printing processes are given in Fig. 1. From the plate making stage on, the production

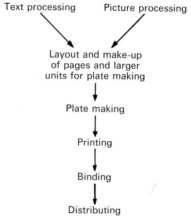

FIG. 1. The production stages of a printing process.

methods of the various processes diverge. The three main types of printing are:

   (i) *Letterpress* in which ink is transferred on to the substrate from raised elements on the plate.

  (ii) *Lithography* in which printing elements are lipophilic and non-printing elements hydrophilic. There are almost no height differences on the plate. The plate is wet continuously by a water-based dampening solution which adheres to the hydrophilic parts and keeps them free from the ink which wets the lipophilic printing elements.

 (iii) *Gravure* in which ink is transferred from etched or engraved depressions in the printing cylinder. The rest of the cylinder is scraped clean from ink by a doctor blade pressed against the etched cylinder.

Most lithography printing is done by transferring ink from the plate on to the substrate via a blanket cylinder in which case the term offset or offset lithography is used. This and some other general features of the three main

TABLE I

Features of the three main printing methods

	Letterpress	Lithography	Gravure
Image elements on plate	Raised	No height differences	Etched or engraved
Term of process in direct transfer from plate to substrate	(Direct) *letterpress*	Direct lithography	(Direct) *gravure*
Term of process in transfer via rubber blanket cylinder (offset transfer)	Indirect letterpress "letterset"	*Offset* (lithography)	Gravure offset or indirect gravure
Ink viscosity (poise) (Pa s)	10–1000 1–100	50–2000 5–200	0·1–0·5 0·01–0·05
Ink film thickness on plate (μm)	2–10	2–10	2–40
Ink vehicle and the principal drying mechanisms	Mineral and linseed oils with synthetic resins; setting by absorption drying by oxidation and evaporation		Toluene or similar solvent with synthetic resins; drying by evaporation
Approximate production share in the U.S.A. (per cent)	50	35	10

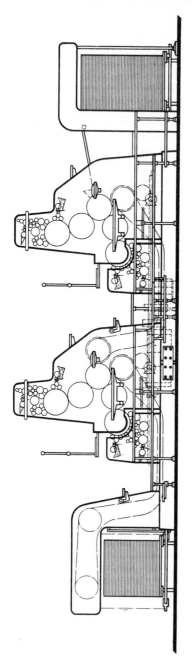

Fig. 2. A four-colour offset press.

processes are given in Table I. More detailed information is given in the paper by Pace (1974) and in handbooks (e.g. Strauss, 1967).

Printing processes duplicate text and picture information into a large number of copies by transferring black and coloured inks on to substrates such as paper. The information content of a print depends on the optical contrast between image and non-image areas and on their frequency. The latter is controlled by the screen ruling of the halftone pictures and on the type, size and spacing of letters and figures in the text. Optical contrast depends on the optical properties of the ink and substrate and also on the physical structure of the ink film. An even ink film is optically most effective, and thus the aims of printing are much the same as those of other coating processes. This paper deals with the most important stage of printing, i.e. ink transfer, and the limitations and capabilities of printing as a process when thin, even and positionally controlled ink films are sought.

B. Technology of printing

Ink is fed on to the plate containing the image and non-image areas in a printing press through an ink distribution system, which is a multi-roller unit in letterpress and lithography presses (Fig. 2) and a tank and doctor blade unit in gravure (Fig. 3). The latter is an almost ideal distribution

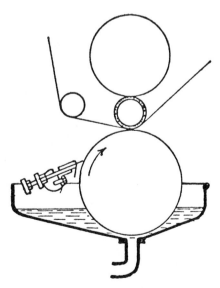

FIG. 3. Gravure printing. Ink transfer on web takes place between a rubber-covered impression cylinder and a printing cylinder. The ink distribution system in the diagram consists of an ink duct in which the printing cylinder revolves.

system in which ink is metered dot by dot on to a substrate by the varying depths of the etched or engraved screen cells on the cylinder. Thus in gravure the only ink transfer problem is to make the ink transfer out of the cells on to the substrate.

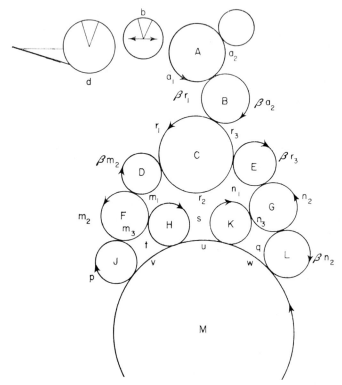

FIG. 4. An offset and letterpress type ink distribution system. Ink flows from the duct (in the far left) to the plate cylinder (M) by splitting in the nips between rollers. The flow can be mathematically simulated by relating the ink film thicknesses on the rollers with equations such as $a_1 + \beta r_1 = (1 + \beta)a_2$ for rollers A and B.

In the letterpress and lithography presses, the ink distribution system evens out the ink film thickness in the machine and cross directions. Differences in cross direction ink requirement can be compensated for by cross machine regulation of the ink feed from the duct to the duct roller (Fig. 4). Machine direction variations of the printing area tend to cause disturbances called ghosting and starvation (periodical variation of print density on the substrate in machine direction) which can partly be eliminated by many plate rollers (J, H, K and L in Fig. 4) and by many distribution rollers with varying diameters as indicated by Mill (1961), Wirtz (1964) and Ruder (1965). A large number of roller nips is advantageous because it permits a high ink film thickness on the duct roller and consequently results in low relative sensitivity to errors in ink feed from the duct and helps to break down ink thixotropy. A large number of nips, however, decreases the response rate of an ink distribution system as shown by Mill (1961).

Ink transfer from nip to nip in an ink distribution system is linear whereas the transfer from plate to substrate is non-linear due to the coverage and penetration phenomena which are discussed in the following sections.

In the text, diagrams and tables the following nomenclature is used:

x = amount of ink on plate before transfer (g/m^2)

y = amount of ink transferred to substrate (g/m^2)

k = smoothness parameter (m^2/g)

A_0 = smoothness parameter depicting an initial value of the coverage function A at $x = 0$

b = immobilization parameter (g/m^2)

f = splitting parameter

A = coverage functions, i.e. fractional area of paper covered by ink

B = immobilization function

D = density of solid print

D_∞ = density of solid print at infinite ink film thickness

m = parameter in the solid print density equation depicting the steepness of the curve

D_r = density of halftone print

F = effective printing area of a halftone which depends on the real printing area and on optical effects in the ink layer and substrate

s = a measure of the filling-in of halftone dots

x' = depth of penetration of ink into paper at the moment of ink transfer

x_0 = ink film thickness at the flattened fraction of paper at the moment of ink transfer (μm)

$\phi(x')$ = roughness distribution of a substrate

II. INK TRANSFER THEORY

The transfer of ink from solid printing surfaces on to paper is generally expressed as an equation between the amount of ink transferred (y) and the amount of ink on the plate before impression (x). These quantities are measured as mass per unit area (g/m^2) or as ink film thicknesses (μm). The basic idea of ink transfer was presented by Walker and Fetsko (1955) in the form of the following equation:

$$y = A(bB + f(x - bB))$$

in which the coverage function $A = 1 - e^{-kx}$, and the immobilization function $B = 1 - e^{-x/b}$ and k, b and f are parameters.

Verbally this means that the amount of ink transferred is the sum of a constant amount of ink immobilized (b) and a constant fraction (f) split from the remaining amount of free ink ($x-b$). At low ink film thicknesses this only occurs in the fractional area A of ink-paper contact. Further at small ink amounts immobilization is smaller than b by a fraction B which tends to unity when the ink amount is increased.

Basically similar models have been presented since by many researchers by modifying the coverage function or by introducing new features into the model. Cropper (1972) recently made an extensive review comparing the different ink transfer models proposed. In this paper only the newest models, modifications proposed by Karttunen (1970) and by Karttunen et al. (1971) —not mentioned in Cropper's review—are examined.

By introducing a new parameter A_0, called the flattened fraction of paper surface, the following modification of the basic model can be obtained:

$$y = (A - A_0)bB + f[Ax - (A - A_0)bB]$$

which is the same equation as

$$y = Afx + (A - A_0)bB(1 - f)$$

in which $A = 1 - (1 - A_0)e^{-kx}$ and $B = 1 - e^{-x/b}$. Computing the four parameters from the experimental test printing data is slightly more complicated than computing the three of the basic model, but the resulting accuracy of the fit is better as is shown by Karttunen et al. (1971) (Fig. 5). The significance of the two smoothness parameters k and A_0 in the coverage function is discussed in the next section.

The models presented above are only applicable in one-colour letterpress and lithography printing. Joyce and Fuchs (1966) have proposed principally similar models for gravure and some attempts have been made by Karttunen and Oittinen (1972) to construct ink transfer models for halftone printing and two-colour wet-on-wet printing.

A. Coverage

Ink is transferred from a plate to a substrate within a very short contact time in a printing nip. In this dwell time, the ink, forced by printing pressure, penetrates into the depressions and pores of the substrate surface. There is not time for the liquid phase of ink, called the vehicle, to separate from the pigment because the pressure is only exerted for about 1 to 5 milliseconds. The movements of the ink pigment after impression are slight. However, optical changes take place in the prints after printing due to solvent evaporation or vehicle absorption into the substrate. Evaporation and absorption of the liquid components of an ink greatly hamper the possibility of predicting the resulting print properties beforehand with the aid of ink transfer data.

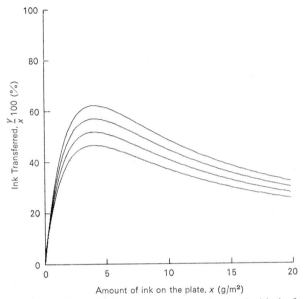

FIG. 5. An example of the form of ink transfer curves as simulated with the four parameter model of Karttunen *et al.* (1971). $f = 0.1$, $k = 0.5$, $b = 5$, A_0 varies from 0 on the uppermost curve to 0.3 on the bottom curve. The relative ink transfer $y/x(\%)$ is presented as a function of the amount x (g/m²) of ink on the plate before transfer.

Substrates such as paper are relatively rough and porous and their surfaces can be described by a roughness distribution, i.e. the distribution of surface elements at varying vertical distances from a given reference level. In his studies Hsu (1962; 1963) used reflectance measurements in the determination of roughness distributions of papers by first making prints varying the ink film thickness on the plate and then measuring the paper coverages A. The types of models he found to depict the distributions were as follows:

(i) For ink transfer between two flat surfaces (Hsu, 1962)

$$\phi(x') = (1 - A)\frac{\mathrm{d}A}{\mathrm{d}x} = \frac{nqx^{n-1}}{(1 + qx^n)^3}$$

in which q and n are parameters.

(ii) For ink transfer between two cylinders (Hsu, 1963) he proposed a normal logarithmic distribution with standard deviation and median of x' as parameters. In this case Hsu assumed x' to be equal to x, and thus $\phi(x') = dA/dx$.

In Hsu's studies the mean depths of roughness of normal printing papers were in the range of about 0·5 to 2 μm. Other types of measurements such as air-leak methods have given similar results. It is self-evident that roughness distributions of surfaces tend to be skew so that the deepest recesses are many times deeper than the mean depths. The deepest surface recesses of paper made of wood fibres with native diameters of about 20–50 μm are about as deep as these diameters.

According to the models of Hsu the coverage function A and consequently the roughness distribution tends to zero when x' tends to zero. This is literally true, but if we think of a unit area of paper under nip pressure, we must accept that a considerable fraction may be flattened to form an initial value A_0 for the coverage function A. This is particularly the case when the ink amount x is close to zero as noted by Karttunen (1970). Figure 6 shows the relationship between x, the amount of ink on the plate and the roughness distribution in the case of the models discussed above. Although it is impossible to determine a roughness distribution without certain assumptions, the dependance of the penetration depth x' and coverage A on the amount of ink on plate x, is theoretically important. The equations in Fig. 6 are based on volume identity of x.

The size distribution of roughness, i.e. the frequency distribution of the sizes of the surface recesses, is also important. In coated papers most surface recesses are so small-sized that their effect on printing is minor or negligible whereas uncoated papers exhibit relatively more large-sized recesses as measured by Brecht and Rothamel (1969). Recesses larger than the halftone screen units (100–200 μm) in particular have been found to be troublesome, because they not only disturb the printed dot structure but also cause complete missing of screen dots. Size distributions of roughness have been mainly studied by the optical contact principle in which a very illustrative picture of contact and non-contact areas of paper surface under pressure is obtained.

Paper as a substrate is generally wet very well by printing inks. The wettability of other substrates such as metal foils, plastic films and synthetic papers varies greatly. Especially polyolefine films with no surface treatment have too low a surface energy to be printed without difficulties. Sometimes severe mottling is observed when printing on a surface already once printed as noted by Carlsson and Lindberg (1971). Mottling is large scale random variation of print density which is understood to be due to surface tension effects. Interactions between ink and substrate additives can also occur causing poor adhesion, slow drying etc.

Real coverage difficulties resulting from poor wettability are rare because considerable pressure is used in printing. The reasons for poor coverage, familiar from our daily newspaper, are mainly caused by the roughness of the printing plates and the substrates which are so prominent that the ink film thickness simply is not high enough to fill the recess volume of substrate

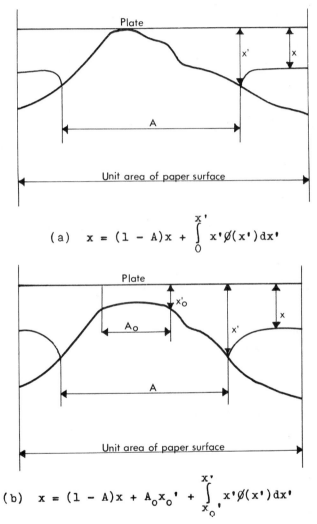

$$(a) \quad x = (1 - A)x + \int_0^{x'} x'\phi(x')dx'$$

$$(b) \quad x = (1 - A)x + A_0x_0' + \int_{x_0'}^{x'} x'\phi(x')dx'$$

FIG. 6. Two hypotheses of the deformation and coverage of a paper surface when contacted with a layer of ink of thickness x, (a) according to Hsu (1961), (b) according to Karttunen (1970) and Karttunen *et al.* (1971).

in printing. The recess and pore volume of porous substrates such as paper depends on pressure in the printing nip. By increasing pressure (and its dwell time, i.e. by decreasing speed), coverage is completed at lower ink film thicknesses, i.e. the smoothness parameters k and A_0 increase.

B. Immobilization

As mentioned before, the ink transfer model of Walker and Fetsko (1955) is linear at large ink film thicknesses at which the coverage and immobilization functions A and B tend to unity:

$$y = b+f(x-b) = b(1-f)+fx$$

This equation shows that there is an intercept on the y-axis and that a constant amount of ink (b) is immobilized by the substrate at large ink film thicknesses. At low ink film thicknesses the case is less straightforward because b is multiplied by the immobilization function B. Rough and porous uncoated papers give b-values up to 5 or 10 µm; coated papers less than 1 µm and films and foils practically zero.

Immobilization increases with increasing printing pressure. With rough papers in particular this is a disadvantage because full coverage is attempted by increasing printing pressure which, however, at the same time increases immobilization. The resulting ink films are rather uneven, if (say) 5 µm has transferred by immobilization and only 2 µm by splitting. Such very uneven ink films are not optically effective as mathematically shown by Tollenaar and Ernst (1971).

The same situation applies to ink viscosity. By decreasing the ink viscosity, coverage improves (cf. the smoothness parameters in Fig. 7) but at the same time immobilization increases.

Penetration of fluid inks into a substrate wetted as easily as paper, makes it impossible to obtain perfectly even films. Smooth and closed surfaces (films and foils) on to which ink is transferred by splitting alone can be expected to give much more even ink films. Unfortunately such surfaces often cause setting and drying problems due to low absorption and mottling and poor adhesion due to wettability difficulties.

C. Splitting

Coverage and immobilization govern ink transfer at the inlet and middle parts of the printing nip. At the outlet the pressure in the ink film gradually turns negative. Even though this happens very quickly—in fractions of milliseconds—cavitation generally occurs in the "free ink layer" $x-bB$.

Cavitation bubbles increase further away from the nip centre and result in filaments which finally break and end the splitting (Fig. 8). The ink transfer phenomena at the outlet region of the nip are depicted by the splitting parameter f which measures the proportion of the free ink layer transferring on to the substrate. The most important variables of f are printing speed and ink viscosity.

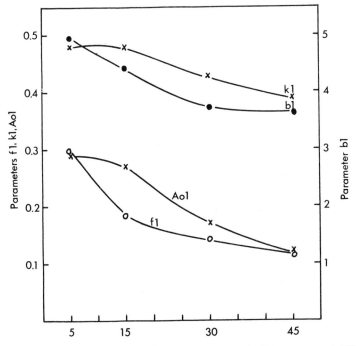

FIG. 7. Ink transfer parameters from the four parameter model of Karttunen *et al.* (1971) as a function of ink viscosity (P).

The splitting parameter f varies from about 0·1 to 0·5 so that the lower bound corresponds to high speed and viscous ink, the conditions for the upper bound being the reverse. A plausible explanation is that a higher speed increases the pull of ink (splitting resistance) which drags the immobilized ink layer more efficiently from the recesses to take part in splitting as compared to a lower speed, thus resulting in a smaller amount transferred. On the other hand a viscous ink has an inherently higher pull than a fluid ink and thus its splitting constant is lower as an indication of a higher ink film thickness in splitting. This reasoning is analogous to the so-called trapping considerations in multicolour wet-on-wet printing as studied by Karttunen and Oittinen (1972).

The directions of the effects of the printing conditions on ink transfer are presented in Table II in which + signs denote an increasing and − signs a decreasing effect. The most pronounced effects are marked with two signs, and "no effect" with zero.

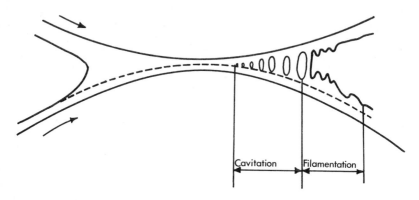

FIG. 8. A printing nip. Two stages of ink splitting at the outgoing side of the nip, cavitation and filamentation, are indicated.

With viscous inks, which are non-newtonian liquids, viscosity is not the only variable to govern the splitting resistance, but at present no exact knowledge is available of the effects of rheological properties other than viscosity or surface energy properties. An example of splitting constant as a function of viscosity is given in Fig. 7. The form of the curve tends to indicate that there is a viscosity limit beyond which f becomes negative.

TABLE II

The dependence of ink transfer on printing conditions

	Ink transfer parameters					Coverage	Total amount transferred
	A_0	k	b	f	A		$y = y(x)$
Printing pressure	+ +	+ +	+ +	0	+ +		+ +
Printing speed	−	−	−	− −	− −		− −
Ink viscosity	+	−	−	− −	−		− −

III. PRINT DENSITY

The optical result of printing is print density which is defined as the negative logarithm of the ratio of the reflective and incident light intensity in density measurement. Density of solid print depends on the ink film thickness on the substrate, on its thickness distribution, and on the optical properties of the ink and the substrate. The exact forms of the ink film thickness distributions are so far unknown and therefore density of solid print is generally expressed, as first done by Tollenaar and Ernst (1967), as a function of the amount of ink on the paper (y) with the aid of only two parameters D_∞ and m:

$$D = D_\infty(1 - e^{-my})$$

The density of halftone prints can be expressed as a function of the density of the solid print D and the effective printing area with an additional filling-in parameter s as proposed by Schirmer and Tollenaar (1971):

$$D_r = -\log_{10}[1 - F^{1-(D/D_\infty)^s}(1 - 10^{-D})]$$

To be able to control a printing process from the ink feed to the final print quality, the print density models such as those presented above should be connected with the ink transfer models. However, the existing theories of ink transfer and print density are very elementary as the effects of ink film thickness distributions are largely unknown.

IV. LIMITATIONS AND CAPABILITIES OF PRINTING

Fluid inks are used in the main printing processes: the higher the speed the lower the ink viscosity. Electrostatic printing is done with solid powder inks, but so far these methods are not used in mass production printing.

A fluid ink penetrates into a substrate because a high pressure is exerted to obtain complete coverage of surface at ink film thicknesses as low as possible. Therefore on rough and porous surfaces, the resulting ink films are rather uneven. More even films can be achieved on smooth and closed surfaces. Ink film thicknesses on the substrate are in the range of 1 to 10 µm; the lower limit corresponds to very smooth surfaces. In screen printing even thicker films can be applied. The substrate is generally in sheet or web form even though very complicated forms like tubes, bottles, etc. can be printed with special methods.

The main capabilities of printing are high speed and positional accuracy. Web speeds up to 10 m/s are used in newspaper and magazine printing. Continuous films or curves, lines, dot structures of any form and size (down to units of about 50 µm) can be printed with a positional accuracy of about 0·5 mm or even more. Various substrate/ink combinations can be used. Paper is the most common substrate but metal foils and sheets, plastic films, textiles, wood, leather, glass, rubber etc. can be printed with special methods. Inks range from thin water- and solvent-based fluids to stiff pastes of thousands of poises (hundreds of Pa s) in viscosity.

Printing of many ink layers on each other, wet-on-dry or wet-on-wet, is every day practice. Coating is performed by printing machines, varnishing by offset lithography and pigment coating by gravure. On the other hand, in paper coating technology, methods relatively close to those in printing are used in printing paper and board production.

In the future, printing processes will naturally achieve better performance than today. More even ink films on relatively rough surfaces are transferred with less pressure by using for example static electricity or jet techniques. These methods also diminish positional ink film thickness variations. Further improvements can be obtained by computer controlled printing. Small scale jet printers as computer output devices have already been realized as reviewed by Kamphoefner (1972) and such installations will additionally eliminate manual reproduction and plate making stages.

REFERENCES

Brecht, W. and Rothamel, H.-J. (1969). *Das Papier* **23**, 6, 326.
Carlsson, G. E. Lindberg, B. (1971). *In* "Advances in Printing Science and Technology. (W. H. Banks, ed.) Vol. 6, p. 283. Pergamon Press, Oxford.
Cropper, M. (1972). *J. Oil Colour Chem. Ass.* **55**, 2, 128.
Hsu, B. (1962). *Br. J. Appl. Phys.* **13**, 4, 155.
Hsu, B. (1963). *Br. J. Appl. Phys.* **14**, 5, 301.
Joyce, E. and Fuchs, G. (1966). Proc. Tech. Ass. Graphic Arts, 291.
Kamphoefner, F. J. (1972). I.E.E.E. Transactions on Electron Devices **ED-19**, 4, 584.
Karttunen, S. (1970). *Paperi Puu*, **52**, 4a, 159.
Karttunen, S., Kautto, H. and Oittinen, P. The 11th IARIGAI Conference 1971, to be published in Advances in Printing Science Technology, P. Keppler Verlag, Heusenstamm, 1973.
Karttunen, S. and Oittinen, P. (1972). *Graphic Arts in Finland*, **1**, 1, 9.
Mill, C. C. (1961). *In* "Advances in Printing Science and Technology" (W. H. Banks, ed.) Vol. 1, p. 183. Pergamon Press, Oxford.
Pace, S. A. These Proceedings, Chap. 25.
Ruder, R. (1965). Dissertation at the Technical University of Karl-Marx-Stadt.
Schirmer, K.-H. and Tollenaar, D. The 11th IARIGAI Conference, 1971, to be published in "Advances in Printing Science and Technology", P. Keppler Verlag, Heusenstamm, 1973.
Strauss, V. (1967). *In* "Printing Industry". Printing Industries of America, U.S.A.
Tollenaar, D. and Ernst, P. A. H. (1967). *In* "Advances in Printing Science and Technology" (W. H. Banks, ed.) vol. 4, p. 445. Pergamon Press, Oxford.

Tollenaar, D. and Ernst, P. A. H. (1971). *In* "Advances in Printing Science and Technology" (W. H. Banks, ed.) vol. 6, p. 139. Pergamon Press, Oxford.

Walker, W. C. and Fetsko, J. M. (1955). *Am. Ink Mkr.* **33**, 12, 38.

Wirtz, B. (1964). Proc. Tech. Ass. Graphic Arts, 102. (Dissertation at the Technical University of Darmstadt, 1963.)

The basic principles of printing processes

S. A. PACE

The Kynoch Press, PO Box 216, Witton, Birmingham B6 7BA, England

I. INTRODUCTION

This paper outlines the basic principles involved in the five most important printing processes, namely letterpress, offset lithography, gravure, flexography and screen process. The characteristics of each process are explained, and throughout, the use of printing terms is kept to the minimum.

II. LETTERPRESS

Letterpress is a relief printing process. The image or printing area of a letterpress plate or type is raised above the non-image areas. The inking rollers of the printing machine contact the raised areas, transferring a thick greasy ink onto them. To make a print, the paper must be brought into contact with the image area under pressure. This pressure forces those raised areas into the paper, indenting the surface.

At the same time the ink is squeezed out to the edges of the image areas, causing a slight distortion of the outline. This, in the printing industry, is termed "squash". Under magnification, the edges of the image show a heavy rim of ink slightly separated from the main body of the characters. This is typical of any relief printing process. Ink on the raised surfaces of the plate is squeezed out to the edges when the plate is pressed against the paper to make the print. This phenomenon is influenced by a number of factors.

Firstly, the surface of the paper affects the clarity of the outline, e.g. on a rough cartridge paper, the rim of ink is less clearly defined than that on a smooth glazed paper. Secondly, it can also be influenced by the amount of pressure used; thirdly, by the amount of ink applied and fourthly by the type of material used to form the image area. The thickness of the ink film being transferred to the surface of the paper will also vary according to the same factors. This film thickness in the letterpress process can be between 1 and 10 μm, but for the normal printing of type and of halftone illustrations on a reasonable quality paper, the thickness of ink film would be 2 to 4 μm. The pressure required to transfer the ink from the image to the paper is about 1000 lbf/in² (6·9 MPa). In general printing, letterpress is gradually being ousted from its predominant position by offset lithography due to the relief process's lower printing speeds and its need for high quality papers to give good tonal reproduction.

III. FLEXOGRAPHY

Flexography is another relief printing process, differing from letterpress in two main respects. Firstly, the image area is carried on a flexible rubber plate and secondly, the inks are very thin and fluid and dry rapidly on evaporation. A rubber-covered roller carries ink up from a duct, transferring it to a second metal roller called a transfer roller. This metal roller is engraved or etched with an all-over pattern of small cells which act as ink reservoirs. The amount of ink carried to the rubber image plate by the second roller is controlled by the number and size of these cells. An impression cylinder brings the printing stock into contact with the raised inked areas of the rubber plates which are wrapped round a plate cylinder. Rubber distorts under pressure and flexographic inks being much more fluid than letterpress, the squash or distortion is very pronounced. A very light or "kiss" impression pressure is essential to minimize distortion and because of this tendency, it is difficult to print fine detail. Flexography is excellent, however, for bold, solid line work and for the use of bright colours. The fluid solvent-based inks have a valuable property— they dry very rapidly by evaporation. For this reason Flexography is widely used for printing on non-absorbent materials such as plastic films and metal foils which are commonly used in packaging.

IV. OFFSET LITHOGRAPHY

Offset lithography is a planographic process, the image and non-image areas of the printing plate are in the same plane—there are no raised areas or indentations. In this process, a printing plate is used on which the image and non-image areas have been defined chemically, the non-image area being so treated that it will attract only moisture while the image area will attract only a thick greasy ink. In offset lithography, the inked image is first transferred onto a rubber blanket and then transferred from this blanket onto paper. The plate which is wrapped round a cylinder first passes under a series of dampening rollers which apply a thin film of moisture over the whole of the plate before it is passed under the inking rollers. The inked image is transferred to the rubber blanket cylinder and then onto the paper which is brought into contact with the blanket by an impression cylinder. Offset lithography is a versatile process with a very wide range of applications but does not effectively print on non-absorbent materials. Under magnification, the printed image clearly indicates the significant difference between the planographic and relief processes. The offset lithographic process is characterized by uniform ink coverage and there is hardly any distortion of the character. There is no rim of ink round the edge of the image and there is no squash in the centre or at the rim of the characters because the rubber blanket used on the offset cylinder conforms well to the paper surface.

Halftone dots are printed very uniformly and are fairly well defined on smooth coated papers with probably only a slightly fuzzy edge on uncoated stocks. Because of the absence of squash it is possible to print finer halftone screens on rougher papers than by letterpress. In offset lithography the thickness of the ink film is governed by the rubber blanket offset cylinder and the range is very small when compared with letterpress, being between 1·5 and 3 μm. This factor requires the ink to be heavily pigmented to give opacity.

V. Gravure

The gravure process is where the image areas are recessed below the level of the non-printing areas. The image area is etched into the surface of a copper-surfaced cylinder and after ink is applied, the surplus ink on the surface of the copper is removed by a very thin steel "doctor" blade, leaving ink only in the recessed areas. The recessed image area or areas are divided up into small cells whose walls provide a structure that supports the "doctor" blade. Ink is transferred direct from the cells to the printing stock under pressure applied by a rubber-covered impression roller. Production costs of producing the gravure cylinder which contains the image are very high, but it has a long life and this process is widely used for long run colour work such as newspaper colour supplements, magazines and packaging printing. As with flexography, solvent-based quick-drying inks are used which make it ideal for non-absorbent materials such as plastics, films and foils, but unlike flexography, gravure is capable of printing high quality fine detailed illustrations.

Under magnification the cell pattern of the screened cylinder is identified by the saw tooth edge to the printed characters.

VI. SCREEN PROCESS PRINTING

In this process, ink is transferred to the paper through a stencil supported on a fine screen fabric of silk, synthetic fibres or metal. Image and non-image areas are separated by blocking the pores of the screen mesh in the non-image areas so that the screen remains porous only on the image pattern. A squeegee forces the ink through the open pores of the screen onto the material to be printed. Screen process printing is very versatile, having the ability to print on almost any material, glass, metal, leather, plastics, textiles and many others, and the surface to be printed can be flat, curved or cylindrical. The main characteristic of screen printing is the thickness of the ink film which can be applied. At an average of fifty μm, this is ten times the thickness of the letterpress ink film. The print is patterned with the mesh structure of the screen fabric. Another characteristic is that the mesh structure of the screen fabric affects the image giving an irregular zigzag outline.

VII. SUMMARY

The descriptions given are both basic and brief and no attempt has been made to cover the many implications of the five processes. Offset lithography is probably the most versatile process, but has limitations imposed by the thinness of the ink film. While great advances have been made in ink technology, lithography cannot match the density or opacity of letterpress.

BIBLIOGRAPHY

PIRA Teaching Guides.

The application of screen printing to the deposition of metals and insulants

R. G. LOASBY

United Kingdom Atomic Energy Authority, Atomic Weapons Research Establishment, Aldermaston, Berkshire, England

I. INTRODUCTION

In its simplest terms, screen printing technology is based on the formation of patterns through the use of woven cloth screens and viscous pastes. Some of the openings in the screen are blocked using a photosensitive filler emulsion and when the paste is forced through the holes remaining onto a substrate a replica of the pattern on the screen is obtained. Virtually any material that can be formulated into a viscous paste may be used: this includes liquids and solids, the latter being carried as powders held in suspension in a paste medium.

As applied to the deposition of metals and insulants, the method is an elevated temperature process relying on the sintering of particles together to form either conductive paths or insulating films. The metals used are mainly noble as these have good electrical conductivity and can be air-fired at low cost. Using powder in the readily obtainable 1–10 μm diameter range, the sintering range becomes 700–1000 °C. This high temperature aspect of the process brings with it both advantages and disadvantages; room temperature stability will be high, after manufacture, but materials compatibility problems are accentuated during manufacture.

The high temperature limits the choice of substrate to some kind of ceramic: usually alumina of 85–98% purity is used as a compromise on cost and thermal conductivity. Nowadays the commercial metallization pastes are designed to bond to the glass phases in this type of material and hence compositions outside this range of purity are rarely used. The common exception to this is in microwave applications which demand a higher purity level.

The materials that are to form the metallization patterns are made up in the form of viscous pastes, somewhat like petroleum jelly in consistency, for application to the substrate. The pastes contain four principal components:

(i) the powdered metal;
(ii) a glass, again in particulate form, which is present to provide an agent for bonding the metal to the substrate;
(iii) an organic vehicle to carry the usable components into position;
(iv) a diluent to control the viscosity of the whole.

These last two components should have a low temperature coefficient of viscosity and be constituted such that they burn off in air without residue. These are minor constraints and hence there is a very wide choice of materials for this application. The particulate size of the powders used is about 5 μm, as small as possible in the case of the metal to lower the sintering temperature.

Insulant materials are similar with the exclusion of the metal component. Resistive pastes usually contain non-noble metals and oxides, to give them high resistivities and low temperature coefficients of resistance. This somewhat unavoidable compositional structure has considerable thermodynamic instability at the firing temperature, making such pastes critically sensitive to firing conditions. With hundreds of combinations of these materials available, it is clear that some of the problems which arise in using the process are associated with compatibility.

The main component of the printing equipment is the screen itself which usually consists of a woven mesh, of stainless steel or nylon, mounted under tension on a metal frame. The circuit pattern is formed photographically on the mesh using an ultraviolet-sensitive filler emulsion. The paste is then deposited on the printing screen, the screen is suspended about 0·5 mm above the substrate and a scraper blade, under pressure, traverses the screen, bringing it into contact with the substrate as it does so. The natural elasticity of the screen material enables it to peel cleanly off the substrate leaving the deposit of paste in position. The scraping step is often carried out by hand but a range of commercial machines is available, enabling the print to be achieved in a controlled manner. Such a machine consists essentially of a substrate holder and scraper blade traverse mechanism, these components being arranged with some precision within a rigid structure such that their position and movement with respect to each other are accurately defined and controllable. The substrate holder has three-dimensional position adjustments and a central vacuum chuck which is often fitted with internal illumination for pattern registration purposes. It is carried usually on a pair of linear bearings for movement from the loading position to the working position beneath the screen. The scraper blade traverse mechanism can be quite complex as control of traverse speed, blade pressure and position is necessary to ensure uniformity of paste deposition. Pneumatic-hydraulic combinations or mechanically spring-loaded systems are generally used, the former giving a greater range of controllable pressure and speed values. Again this mechanism is mounted on a pair of linear bearings: the three positional planes for scraper, screen and substrate must be maintained parallel and in a fixed position throughout the printing sequence, so the whole machine structure has to be rigid. There are many variables in the deposition sequence, which have been discussed for example by Savage (1969), but the main factors associated with the machine are rigidity, mechanical precision and uniformity of speed and pressure at the scraper blade-screen interface. To achieve a dimensional precision of greater than 250 μm, nothing is critical and a complete equipment need not cost more than £100. At the same time, for the most precise work where tolerances of less than 25 μm may be required, a machine may cost £3000 and a screen £50.

The firing of the deposited paste produces, in sequence, the burn-off of the organic vehicle at 200–500 °C and then the sintering of the metal particles and the melting of the glass particles at temperatures in the range 700–900 °C. The glass forms a mechanical key on the rough underside of the metal and wets the glass phase in the alumina. A continuous bond is thus established between the metal and the substrate. Although the details of the firing step are most critical for resistors, care still has to be taken with conductors as the alumina can readily and rapidly take up the bonding glass, resulting in a degradation of strength if the arrays are overfired. The firing step, and its complications, are clearly related to the materials used for producing patterns, rather than to the printing process itself, so that it is useful to discuss the whole process in terms of these two aspects, pattern deposition and materials. They overlap a little but this can be ignored for now.

II. PATTERN DEPOSITION

On the deposition step, it is clear that the printing sequence is very crude and has plenty of scope for losing dimensional control of the pattern. Firstly, the print will be made on a non-flat surface if cost is to be minimized: available alumina substrates are usually warped, often with dips and high areas of amplitude 10–20 μm spread over distances of 5–10 mm and this can affect the amount of paste deposited unless care is taken to ensure that the screen follows all the undulations in the substrate surface. This is an easy condition to satisfy with the mesh material usually used, but this material brings with it a number of disadvantages. It is a woven material with many crossover points and without a flat surface. The stresses which arise due to friction at the crossovers will gradually relax in use, with consequent loss of pattern size and mesh tension, so steps must be taken to relieve these stresses before developing the pattern. This is not very difficult either and it is possible to hold pattern dimensions on a screen to a few micrometres over its life.

The complex shape of the mesh, as seen by the paste printed through it, results in poor line edge definition. A good quality screen filler emulsion will give straight edges, which tends to improve definition, but it is difficult to obtain edge errors of less than 10 µm by attention to the emulsion. A high build-up of emulsion, which also decreases the edge effect, introduces a more significant problem of deposit thickness control as the suspended mesh will bend down under the scraper pressure. The result of this is a variation of deposit thickness dependent on the unsupported area; combined with substrate bow, this can be a serious restriction on the control of resistor values.

A partial answer to this predicament is to use one of a number of different types of solid metal screen which have been developed over the last few years. The most common of these has the pattern chemically milled in a 25–75 µm thick sheet of metal. The pattern is taken halfway through the sheet and an array of feeder holes is formed on the reverse side. This results in a screen which is altogether cleaner than the mesh type and gives a better pattern but the price of this improvement is a loss of mechanical flexibility of the screen. Such a screen demands flatter substrates, which have to be ground or selected, and the loss of the natural peeling action of the screen off the substrate gives some plugging of the screen holes leading to a certain amount of porosity in the print. The upshot of this situation on screens at the moment is that for the majority of work where definition is not crucial, mesh screens are used—and these nowadays can be faced with metal to give long life—and for those particularly demanding applications where fine lines and spacings are needed, for example in printing integrated circuit cell patterns, the more rigid solid screens are used. A more detailed discussion on the properties of the printing screen has been presented by Savage *et al.* (1969) and Caronis (1967).

By and large the printing process has developed to the point where it is under control and dimensionally it is capable of meeting most of the requirements. Circuit patterns with line widths down to 125 µm over areas up to 15 cm², with short feeder lines for active device connection down to 25 µm widths, all with a width tolerance of ± 10 µm and a positional tolerance of ± 25 µm, have been achieved. At the limit there is a fall-off in yield: around 25 µm line width of conductors more effort is necessary, and hence more cost, than can be justified in any but the most sophisticated applications.

III. The materials used in the context of electronic circuitry

The materials used in thick film have developed, and are developing very quickly. Generally they have lagged behind the other aspects of the process as they are more complex and require much more research, but at the same time the materials have a high value as circuit elements and hence can be relatively expensive without limiting their use. In the past two years conductor pastes have improved to within a factor of two of the bulk conductivity, the hardness has become controllable for bonding purposes and the flow at the edges is now within the error due to the printing process. A factor of two in values is often unimportant for conductors and insulators but resistor materials present a constant problem as there are more critical demands made on their electrical properties. One per cent tolerance as-fired, and 10 p.p.m./degC TCR is not an unusual requirement and this cannot be achieved at the present time. Less than 5% and 50 p.p.m./degC is currently available, and both these figures are a factor of 10 better than those available three years ago. It is reasonable to expect that the first figures will be achieved in two to three years' time.

The other main material used, the insulant, has developed greatly in the last three years, much to the advantage of thick film technology as a whole. Mention was made earlier of the presence of glass phases in the substrate, the metallization pastes and, of course, the insulants. Now, using the glasses normally available four to five years ago, it was not possible to utilize more than a single layer of insulant as all the glass phases interacted, usually to form lower melting point mixtures, resulting in a loss of dimensional control, electrical continuity and layer-to-layer insulation as the component materials moved out of position. Two kinds of glass mixture have been developed to overcome this problem, both designed to limit interaction to the bonding surfaces, that is, to allow just sufficient interaction to achieve a bond and no more. There are devitrifying glasses, which precipitate a high melting point crystalline phase on firing, making the layer more rigid, and loaded glasses, which contain ceramic powder and rely on a high viscosity to limit interaction and flow. The difference between the performance of the two types is only important at the smallest dimension, the loaded glass moving less on an edge due to the absence of the flow which occurs prior to the recrystallizing process in the devitrifying type. With both these materials the compatibility limitations are largely removed and it becomes possible to refire a number of times and produce multilayered structures. An example of such a circuit which has eight layers of metal and insulant on an alumina substrate is shown in Fig. 1.

Summarizing the materials situation at present, there are resistor materials

Fig. 1. This circuit, bearing 8 alternating layers of gold conductors and glass insulant, is 4 inches (10 cm) square and is typical of modern screen printed circuitry.

available that are capable of covering the range 1 Ω to 10 MΩ per square with as-fired tolerances in the 5% to 20% range and temperature coefficients of between 50 and 500 p.p.m./degC, insulant materials capable of 99·99% crossover yields and with permitivity values in the range 10 to 1000 and conductor materials with resistance values from 2 to 100 mΩ per square.

REFERENCES

Coronis, H. L. (1967). Proc. 2nd Symposium of Hybrid Microelectronics, Solid State Technology Book Division.
Savage, J. (1969). *Thin Solid Films* **4**, 137.
Savage, J., Houghton, D. G. and Willis, D. (1969). Proc. Conf. Multilayer Interconnection Technology, Paper 6 Int. Soc. Hybrid Microelectronics (U.K.).

The basic principles of air and airless spraying

J. MUIRHEAD

The DeVilbiss Co. Ltd., Ringwood Road, Bournemouth, Hampshire, England

I. INTRODUCTION

Where a surface coating has been designed to produce certain results, it has been formulated with the understanding that certain rules of procedure will be observed, and in terms of spray application these are the rules which govern viscosity, correct solvents, film thickness, atomizing pressures, stoving times and temperature.

If surface coating processes are judged in terms of their relative efficiency, the choice of systems can in many cases present one or two anomalies because of the extent to which the systems overlap each other in their relative merits. Such considerations are also further complicated by the several variables which can influence the choice. These variables are often directly concerned with parameters involving production methods, fabrication requirements of the article itself, coating material specifications, material handling and existing finishing-shop limitations.

II. ATOMIZATION

Approximately 90 years ago it was established by John Rayleigh, the British physicist working in conjunction with Sir William Ramsey, that a free stream or jet of liquid would atomize whenever it exceeded a certain speed. It was also established that this critical speed, the speed at which atomization took place, was closely related to the viscosity and surface tension of the liquid.

The higher the viscosity of the liquid, the higher its critical speed. It therefore follows that any atomizing device must be primarily capable of projecting a liquid into a free stream or jet with a velocity sufficient to overcome the viscosity of that liquid and thereby cause it to disintegrate into a stream of small droplets.

III. CONVENTIONAL SPRAYING

Based on the principle of atomization, the first attempted use of a spray gun was as early as 1869 and the first major application of paint by this system was carried out in 1893. The first successful spray gun designed for general use in industry was placed on the market in 1909.

The principles then embodied are still the basis of today's designs in the sphere of conventional spray application by compressed air atomization of the coating material.

The spray gun directs the liquid into a free stream by striking it with a high speed jet of compressed air. The speed, or energy content, of the air stream is partially transferred to the relatively slow moving liquid. If this transfer of energy is sufficiently great, the liquid is accelerated to a speed above its critical point and atomization will result.

Since air is very light in weight, a comparatively large volume moving at high speed is required to impart a portion of its speed to a heavy, viscous material such as paint. Thus, each part of paint which leaves the spray gun is accompanied by a high proportion of high speed air, the droplets in this stream moving with and propelled by the expanding stream of atomizing air, the correct balance between air and fluid being essential and of vital importance.

A method must therefore be used to meter the coating material into the air stream and this is achieved by the use of a nozzle or fluid tip, the quantity to be passed being governed by the size and design of the orifice in the fluid tip, the type and viscosity of the liquid, and maintenance of the correct balance between the air pressure exerted on the fluid stream, and the fluid flow.

It is therefore established by the foregoing that in order to atomize a given quantity of paint, a definite volume of air is necessary at a suitable pressure, the volume of air being determined by the particular type of air cap (Fig. 1), the number of air cap holes present, and their size and position relative to the orifice of the fluid nozzle. Air caps and fluid nozzles are complementary and we shall examine this relationship in due course.

An air cap's ability to provide correct atomization conditions is dependent on:

(i) Its consumption in terms of air volume.

(ii) Position of the fluid nozzle face relative to the air cap face.

(iii) Air cap hole convergence where atomization occurs.

Approximately 60% of the volume of air passed by an air cap is utilized in atomization and 40% to pattern formation. From the instant that compressed air is released from the air cap, it starts to expand. As it expands it

SIMPLE AIR CAP TYPES

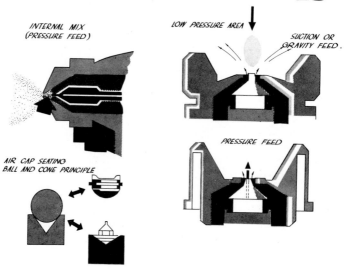

FIG. 1. Simple air cap types.

loses velocity, pressure and atomization potential—in other words "horsepower". Therefore to utilize this potential horsepower, the act of atomization should take place as close to the air cap face as possible, having due regard to the turbulence which is set up at the meeting point of air and fluid. If this turbulence is not controlled, a build-up of paint in the air cap orifice will result, thus causing a noticeable drop in spray gun performance together with distortion of the spray pattern. The main nucleus of the jet thus formed on emission from the air cap takes on the form of a cone. This nucleus has been found to extend for a length equal to about $3\frac{1}{2}$–4 nozzle diameters. In this region the paint undergoes the most vigorous atomization. In the remainder of the jet, the air stream is highly turbulent, promoting rapid mixing of the paint particles with the air. With increasing distance from the nozzle, the air velocity falls sharply, to reach a value of about 20–35 m/s at a distance of 150–200 mm from the nozzle, against an initial critical velocity of up to 313 m/s. The optimum distance between the spray gun head and the surface to be coated can be taken as 200–300 mm.

IV. Deposition of Paint Film

Having discussed the basic principles of atomization by compressed air, let us examine in broad outline what happens in the process of applying a surface coating by conventional compressed air spraying.

The solids-constituent parts of the coating material in the form of gums, resins and pigments are broken up by the action of the air, into tiny globules. This process takes place between the gun and the surface being sprayed, and the air stream carries these tiny globules or spheres on to the work surface. Simultaneously the process of vaporization is also taking place. As we are aware, paint has thinner or reducer added at the time of manufacture and it is often necessary to augment the thinner content in preparing the paint for application. This thinner or reducer is of course in liquid form, and as the material flows through the spray gun, the air stream also acts on the thinning medium causing a vapour or gas to be created. A portion of this gas, sometimes up to 20%, escapes from the spray pattern in its progress towards the work surface. When we remember that a very important function of the thinner or reducer is to "flow out" or level the globules or spheres in the resultant paint film we create, we realize that an undue loss of thinner or reducer due to excessive vaporization will seriously curtail the normal flow-out characteristics of the paint film and result in a low quality finish with a dry or sandy appearance.

One of the most important factors in spray finishing now becomes apparent. Too high an atomizing air pressure will result in excessive vaporization in the spray cone and conversely, too low a pressure will result in a large droplet size with consequent runs or sags on the surface.

As the air flows through the spray gun (Fig. 2) via the trigger operated valve, a second valve is encountered which controls the amount of air fed to auxiliary holes in the horns of the air cap. With this spreader adjustment valve closed, the air flows only through the centre hole and atomizing holes of the air cap, thus producing a round pattern. With the spreader valve open, this same flow of air prevails and an additional air stream now flows through the baffle and into the horns of the cap thus spreading out the round spray and producing a fan spray pattern.

By adjusting the spreader valve, a round or fan spray pattern of various widths can be established according to the work in hand. It must be noted that for use with suction or gravity feed, the realignment of fluid nozzle and air cap is essential to allow the fluid tip to protrude slightly beyond the air cap, thus creating a vacuum at this point allowing atmospheric pressure to force the material from the container to the spray head of the gun.

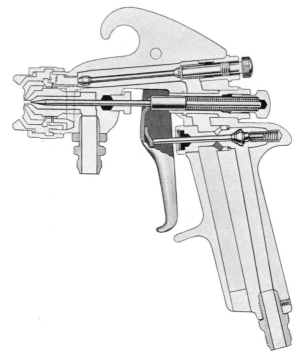

Fig. 2. Air spray gun.

Modern developments and improvements in spray gun design combined with the correct choice of fluid nozzle and air cap have overcome many of the problems of accurately maintaining the correct balance between air and fluid in order to obtain a spray pattern which is uniform in shape and density and gives a well atomized coating throughout its entire area. The controls of this balance can be affected by:

(i) Reduction of the material viscosity within defined limits, resulting in an increase in the quantity of paint being passed through the fluid nozzle and vice versa.

(ii) Regulating the spread of the fluid being atomized by means of the spray width control, so that a greater spread is obtained by the elongated shape of the fan.

   (iii) Control of both air and fluid pressure. Where material is being supplied to the gun by pressure feed supply, two independent adjustments are possible—the air pressure on the pressure feed tank can be varied to supply more or less material, and the atomizing air pressure can be increased or decreased while maintaining constant fluid pressure.

   (iv) The ratio of fluid to air can also be varied by using the fluid needle adjustment on the spray gun. This method however has the disadvantage of exerting additional tension on the trigger of the gun and can induce operator fatigue.

A further variation can be utilized in the form of an internal mix air cap, in which case the air and liquid are mixed inside the cap before being expelled. The use of internal mix air caps permits lower air pressures to be used and requires the application of balanced air and fluid pressures in operation.

This latter type of cap is often used to reduce fog or spray mist when operating in confined areas of the workpiece.

V. Hot spray application

The basic worth of any system depositing a surface coating is the efficiency at which the material is deposited on the article in relation to the material usage. The difference between the theoretical efficiency and actual efficiency is waste, which can be in the form of evaporation, settling out, hardening, and —in the case of systems employing spray guns—paint overspray. The recognition of this latter fact has resulted in the development of many systems of paint application, all designed to minimize paint losses.

Originally, paint heating was introduced as a method whereby viscosities could be reduced by heat thus avoiding the need for solvent. Claims were therefore made for considerable savings and, whilst these were justified, sight was lost of the fact that the material being applied was almost pure paint and not a mixture of paint and solvent and therefore any overspray was a more expensive commodity. The system of applying materials at elevated temperatures was originally based on the theory that high film builds could be applied in one coat. For this purpose paint manufacturers produced special high solids content materials which did of course provide heavier coats in one application, and thus saved time, but in many cases put more paint on the article than was either needed or required. A more advantageous method of utilizing paint heating is to reduce the viscosity of the material to as low as possible, consistent with the dry film build required in order that a low atomizing pressure can be used. By using conventional materials and not the special high solids content type, thinning them to a normal cold spray viscosity of approximately 20–25 seconds No. 4 Ford Cup (60–75 cSt) with a fast-evaporating low-boiling solvent, and then passing this mixed material through the paint heating system to still further reduce its viscosity, atomizing pressures as low as 15–25 lbf/in^2 (103–172 kPa) can be utilized.

At first sight it would appear that applying paint at a lower viscosity through the paint heater will more readily give rise to runs and sags, but this is not the case, since the unwanted solvents are evaporated by the effects of heat inside a sealed unit instead of by air pressure. In the heat exchanger the added solvents in the paint reach nearly their boiling point so that when they emerge from the nozzle they expand and evaporate without the need for compressed air.

VI. Airless spray application

Reference was made at the beginning of this paper to the fact that the atomization of any material was dependent on a critical speed being achieved to permit the break-up of the material into tiny droplets. Any paint spraying method has therefore primarily to induce this high velocity in the material flow.

Atomization by the airless method is achieved by forcing material at high pressure through an accurately designed small orifice in the spray cup. A combination of high velocity and the rapid expansion of the paint upon passing through the fine orifice into the atmosphere, causes the material to break up or atomize into a spray of fine particles. The pressures applied to the material are provided by a fluid pump having a pressure ratio of up to 30:1 thus providing final fluid pressures of up to 3000 lbf/in^2 (21 MPa). The pressurized paint is fed to the spray gun via a spirally wound stainless steel reinforced P.T.F.E. or nylon hose, specially designed to withstand pressures in excess of the possible maximum developed.

The airless spray gun (Fig. 3) is similar in appearance to the conventional gun with the exception of having only a fluid inlet and no atomizing air inlet. The important components of the spray gun are the fluid tip and spray cap, the latter normally made from tungsten carbide to withstand the high abrasive wear caused by the coarse particles in the paint. These caps are specially shaped to give varying fan patterns and are normally available in a large variety of sizes ranging from 75–700 µm diameter. The fan pattern is predetermined by the angle at which the spray cap slots are machined in the respective caps.

Unlike the process of conventional spraying, there is no carrier or air stream to convey the atomized particles to the work surface on leaving the spray gun. The atomized particles of material have sufficient forward velocity created by the high pressure exerted on the material and on being released from the tip of the spray cap rapid expansion takes place. In addition to this hydraulic mechanism, atomization in airless spraying equipment is assisted by two additional factors. Firstly, the effect of the resistance of the atmospheric air surrounding the spray cap to the high speed jet of material, and secondly by the rapid evaporation of the volatile solvents present in the coating material.

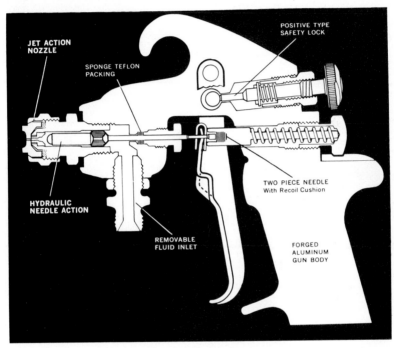

POSITIVE TYPE
SAFETY LOCK

JET ACTION
NOZZLE

SPONGE TEFLON
PACKING

TWO PIECE NEEDLE
With Recoil Cushion

HYDRAULIC
NEEDLE ACTION

REMOVABLE
FLUID INLET

FORGED
ALUMINUM
GUN BODY

FIG. 3. Airless spray gun.

The material is applied much wetter than with conventional spraying since the solvents are not subjected to the same degree of vaporization as with conventional spray methods, and for this reason some materials must be modified in their solvent content, using faster evaporating solvents than with conventional spraying. For instance, if a material has been formulated to offset high atomization air, a large percentage of the thinners would be of the slow acting type. These would be employed to prevent the material from being applied too dry. If this same material were used with airless spray methods it would be impossible to obtain a sufficient coating without sags.

Pigment size must also be considered if airless atomization is to be used, and finely ground pigments must be used to avoid clogging. Similarly the cell structures of some materials have cohesion characteristics which do not permit the necessary function of tearing apart when suddenly released from the high pressure.

It will be noted from this that certain limitations exist in the use of airless atomization, these limits being largely dictated by the suitability of the surface coating materials.

In conclusion it must be said that the subject of the basic principles of spraying is far broader than this presentation could possibly cover. It has been my intention to cover it from its more important angles, pointing out the essentials, and pointing out also some of the controls which should be exercised. I hope that this has in some way been of assistance.

Use of spray applied coatings in colour television picture tubes

D. DE GRAAF

Acheson Colloiden B. V., Scheemda, Holland

I. INTRODUCTION

The glass envelope of a colour television tube consists of three parts—the screen, the cone and the neck. The screen carries a fluorescent coating and an electron gun is inserted in the neck. Between the screen and the cone a shadow mask is present for the focusing of the electron beams on the screen (Fig. 1). In colour tubes, unlike black and white tubes, the whole inside of the cone is coated with a conductive layer.

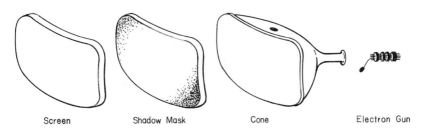

Screen   Shadow Mask   Cone   Electron Gun

FIG. 1. The four parts of a colour television picture tube.

The main purpose of the coating is to serve as a ray-focusing anode preventing distortion of the image by stray electrons reflecting back into the beam. It also establishes an electrical contact between the anode contact and the electron gun system (Fig. 2). Because of the high voltage used in colour tubes, heat will be generated. This heat should be distributed over the entire interior of the cone to avoid accumulation of heat on the shadow mask and thus causing distortion of the picture. To be sure of a good electrical contact between the anode contact and the electron gun system, the anode contact and neck are coated first in the same way as for black and white tubes.

After a short time of air drying, the inside is coated either by spray or sponge application. The coatings are applied before assembly of the screen, shadow mask and cone. As a conductive material, water based graphite dispersions, with sodium or potassium silicate as a binder, are used.

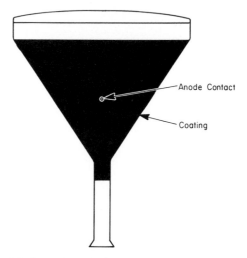

Fig. 2. The internal coating in a colour television picture tube.

II. The internal coating

A. The application

The glass cone is thoroughly cleaned with solutions of hydrofluoric acid and sodium carbonate in an ultrasonic cleaning machine. Then the cone is dried in a warm air stream. During this cleaning process the interior wall of the cone is etched to some extent by the hydrofluoric acid to improve the adhesion of the applied coating.

The coating between the anode contact and the neck is applied by brush on the cone, preheated to 40 °C. After air or infra red drying, the rim of the cone is screened with adhesive tape and the cone is then placed in a holding device, with the neck pointing downwards. In this position a coating is applied by spray. During spraying the cone is rotated either by hand or automatically. Then the tape is removed and the cone is placed on the transport system where warm air is blown over the coating to speed-up drying

(Fig. 3). Afterwards the coating is baked in a 30 metre long oven at a temperature of 420–450 °C for about 20 minutes. After curing of the coating the screen is fritted on the cone. Sometimes the fritting and baking for the coating are done at the same time.

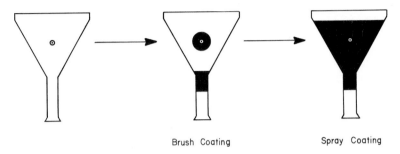

Brush Coating   Spray Coating

FIG. 3. Sequence of the application of the two coatings in a colour television picture tube.

Some tube manufacturers prefer application by sponge. This is done to avoid soiling of the rim of the cone by graphite, which can cause poor vacuum tightness of the seam between screen and cone. For this application a sponge, soaked with the graphite dispersion, is pressed against the interior wall and by rotating the cone and moving the sponge downwards, the cone is coated.

Before use, the dispersion should be well homogenized with a stirrer or a rolling equipment to redisperse the sediment. It is recommended to sieve the dispersion through a nylon gauze, preventing dried-out particles in the coating.

B. Required properties

1. High vacuum stability under electron bombardment

To guarantee long life of the television tube, the coating should not give off gaseous products under high vacuum and electron bombardment. Gaseous substances will cause a drop in emission. Organic materials, still present in the wet coating, must decompose during the curing cycle. Only inorganic, stable components, should be left. These organic materials, such as dispersants and wetting agents, are unavoidable in the production of stable dispersions. The amount of organic substances must be kept as low as possible. The degassing properties and temperature stability of the coating can be checked by a thermogravimetric analysis during which a small sample is heated at a constant rate. The loss of weight of the coating is measured.

2. No coating defects

The coating should not show cracks or blisters. This is particularly important for the area around the anode contact where two coatings are applied over each other. Irregularities around the anode contact will collect the dispersion, resulting in a thick coating after drying. The type of graphite and silicate should be selected carefully to avoid cracking. To avoid blisters around the anode contact, the brush applied coating should be completely air dry before the spray coating is applied.

Blisters on places other than the anode contact can be caused by poor degassing properties.

3. Good adhesion and cohesion

The coating should adhere well to the glass substrate. Adhesion is obtained by using potassium or sodium silicate as a binder, since these do not decompose during curing. The cohesion must be such that no loose graphite particles are present, as these can affect the electrical functioning of the tube. The cohesion can be tested by a tape test in which a standard adhesive tape is applied to the coating. The tape is then pulled away and examined for graphite particles.

4. High scratch hardness

The contact between the graphite coating and the electron gun system is made by three contact springs. The coating should not be too soft causing detaching of graphite particles. Some tube manufacturers prefer a product with a higher silicate content as a coating material on the anode contact and neck to increase the hardness and scratch resistance of the neck coating.

5. Covering power

To speed up production the material should cover in one spray. Therefore a fine particle size graphite has to be used.

6. Sedimentation

The dispersion should show no sedimentation because large containers are used for feeding the spray gun. Sedimentation would lead to variations in coating thickness, resulting in variations in coating characteristics.

7. Good workability

The rheological properties of the dispersion should allow easy application by spray gun.

8. *Low electrical resistance*

The resistance of the coating is of minor importance because of the high voltage used in the tube. As the silicate is an insulator, compared to graphite, somewhere a compromise has to be found between adhesion, cohesion, scratch hardness and conductivity.

The basic principles of flame spraying

C. W. SMITH

Metco Ltd., Chobham, Woking, England

I. INTRODUCTION

This paper describes two flame spraying processes—one utilizing material to be deposited in wire form and designated "wire combustion spray", and the second utilizing material in powder form and designated as "powder combustion spray". Both processes use a combustion flame as a heat source.

II. THE WIRE COMBUSTION SPRAY PROCESS

Figure 1 illustrates how this process can readily be subdivided into the following:

A. *The process occurring at the source*

As illustrated in Fig. 1 the source comprises a nozzle, through the centre of which wire is passed into an oxy-fuel flame. An annulus is formed around the outside of the nozzle, and the inside of an air cap, through which compressed air is fed. The wire tip is continuously heated to its melting point, and then broken down into particles by the stream of compressed air.

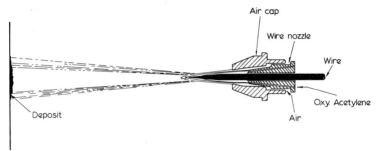

FIG. 1. A schematic of the wire combustion spray process.

B. The process occurring in transport

Molten particles are accelerated by virtue of the air stream towards the substrate. In so doing they are both cooled to a plastic or semi-molten condition, and a certain amount of oxidation occurs on the particle surfaces.

C. The process occurring on impact with the substrate

Upon impact with the substrate the particles flatten and bond to the former. Subsequent particles also flatten, and bond to those already adhered to the substrate, thereby fabricating a coating.

III. THE POWDER COMBUSTION SPRAY PROCESS

Figure 2 illustrates how this process can readily be subdivided into the following:

A. The process occurring at the source

The source essentially comprises a nozzle, through the centre of which material in powder form is conveyed by being suspended in a carrier or aspirating gas. The material passes out of the nozzle into an oxy-fuel flame, and is thereby raised to a temperature approaching its melting point.

B. The process occurring in transport

The flame velocity accelerates the semi-molten particles towards the substrate. In so doing there will be a certain amount of oxidation on the particle surfaces. Compressed air may be used further to accelerate particles, thereby producing higher particle velocities.

C. The process occurring at the substrate

The process occurring at the substrate is essentially the same as that described in the wire combustion spray process.

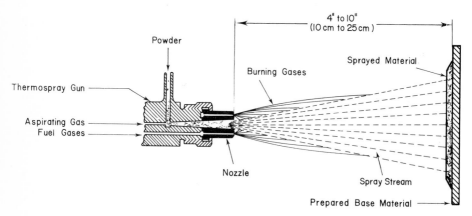

Fig. 2. A schematic of the powder combustion spray process.

IV. SCOPE AND LIMITATIONS OF FLAME SPRAYING

The following indicates the scope and limitations of both the coating and the substrate, and is applicable to coatings produced by both the wire and powder processes.

A. Those pertaining to the coating

1. *Materials*

(a) Wire combustion spray. Any material can be sprayed which can be produced in wire form, will melt in the environment of an oxy-fuel flame, and does not decompose on melting. Of materials commonly sprayed, molybdenum has the highest melting point, this being approximately 2600 °C.

(b) Powder combustion spray. Materials can be deposited which are available in a suitable powder form, will melt in the environment of an oxy-fuel flame, and will not decompose on melting. These powders normally exist in the following categories:

(i) Chemical elements. For example, copper and aluminium.

(ii) Chemical compounds or metallurgical alloys. For example, aluminium oxide and aluminium bronze alloy.

(iii) Composites. In this group each discreet powder particle will be made up of two or more different materials, physically bonded together, but chemically and metallurgically separate. During deposition by flame spraying, such materials may chemically and/or metallurgically combine. For example, composite powders or nickel and aluminium produce coatings containing nickel aluminide.

(iv) Physical blend. These materials are produced by blending together two or more powders of the above types.

2. Structure

Figure 3 is a microphotograph showing the structure of a particular flame sprayed coating system. This coating system is sprayed on to a steel substrate, and is in two parts, as follows:

(a) Nickel aluminide underlayer. This coating is produced from a composite powder of nickel and aluminium, by the powder spray process. The nickel and aluminium constituents of the powder react in the spraying process to produce nickel aluminide in the coating, which is therefore made up of nickel, aluminium, nickel aluminide and oxides of these materials. The coating produced from a composite powder of nickel and aluminium is used in this coating system, as it frequently is in practice, as a bond coat for subsequent coatings of other materials. For reasons to be discussed later, composites of nickel and aluminium tend to produce high bond strengths by comparison with other flame sprayed coatings.

(b) 0·8% Carbon steel top coating. This is a coating produced by the wire spray process. This coating is frequently used in practice for its good wear-resistant properties.

Figure 3 illustrates clearly the laminar structure of flame sprayed coatings, showing how particles flatten on impact with the substrate. The black areas indicate porosity, which is typically between 5% and 15% by volume. However, for certain industrial applications a high level of porosity is advantageous, in which case porosity levels may rise to 30% or even 40%.

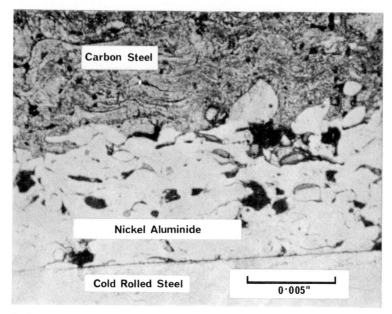

Fig. 3. Cross section of a 0·8% carbon steel coating, deposited onto steel utilizing a nickel aluminium composite bond coat.

The as-sprayed surface finish of coatings produced by either the wire or powder processes would typically be between 250 and 600 microinches (6–15 μm) c.l.a. This, however, is very much a function of the material being sprayed, the deposition parameters, and in the case of powder spray, the size of powder being used.

3. *Deposition rate.*

This will vary depending upon the material being deposited. For example, certain oxide ceramic coatings produced by the powder process would be deposited at approximately 2 lb per hour (0·25 g/s), whereas zinc coatings produced by the wire process would be deposited at between 60 and 70 lb per hour (7·6–8·8 g/s).

4. *Thickness*

The thickness to which flame sprayed coatings can be applied varies, depending upon the material being sprayed.

There is a tendency for all coatings to shrink due to the rapid quenching of semi-molten particles. Certain materials will have a greater tendency to shrink than others. These differences, coupled with different bond strengths, will define what this thickness limit will be in practice. Certain materials are limited to a thickness of 0·020–0·030 inch (500–750 µm), whereas others can be sprayed up to thicknesses of $\frac{1}{2}$ inch (1·3 cm) or so. Typically, in industry, sprayed coatings are used in thicknesses ranging from 0·005–0·100 inch (125–2500 µm).

5. *Special properties*

In practice sprayed coatings are used to resist corrosion, oxidation, cavitation, wear, or any combination of these. Additionally, many special-purpose applications call for coatings offering other properties. Compared with like materials in the "solid form", sprayed coatings often have very different characteristics, due to their unique structure. For example, because of their porosity, sprayed coatings tend to retain lubricants, thereby providing desirable properties for bearing surfaces, etc.

B. *Those pertaining to the substrate*

1. *Material*

Generally speaking, flame sprayed coatings are deposited onto metals. However, they can be applied to other materials, such as glass, plastics, etc. In general, bonding to metallic surfaces is superior to that of non-metallics, although there are exceptions.

2. *Temperature limitations*

Generally speaking, these are defined by the substrate itself. The temperature experienced by the substrate can to a very large extent be controlled by the particular spraying operation, i.e. workpiece and spraying head traverse rates, etc.

3. *Pre-treatment*

In order to maximize bond strengths, substrate surfaces would normally be grit blasted, using either angular steel or aluminium oxide blasting media.

V. ADHESION TO THE SUBSTRATE

Factors which influence this are:

A. *Those pertaining to the material being sprayed*

The prime factor is the energy content of any given particle. The higher the energy level (both thermal and kinetic) then the higher will tend to be the bond strength.

Certain materials will have an inherently high bond strength due to their ability to form an alloy bond layer with the substrate. Materials which tend to do this have been specially developed, and would typically comprise a composite of nickel and aluminium. Such a material is shown in coating form in Fig. 3. When this material is flame sprayed, the nickel and aluminium react exothermically, thereby providing a higher particle energy level. For this reason this type of material is frequently used as a bond coat. Other factors which influence the energy content of a particle are: particle mass, particle velocity, latent heat of fusion, specific heat and melting point.

B. Those pertaining to the substrate

Properties of the substrate will play a large part in the formation of any alloy bond layer, thereby influencing the adhesion of the coating. Substrate properties which influence this are melting point, thermal conductivity, specific heat, pre-heat temperature and nature of surface preparation.

BIBLIOGRAPHY

Apps, R. L. (1970). "The significance of surface preparation and coating thickness upon the bond strength of flame sprayed aluminium and zinc coatings on mild steel".*

Brown, R. L. (1970). "The flame spray forming of silicon for the manufacture of silicon nitride ceramic".*

Buza, J., Damits, M. and Kniewald, D. (1970). "Theoretical quality evaluation of surfaces treated by abrasive blasting".*

Endoh, M. (1970). "On adhesive strength of aluminium oxide sprayed coatings on steel substrates".*

Fletcher, R. K. (1962). "Metal spraying and how it is used". British Welding Journal, July, 428.

Fletcher, R. K. (1963). "Some methods and equipment used in the prevention of corrosion of steel pipes and structures". Glos Ursusa Zam, Poland, 524–524–II VII 1963–100.

Hasui, A. and Kitahara, S. (1970). "On the relation between properties of coating and spraying angle".*

Hermanek, F. J. (1970). "Properties of self-bonding materials".*

Hirose, S. (1970). "Mechanism of erosion by shot or grit blasting".*

Ingham, H. S. and de Gasero, C. (1970). "New optical measuring techniques for particle size of spray powder".*

Knotek, O. and Steine, H. (1970). "On reactions in and on layers of sprayed and fused metal powders".*

Matty, A. and Becker, K. (1956). "An experimental investigation of the metal spraying process". Electroplating and Metal finishing, April, 143–145.

Rammelts, G. J. (1963). "The properties of flame sprayed surface coatings". Industrial Finishing, February.

Shephard, A. P. (1962). "Second generation of flame sprayed hard surfaces". *Machinery*, February.
Steffens, H. D. (1970). "Production and properties of homogeneous deposits of special materials with high oxygen affinity".*
Sutton Smith, C. (1970). "New engineering applications for flame spraying".*

* These papers were presented at the 6th International Metal Spraying Conference, 1970. Reprints are obtainable from Institut de Soudure, 9 Boulevard de Chappelle, Paris, France.

Metal sprayed, polymer impregnated surface coatings

T. A. MADEN

Fothergill and Harvey, Ltd., Littleborough, Lancashire, England

ABSTRACT

Fluorocarbon coatings, in particular coatings based on polytetrafluoro-ethylene (PTFE), have a wide range of use throughout industry, due to their non-stick, low slip, high temperature stability, good chemical resistance and non-ageing properties.

However, one of the disadvantages of PTFE coatings is that they exhibit poor mechanical properties and are, therefore, easily damaged in use. In practice, the limited life of unreinforced PTFE coatings is such that they prove to be unsuitable in many industrial applications. It has been found, however, that coatings with far superior wear resistance can be achieved by reinforcing the fluorocarbon film with metals. One such process is carried out by metal spraying suitably prepared metal substrates with stainless steel. The stainless steel coating is purposely deposited in such a way that a porous matrix, nominally 0·002 inch (50 µm) thick, is applied. This matrix is then impregnated with an air sprayed, water based, dispersion of PTFE. The resulting coating is then dried to remove all the water and cured at a temperature of 370 °C.

On subjecting this type of coating to wear, the stainless steel matrix takes the load and so protects the PTFE. After being subjected to abrasion, the peaks of the matrix are exposed and these resist further wear, protecting the relatively large areas of PTFE surrounding them, thereby maintaining the desirable non-stick low friction characteristics of the surface.

The basic principles of detonation coating

R. G. Smith

Union Carbide U.K. Limited, Swindon, Wiltshire, England

I. Introduction

Experiments in the U.S.A. in the early nineteen-fifties on the mechanism and reaction products of mixtures of explosive gases in open-ended tubes, demonstrated that detonations produced by igniting such mixtures could be harnessed to produce a high velocity, high temperature gas stream.

It became apparent that if the energy of detonation could be controlled and powder particles suspended in the gas mixture prior to detonation, the high velocity gas stream which accompanies a detonation could be harnessed to accelerate and heat the powder particles prior to impact on a chosen substrate, thereby producing an extremely dense coating with high bond strength. From these earlier researches, detonation coating—or D-Gun coating—began; a coating technique closely allied to thermospray processes but differing in that a considerable amount of the energy required to fuse a powder particle is generated by the high kinetic energy stored in each particle during acceleration and released as heat on impact with the substrate.

II. BASIC PRINCIPLES

How is a detonation initiated? A most convenient form for gas mixtures is a steel tube closed at one end, approximately 2·5 cm internal diameter and 135 cm long (Fig. 1). The tube is filled with an explosive mixture of gases such as methane, propane with oxygen or air and, in the case of the D-Gun, acetylene with oxygen. By means of an electric discharge or sparking plug,

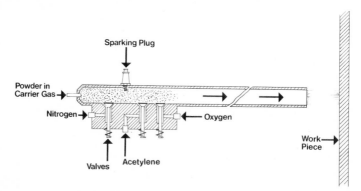

FIG. 1. Schematic diagram of detonation gun.

the mixture is ignited at the closed end. The gases burn for a short period of time, and the flame front accelerates, compressing and heating the unreacted gas zone ahead of it. The velocity of this flame front is of the order of 10–15 m/s. Depending upon the gas mixture being compressed, a critical temperature is reached and self-ignition of the unreacted gas mixture occurs, producing a strong light flash and a detonation or shock wave. Up to this point the gas chemical reaction occurring has been gentle.

The shock wave at the head of the detonation wave front compresses and raises the gas temperature, inducing violent chemical activity and reaction resulting in high gas pressures and temperatures of the order of 4740 K. The detonation wave front travels at constant velocity, for a 1:1 ratio of oxygen/acetylene at 2930 m/s. The detonation velocity is the speed at which the phase boundary, between the completely reacted and unreacted gas mixtures, moves within the tube. When the shock wave issues from the open end of the barrel it compresses the surrounding air outside with a resultant loud bang for each successive detonation. The detonation wave travels in both directions towards the open and closed ends of the tube. Chemical reaction ceases when the detonation front passes through the flame front. The detonation front then continues to travel towards the closed end of the tube in the form of a shock wave only. The change from combustion to detonation gas mixtures differs with respect to the composition of the gas mixture, which is critical with respect to upper and lower percentage limits of certain gases; outside these limits a detonation will not take place at all. The detonation velocity varies with differing gas mixtures, and can be determined by the summation of the velocity of the gaseous combustion products in the detonation phase boundary plus the velocity of sound in the gaseous media.

For a gas mixture of equal volumes of oxygen and acetylene in a tube 135 cm long, the detonation wave front travels the tube length and complete chemical reaction occurs within 0·5 milliseconds.

In very simple terms the energy produced by the dramatic chemical reaction of the mixed gases maintains the shock wave until it passes out of the open end of the tube, leaving a tube full of reaction product gases at high pressure and temperature. They have only one way to go, quickly, out of the open end expanding and accelerating but cooling down rapidly.

How do we utilize this high temperature supersonic gas stream? By injecting and suspending measured quantities of powder particles into the unreacted gas mixture immediately prior to ignition, we have an ideal medium for heating and hurling particles of a chosen material onto a prepared substrate, placed in the path of the tube. As the detonation wave front passes through the suspended powder particles in the tube, they are heated, accelerated and decelerated. Calculations of the particle velocities induced by the chemical reaction zone are of the order of 40 m/s. However, the final high temperature high pressure system set up within the reacted gaseous products in the tube on completion of the detonation, picks up the molten particles and hurls them out of the open end of the tube at supersonic speeds.

Particle diameter size and gas viscosity both affect the ultimate particle velocity; however, experimental evidence shows that particle velocities of the order of 800 m/s are achieved with, for example, tungsten carbide particles 30 μm in diameter.

How do we convert these basic concepts into practical processing?

III. PRACTICAL PROCESSING

As the name implies, the detonation equipment is like a gun. Figure 2 shows a picture of the gun which has a 1·4 metre long water-cooled barrel, approximately 2·5 cm inner diameter. The noise level produced by the detonations is high, approximately 150 decibels (relative to 1 dB as the lowest detectible level). Because of this, the equipment is housed in a double-walled concrete cubicle and remotely controlled from outside at an operator console station.

Consider a complete cycle of one detonation and for the purpose of this discussion, refer back to Fig. 1.

Fig. 2. Detonation gun.

Accurately metered quantities of oxygen and acetylene are introduced via the poppet valves into the combustion chamber, filling the combustion chamber and the open-ended barrel. Powder stored in a pressurized dispenser is accurately metered into a nitrogen carrier gas stream, which transports the powder into the rear of the gun barrel and suspends the powder in the gas mixture. At the same time, nitrogen from a separate source is metered into the combustion chamber to surround the poppet valves to protect them from the hot gas erosion following a detonation. With all valves closed, the gas mixture is ignited by means of a spark plug, fed from a magneto which is synchronized with the poppet valve system. The detonation or shock wave which follows microseconds after ignition, attains a velocity of 2930 m/s. The high velocity gas stream imparts energy to the powder particles which reach an exit velocity of approximately 800 m/s. During the short duration of particle acceleration, heat is also imparted by the hot gas stream and the particles are melted. At the head of the detonation front where the chemical reaction is complete, the gas temperature achieved with a 1:1 oxygen/ acetylene gas mixture is 4740 K. This value decreases linearly with distance along the D-Gun barrel towards the breech dropping to 3000 K approximately. Experimental evidence shows that certain materials are completely melted by the hot gas stream whereas certain high temperature melting point materials are melted only on the outer surface of the particles.

The cycle is completed with the nitrogen valve opening and a purge stream sweeping out the gaseous products of the combustion. Depending upon the type of coating applied, this cycle is repeated 4·3 or 8·6 times per second. This produces a coating structure which is built up of a series of detonations on the prepared substrate placed approximately 7 cm in front of the barrel.

IV. COATING/SUBSTRATE INTERFACE

Powder particles in a molten or plastically deformed state leave the end of the D-Gun barrel at 800 m/s. During the relatively short journey to the prepared substrate, further acceleration due to the adiabatic expansion of the gas occurs but the particles start to cool rapidly.

The particles impact on the substrate surface and flatten or splatter into thin overlapping platelets such that their diameter is many times greater than their thickness. The high kinetic energy of the particles, approximately twenty five times that of the energy in particles produced by conventional flame spraying, is converted to additional heat on impact with the surface. This additional benefit of maintaining a molten state with high velocity at the substrate allows the particles to completely deform and "wet" the surface and closely follow the substrate surface contours, to produce extremely high bond strengths between coating and substrate. The coating is built up by successive detonations or "pops" of powder with interlayers approximately 6 μm thick overlapping each other.

The nature of the bond adhesion is not fully understood. Tensile bond strengths in excess of 700 kgf/cm² (69 MPa) are regularly recorded. It is suggested that the bonding mechanism is not purely mechanical, that considerable "wetting" of the surface by the particles occurs, and that in certain instances, penetration of the surface by the particles may be observed. The possibility of microscopic welding between particles and substrate cannot be ruled out. However, this is not completely verified to date and requires further practical experimentation.

V. Coating

A. Coating materials

The types of coating which can be deposited have been developed primarily for wear resistance specific to certain industries and are capable of operating at elevated temperatures and under varying chemical environments. This has established coating technology around high temperature cermets, ceramics, alloys and, to a lesser degree, pure metals. The predominant materials include tungsten carbide with cobalt binder in varying percentages, mixed tungsten and chromium carbides, chromium carbide with nickel chrome binder and ceramics such as aluminium oxide and mixtures of aluminium oxide/titanium oxide and aluminium oxide/chromium oxide.

B. Deposition rate

Deposition rate in D-Gun coating terms is usually referred to as "coating thickness per pop" or "single detonation". The variation in density of the coating material does affect deposition rate and is closely related to the powder feed rate for each particular coating. However, in general terms a single detonation produces a 2·5 cm diameter coated area, approximately 6 μm thick.

C. Coating structure

The structure of a detonation coating can best be illustrated by Fig. 3. This figure shows a typical cross-section of a tungsten carbide/cobalt D-Gun coating. At the substrate/coating interface, there is no separation or porosity. The amount of visible porosity of the tungsten carbide series of coatings is usually less than half per cent. The laminar structure is well illustrated, showing the cross-section of the platelets formed by the high velocity impact of particles against substrate and particles. The structure of D-Gun coatings is usually very uniform. To give an idea of size, the tungsten monocarbide particles illustrated in Fig. 3 are approximately 6 μm in diameter. The general matrix of the coating from which its hardness can be derived, is a mixture of various tungsten carbides and cobalt-rich binder phases. These are mainly W_2C and a range of tungsten carbides with varying percentages of carbon and cobalt.

D. Coating thickness

The D-Gun process is normally associated with the deposition of thin, wear resistant coatings in the general order of 0·1–0·3 mm in thickness.

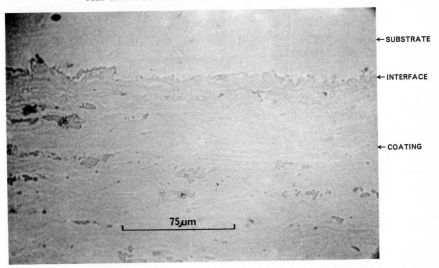

FIG. 3. Metallurgical cross-section of detonation gun coating structure.

Thicknesses up to 0·7 mm on specific applications with certain coatings are perfectly satisfactory. However, above these values, internal stresses are set up within the coating structure.

E. Special properties of D-Gun coatings

The uniform closely packed laminar structure of a D-Gun coating produced by an extremely high velocity hot gas stream, greatly enhances such properties as hardness, density and bond strength. Typical values for tungsten carbide/cobalt coatings are HV 1300 hardness, 0·5% porosity and bond strengths in excess of 700 kgf/cm^2 (69 MPa)

VI. SUBSTRATE

A. Substrate materials

All metallic materials with a surface hardness less than 58–60 Rockwell C, can be coated by the D-Gun process. There is no deformation of these materials as the shock wave is dissipated before the coating powder contacts the substrate. Generally, non-metallic substrates cannot be coated as the high velocity hot gas stream produces surface erosion.

B. Substrate pre-treatment

Prior to coating, the substrate surface is degreased and grit-blasted to clean and increase the surface roughness. This effective increase in surface area improves the mechanical aspect of the coating bond. Tungsten carbide/cobalt coatings can be successfully applied onto clean, non-gritted substrates.

C. Substrate temperature limits

The heat pick-up by the substrate from the powder particles can be readily dissipated by cooling with air or carbon dioxide. The component temperature can be held below 375 K so that no metallurgical or dimensional changes occur in the substrate.

D. Topographic limitations

Each time the gun detonates (and this is at least four times per second) the sound barrier is broken. The limitation imposed is one of "How large can one build a sound proof cubicle to cater for the size, weight and shape of the component to be coated?" At present, parts can be processed up to 10 metres long, 10,000 kilograms in weight and 2 metres in diameter. The D-Gun process is a line-of-sight process; it cannot bend round corners or coat into blind holes. However, all exposed external surfaces, flat or cylindrical, can be coated and the limitation upon internal surfaces, such as bores of cylindrical components, is that a coating can be applied to an internal length equal to the cylinder diameter from each open end.

VII. FINISHING OF COATINGS

The majority of detonation gun coatings are used for wear resistance and the surface finish and dimensions of the coating are usually critical for these applications. In some cases, the dimensional tolerances and the "as-coated" surface finish is quite satisfactory. The "as-coated" surface finish is in the range of 125–250 microinch (3·1–6·3 μm) r.m.s. similar in appearance to rough sand paper. This roughness can be readily improved to between 50–100 microinch (1·3–2·5 μm) r.m.s. by non-dimensional, non-precision techniques such as brush finishing and sanding. When fine surface finishes and close tolerances are required, D-Gun coatings have to be machine-ground. The grinding must be performed with diamond wheels and diamond lapping compounds under accurately controlled conditions.

The surface finishes that can be achieved by grinding are in the range of 5–15 microinch (0·1–0·4 μm) r.m.s. Further improvements to this surface finish can be obtained by diamond lapping techniques and, in this manner, surface finishes in the range of 0·5–4 microinch (0·01–0·1 μm) r.m.s. can readily be achieved with most D-Gun coatings as illustrated in Fig. 4.

FIG. 4. Detonation coating—finished by grinding and lapping.

ACKNOWLEDGEMENTS

The author wishes to acknowledge and thank Mr. R. G. Rudness, Materials Systems Division, Union Carbide Corporation, U.S.A. for his considerable assistance in compiling this paper.

BIBLIOGRAPHY

Shesternenkov, V. I. (1968). *Poroshkovaya Metallurguya*, 37.

Applications of detonation coatings

J. E. LAND

Union Carbide U.K. Ltd., Swindon, Wiltshire, England

I. DETONATION COATINGS IN THE AIRCRAFT INDUSTRY

Detonation coatings are widely employed in today's industry but in this paper reference will, in the main, be made to the aircraft gas turbine engine. The aircraft gas turbine, as illustrated in Fig. 1, indicates the areas where coatings are employed. These include many operating environments—below zero temperature at the air intake to above 1000 °C in the turbine section. It also includes many types of wearing action such as sliding, fretting,* rotating, impact and many combinations of these. Additionally, in certain areas, there are corrosion problems, particularly in the turbine sections. You will observe from the illustration that coatings are applied by the detonation gun throughout the engine.

At the front of the engine (the compressor intake) titanium is employed to a large extent. Whilst this material has remarkable strength for its low weight, it does not resist wear, particularly when rubbing against a similar titanium component. Mid-span stiffeners (1 in Fig. 1) are employed in the engine to reduce the vibration of compressor blades (Fig. 2), and on the mating surfaces of these stiffeners, detonation gun applied tungsten carbide is used at a thickness of approximately 0·25 mm and is left in the "as-coated" condition. The wearing action, here, is a combination of sliding and impact. An uncoated compressor blade could fail after approximately 100 hours' operation but can last over 10,000 hours with the detonation gun coating.

A further area in the compressor of the engine is on the shaft diameter (3 in Fig. 1) where the inner track of a ball or roller bearing will be mounted.

---

* Fretting is the wearing action caused by two surfaces rubbing together with a small amount of relative movement.

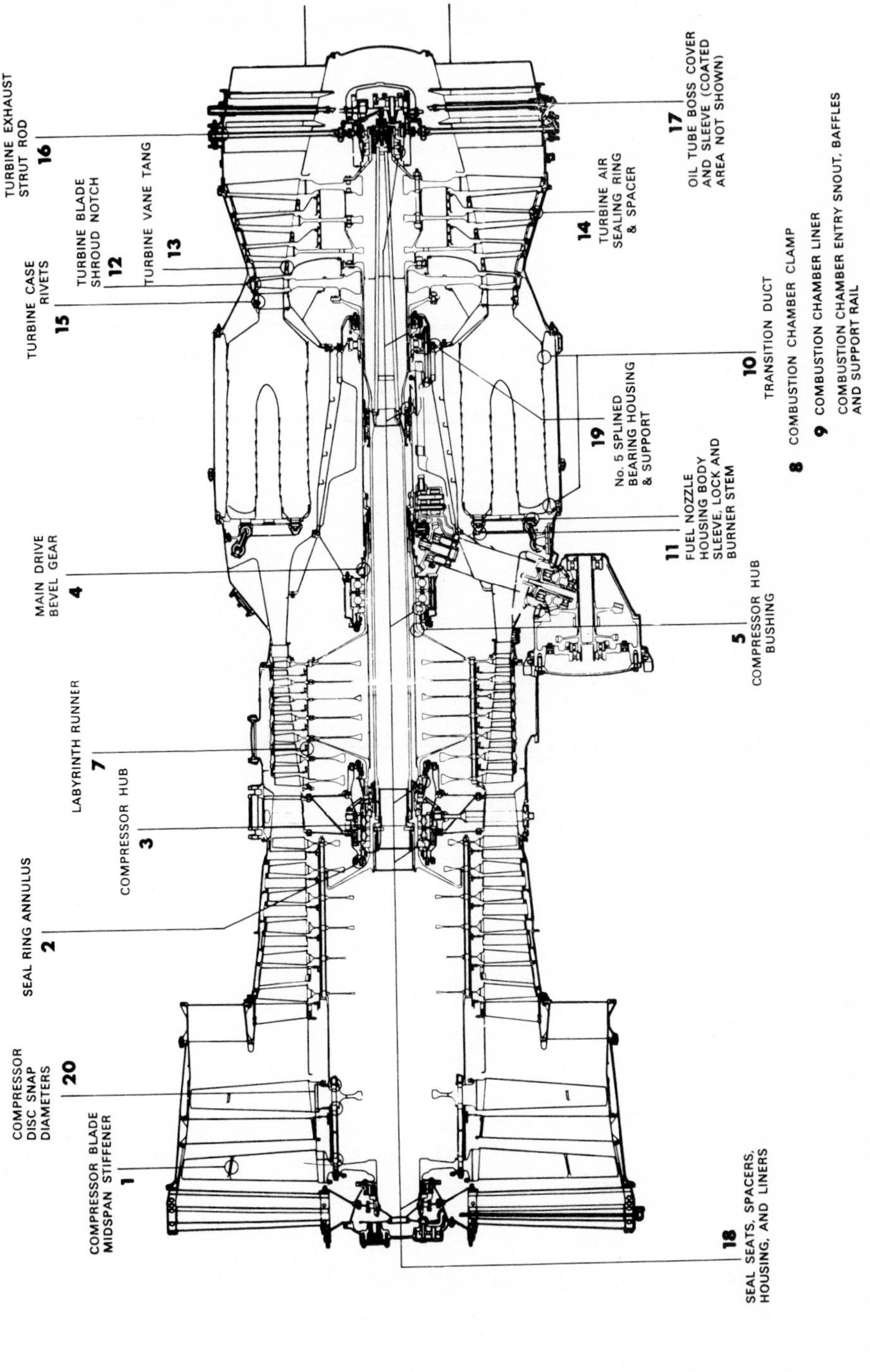

FIG. 1. Typical jet engine coatings applications.

1 COMPRESSOR BLADE MIDSPAN STIFFENER

20 COMPRESSOR DISC SNAP DIAMETERS

2 SEAL RING ANNULUS

3 COMPRESSOR HUB

7 LABYRINTH RUNNER

4 MAIN DRIVE BEVEL GEAR

5 COMPRESSOR HUB BUSHING

11 FUEL NOZZLE HOUSING BODY SLEEVE, LOCK AND BURNER STEM

19 No. 5 SPLINED BEARING HOUSING & SUPPORT

10 TRANSITION DUCT

8 COMBUSTION CHAMBER CLAMP

9 COMBUSTION CHAMBER LINER

COMBUSTION CHAMBER ENTRY SNOUT, BAFFLES AND SUPPORT RAIL

14 TURBINE AIR SEALING RING & SPACER

17 OIL TUBE BOSS COVER AND SLEEVE (COATED AREA NOT SHOWN)

16 TURBINE EXHAUST STRUT ROD

15 TURBINE CASE RIVETS

12 TURBINE BLADE SHROUD NOTCH

13 TURBINE VANE TANG

18 SEAL SEATS, SPACERS, HOUSING, AND LINERS

Under certain engine conditions, the bearing inner track may spin on the highly stressed titanium component, causing extreme wear. Detonation gun tungsten carbide applied to the shaft and diamond ground, eliminates this wear and the need to scrap a high cost component.

Before we turn to the high temperature section of the engine, one further item on the compressor is the small compressor rotor blades. On certain engines, such as those operating in desert or arctic conditions, sand or ice

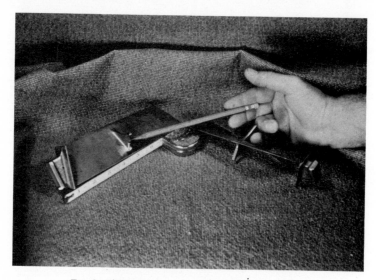

FIG. 2. Titanium compressor blades (scale in inches).

can be ingested into the engine through the intake during landing and take-off manoeuvres. This ingestion causes erosion of the blade material and a subsequent loss of engine performance. Here again, detonation gun-applied tungsten carbide has been shown to increase the life of this compressor blade by a factor of three.

For the high temperature applications, reference will be made to two areas—combustion chambers (8 in Fig. 1) and turbine blades (12 in Fig. 1). At the rear of the combustion chamber (Fig. 3) severe fretting wear occurs due to engine vibration. The engine must be designed to expand both axially and radially and it is for this reason that the combustion chamber components must be free to move. Having designed freedom of movement into the engine, a coating is needed to eliminate the wear that occurs. Detonation gun-applied chromium carbide is employed. You will observe a change from tungsten to chromium carbide. The high temperatures in this area of the

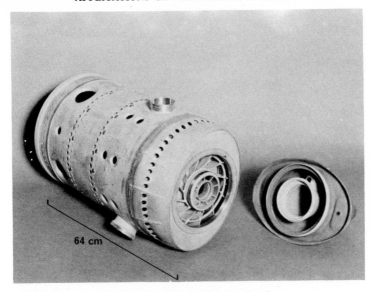

FIG. 3. Gas turbine combustion chamber.

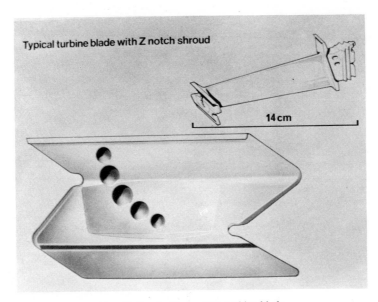

FIG. 4. Gas turbine engine, turbine blade.

engine, between 700–800 °C, necessitate a coating which will not oxidize and therefore we use chromium carbide. The rear end of the combustion chamber and the clamping ring which locates it to the mating component, are coated to a thickness of 0·12 mm. Engines which have experienced failures in less than 4000 hours, now, with a coating, last in excess of 28,000 hours.

The final area on the aircraft gas turbine engine referred to is the turbine blade (Fig. 4) where wear occurs at the shroud surfaces (12 in Fig. 1). Shrouds are employed to reduce the blade vibration and at temperatures in the region of 700–950 °C, the two mating surfaces of the blade material would wear

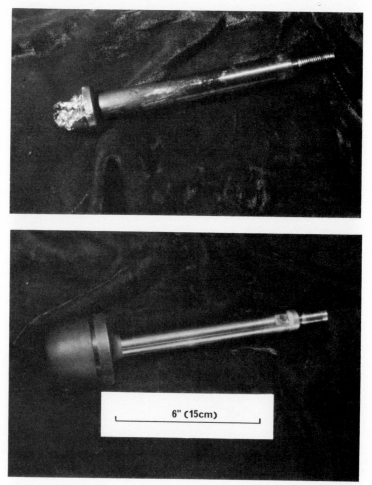

6" (15cm)

FIG. 5. Sulphuric acid pressure control valve (*a*) in worn condition, (*b*) in coated condition.

due to a combination of impact and sliding. In addition to eliminating wear, the coating must also be resistant to corrosion from the products of combustion, i.e. salt and sulphidation. In this area, detonation gun-applied chromium carbide has been shown to improve the life of turbine blades by an appreciable factor. The coating, mating against itself, is employed in the "as-coated" condition at a thickness of approximately 0·18 mm.

II. Other applications

Considering applications outside the aircraft industry, the pressure control valve illustrated (Fig. 5a) is from a sulphuric acid line and was found to be in this condition after 6 months' operation. The valve is now coated with tungsten carbide applied by the detonation gun to the area shown (Fig. 5b) and after two years, is still operating successfully.

In another type of industrial environment, detonation gun coatings have been employed on the brake drums (Fig. 6) of a 10 Mg (ten megagram) crane. The uncoated brake drum could fail in three months and in certain instances, in as little as 7 days. The item is now coated by the detonation gun with a tungsten carbide coating and after 16 months' operation, no wear has been detected on the wearing surfaces of the brake drum. In addition to reducing the wear, the coating has allowed the use of a more ductile steel, thereby

Fig. 6. Steel industry crane, brake drum.

eliminating the cracking caused by severe overheating on the previously-employed conventional cast iron components.

Finally, the foregoing examples show how detonation coatings with their characteristic high bond strength and low porosity, can be employed to overcome many types of wear, including a combination of wear and corrosion.

Arc plasma spraying

A. R. Moss and W. J. Young

RARDE, Fort Halstead, Kent, England

I. Introduction

There are several methods of spraying metals and ceramics which are conveniently grouped under the generic term "arc plasma spraying"; the commercial expression "plasma arc" is to be deprecated. In these various processes the thermal energy of an electric arc and its associated plasma jet is utilized in melting and projecting material as a high velocity spray onto a workpiece, where it solidifies as a coating or deposit. The simple process used as a basis for discussion in this paper, together with items and section of interest is illustrated in Fig. 1. The processes have proved invaluable in advanced technologies such as aerospace and nuclear engineering. Increasing advantage of their cost-effective potential is also being taken in the less exotic industries. Arc plasma devices in general are used for applications requiring the use of a heat source having a very high rate of heat transfer and a temperature significantly higher than that obtainable from combustion reactions like oxygen-acetylene. Unlike chemically produced flames, the composition of plasma or "electric flames" can be varied at will. Contamination can be minimized by using inert gases.

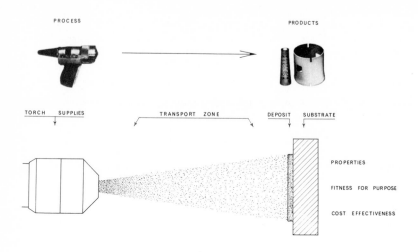

FIG. 1. Arc plasma spraying.

II. PROCESSES

The details of the various types of equipment and the electrical power requirements have been sufficiently well described in the literature and in manufacturers' publications to avoid repetition (Ballard, 1963; Moss and Young,
1964; Gross *et al.*, 1968; Marchandise, 1970; Bidmead and Hutchings, 1972).
The commonly used arc plasma spraying process illustrated in Fig. 1 typically
has an electrical capacity ranging from 10 kW to 50 kW. In this process a
constricted arc passes from the central electrode in the torch and through the
nozzle which forms the anode. The spraying material is generally in powder
form and is fed into the plasma where it melts and gains velocity. Sometimes
wires or rods are used instead of powder, in which case, if they are electrically
conductive, they may constitute a second anode. This mode of operation
improves the thermal efficiency. In a further modification an arc may additionally be transferred to a metallic workpiece in order to melt it and so
facilitate fusion penetration or bonding. Other spraying methods include
the use of transpiration or radiatively cooled nozzles, the induction coupled
high frequency system and the twin wire process (Rosenbery, 1966). There
are examples of hybrids comprising two and even three pure processes.
For instance, twin wire arc spraying/plasma torch, and arc/high frequency.
Both manually operated and automatically controlled spraying equipments
are available (Levinstein, 1960; Donovan, 1965; Moss and Young, 1971).

The latter are preferred because of their technical superiority and economic efficiency in spraying.

III. PRODUCTS

Arc plasma spraying is concerned with the production of coatings and similar deposits. These comprise:

  (i) thin high quality impervious coatings up to say 2 mm thickness;
 (ii) thick high quality and relatively void-free deposits;
(iii) deposits of any thickness containing intentional, or unintentional voids—typical of those required for bearing surfaces; and
 (iv) thick deposits typically up to several cms thickness, such as forging blanks, which are to be heat treated and mechanically worked before being put into service (Moss and Young, 1964).

Coatings resistant to heat, wear, abrasion, erosion and corrosion, thermally and electrically conductive coatings, thermal barriers and dielectric coatings, decorative and identification marker films, may be produced from the many metallic and non-metallic materials now readily obtainable. As metals, ceramics and certain plastics can be deposited individually or as mixtures, almost endless permutations of composition and properties are possible. Deposits may be layered or progressively graded from, say, a metal to a ceramic to a plastic. Compound deposits are used for example for wear resistant non-stick surfaces and other tribological applications.

A particular spraying technique is that known as spray forming. In this, self-supporting deposits of the desired thickness are sprayed onto a pattern or mandrel which may be removed or left in position. Such articles may possess advantages unachievable by conventional methods of manufacture. Objects of complex shape have been produced to fine dimensional tolerances, without machining, from hard materials such as ceramics and carbides. An example is shown in Fig. 1. The subject has been detailed by Moss and Young (1964). See also Anon. (1970), Palermo (1963), and Donovan (1966).

A recent analysis of general spraying enquiries in Britain by Moss (1971) showed an 83% interest in coatings and 17% in spray forms; related data are given in Table I.

IV. TORCH: SUPPLIES

All types of arc plasma spraying torches require an electricity supply and a plasma forming gas. There may also be an additional gas requirement to

TABLE I

Distribution of interests in arc plasma spraying, (1971).

Metals 51%		Non-metallic materials 49%	
Spray forms 29%	Coatings 71%	Spray forms 9%	Coatings 91%

fulfil some special function, such as cooling the workpiece, deflecting the hot plasma jet (but not the spray particles (Anon., 1963)) away from the deposit and blanket shrouding to prevent atmospheric attack. The composition of the plasma forming gas is important because it affects the voltage of the arc and this in turn alters the temperature and enthalpy (heat content per unit mass) (Jordan and King, 1965). Normally argon or nitrogen is used but a small addition of hydrogen or helium is often beneficial. Other gases may be used subject to their compatibility with the electrodes and spraying materials. The chemical effects arising from the use of small additions of different gases may also be of importance. In the spraying of carbides, the retention of stoichiometric composition is assured by a small addition of methane. The reducing effect of hydrogen is another example. A significant difference in behaviour exists between monatomic and diatomic gas when highly heated. The enthalpy of monatomic gases is proportional to the temperature increase until ionization occurs. This is not so with diatomic and polyatomic gases which tend to dissociate into their constituent atoms before being ionized. Although considerable energy absorption is necessary for this dissociation to occur, the energy is later released, mainly as heat, during recombination.

A carrier gas is required to transport spray powder particles from the dispenser and to inject them into the plasma. The quantity of gas used to achieve this has to be determined with care as too much dilutes and cools the plasma to such an extent that the powder particles do not melt in the short time available. Argon is usually used because of its low heat content.

An extensive range of elements, compounds and mixtures are currently available for both production spraying operations and experimental trials. Most materials which can be melted and projected as a spray onto a substrate where they resolidify may be used. Apart from the nominal chemical composition and property aspects, the technical factors normally considered when selecting a powder include: flowability, the particle size distribution, shape, porosity and thermal conductivity (Moss and Young, 1964). Mixtures of powders may be sprayed but, unless these have been compounded by specialists, better results are usually obtained by feeding the individual components separately each at its optimum position along the plasma jet.

Gravitational separation of mixtures has been experienced with some vibratory type dispensers. Sometimes wires or rods are used instead of powders. Protagonists of the former stress that wire must be molten before it can be comminuted, whereas with powders it is possible to have unfused particles in the spray beam, which ultimately impairs deposit quality. As all materials cannot be obtained in wire form, a method of overcoming the problem is to use an extruded mixture of powder in a plastic carrier; a mixture of polyethylene and polyisobutylene has been proposed.

V. TRANSPORT ZONE

The transport zone is that region between the torch and the workpiece where material is melted, accelerated and projected as a spray in the most technically efficient and economic manner. Although much of the available data is not directly applicable to all spraying torches, there are certain common factors and trends.

The powder to be sprayed is preferably fed into the plasma at the exit of the nozzle, where the enthalpy is high enough to ensure complete melting in the few milliseconds available. Some superheating is necessary, but loss through vaporization must be under strict control. The projection of unfused or resolidified material onto a substrate results in weak, agglomerated deposits. The impacting velocity of the molten particles needs to be as high as practicable if low porosity deposits and high values of interface adhesion are needed. Unfortunately, with some types of plasma torches, increasing the velocity decreases the time available for melting the powder. The protective plasma does not always embrace the entire transport zone so some spray particles may be exposed to atmospheric contamination. In addition, the high velocity jet exerts an injection effect which entrains the ambient atmosphere (Hasui and Kitahara, 1964). These two factors were not appreciated in the early days of plasma spraying. With manually operated torches the difficulties are surmounted by the use of shrouds and gas blankets. The question does not arise, of course, with controlled atmosphere equipment. When using short spraying distances, overheating of the workpiece by the hot plasma may be reduced by the employment of cooling gas jets.

The chemical and physical characteristics of the transport zone are strongly influenced by the particular torch design, the type and quantity of the plasma forming and carrier gas, and by the electrical power consumption. The interacting complexities between these items in relation to particle heating and projection have been studied by many workers including Mash et al. (1961), Levinstein (1962), Palermo (1963), Kudinov (1965), Marynowski et al.

(1965), Atkins (1967), Okada and Maruo (1968), Nachman and Gheorghiu (1969), and Lemoine and Le Goff (1970).

VI. DEPOSIT: SUBSTRATE

Nearly every characteristic of a substrate or workpiece seems to be of importance in influencing the quality and adherence of a deposit sprayed upon it. Ideally there should be consultation at the drawing board stage so that steps may be taken to minimize potential troubles. All too often the sprayer is confronted with a mass of components and a drawing cryptically marked "spray here". It cannot be overemphasized that the entire spraying procedure needs to be in accord with recognized standards and recommendations, for example, Brit. B.S. 4891:1972 and Brit. C.P. 3012:1972.

Appropriate substrate preparation is an essential prerequisite for the control of adhesion. Spraying should be done immediately after preparation; a few hours later is too late. For satisfactory protective coatings and building-up operations every effort is needed to increase adhesion. On the other hand, for some types of spray forms the minimum effective adhesion is necessary to ensure the unsullied removal of a deposit from the mandrel acting as substrate. When spraying onto unintentionally prepared surfaces, e.g. as-machined, there can be no control over the quality of fortuitously produced adhesion. In everyday spraying practice, an almost infinite range of chemical and physical forms of substrates may be encountered. They include metallic and non-metallic materials in massive foil, cloth-like and even powder form. They vary from plastic teeth to sheets of glass and paper, and from wood to concrete; the size ranges from tiny electric contacts to ships. The shape and contour of the workpiece is of importance. Sprayed deposits are more readily satisfactorily produced on the external surface of a hollow cylinder than on the internal surface, because the residual stress pattern is more favourable. On internal surfaces there is a tendency for the interface to fracture and so permit the deposit to shrink away. This is understandable as sprayed coatings are usually under tensile stress in consequence of the contractional forces produced during solidification of the spray particles. The interface is subjected to a shear stress hence, for a given bond strength, only a certain tensile stress can be tolerated in the coating before failure occurs.

Experience has demonstrated the necessity for cleanliness together with some form of pre-spray roughening treatment of the substrate, in order to achieve adequate adhesion. Ballard (1963) has reviewed the methods developed over the years to facilitate mechanical interlocking-type bonding. They include shot blasting, grooving, turning, knurling and arcing. Under certain circumstances, "self-bonding materials" such as molybdenum produce high bond strengths, even on smooth substrates—why? Little is quantitatively known of the optimum topographical requirements but empiricism has indicated reasonably satisfactory guide lines. For instance, blasting with clean sharp-edged shot gives better adhesion than does the use of worn rounded material of the same composition.

On several occasions, as spraying commenced in a controlled atmosphere chamber, a short induction period was noted, during which time all spray particles projected at various substrates bounced off, i.e. zero adhesion (Moss 1963). After this induction period, spraying proceeded normally with good adhesion being obtained. The implications arising from this observation led to intensified research (RARDE, 1962–1966), some of the results of which are very briefly discussed in Section VII. Considerable evidence indicated that three characteristics of the substrate are predominant in determining bond strength; they are temperature, chemistry and topography.

Most of the elements, compounds and mixtures now available are successfully deposited directly onto a wide range of substrates without the need for special precautions. The few instances where difficulties do occur are usually overcome by using special techniques or by modifying existing materials. One way of circumventing high stress concentrations or the formation of undesirable microstructures, is to use graded layer techniques or an intermediate buffer or transition layer. For example, the formation of brittle carbide at the interface of a tungsten deposit on graphite is avoided by interposing a thin layer of rhenium. The complex subject of interfacial stress and stress patterns has been studied by Marynowski *et al.* (1965).

The occasional decomposition and melting problems can sometimes be overcome by the use of appropriate techniques. Some materials actually undergo purification as a result of outgassing, chemical reduction and vaporization of impurities. MgO, SiO_2 and B_2O_3, as impurities, have been removed from boron. Similarly, high oxygen content of tungsten powder can be reduced by using a hydrogen-containing plasma forming gas. Silicates may be dissociated to form individual oxides on re-solidification; $ZrSiO_4$ transforms to ZrO_2 and SiO_2. A few polymeric compounds decompose if overheated, and certain carbides and borides may lose their stoichiometry. Substances which sublime are best avoided as they tend to form fragile deposits of condensed vapours. The use of materials having a high vapour pressure may result in melting difficulties because of the Leidenfrost phenomenon; i.e. the thick layer of vapour drastically reduces the rate of heat transfer to the particle.

Cohesion problems leading to friable deposits are rarely encountered. When they do it is usually an indication that some aspect of the spraying procedure is grossly incorrect. Lack of cohesion generally results from insufficiently heated spray particles, i.e. they are not molten on impact, or from the use of materials that sublime, such as MgO and BN. In both instances a weak agglomerated mass of particles can be expected. Ideally, the structure of a transverse cross section of a sprayed deposit comprises a continuum of uncontaminated lamellae, formed when successive molten globules flatten out to form strong interparticulate bonds on impact.

The problem of porosity is quite different and less serious than that of lack of cohesion. For most applications a small amount of porosity is no embarrassment if the deposit is mechanically strong, as it usually is; there is satisfactory cohesion between particles. For oil lubricated bearings, some porosity is advantageous as it retains lubricant. When thin coatings contain interconnected porosity, as may happen with wear resistant deposits of alumina on ferritic steels, penetration of water might cause rusting (Calabrese and Pan, 1971). This and similar problems can be overcome by applying phenolic or other sealing media. The cause of porosity is poorly understood. It is mainly, though not exclusively, a gas condition which is strongly influenced by the spraying parameters, and the size distribution and chemistry (e.g. Svirskii *et al.*, 1971) of the powder particles used. Increasing the spraying velocity, preheating and using moisture-free powders significantly alleviates, but does not completely solve, the problem. Moss (1963) used a specially designed swinging rotary furnace to outgas powders. The treated particles were transferred to an automatic controlled atmosphere spraying unit, such as that shown in Fig. 2, and used without being exposed to atmospheric oxygen and nitrogen. Oversprayed particles may be recycled.

VII. ADHESION: COHESION

In this section it is shown, contrary to popular belief, that mechanical interlocking of a deposit into a substrate is not the prime bonding mechanism in metal spraying; a view supported by Steffens (1965).

A discussion on adhesion is complicated by the fact that, depending on the spraying process used and on the combination of materials involved, one or more basic bonding mechanisms may be operative. Furthermore, it does not necessarily follow that because a high strength bond results from spraying metal A onto substrate metal B, that the latter could satisfactorily be deposited onto metal A (Elyutin *et al.*, 1969). Despite the accumulation of data over the past decade, the paucity of scientifically based information does not permit the formulation of even a general hypothesis of adhesion. Incidentally, adhesion concerns bonding between a deposit and the substrate, whilst cohesion is a particular case of adhesion concerned with bonding within the deposit only.

FIG. 2. Semi-automatic, controlled atmosphere, spray forming unit.

The three main possible mechanisms in spraying are:

  (i) mechanical interlocking,
 (ii) metal/metal bonds—liquid/liquid as in fusion welding,
                          liquid/solid as in soldering and brazing,
                          solid/solid as in solid phase pressure welding,
                          solid/liquid as in hot dip galvanizing,
(iii) chemical bonds, for example spinel formation.

Schoop invented the metal spraying process as a result of observing the strong adherence of squashed lead bullets fired from a pistol at a brick wall. It has since generally been thought that mechanical interlocking was the key to high values of adhesion in spraying. His success was probably due to the fact that common building bricks are porous and the faces contain many re-entrant cavities. The explanation is also applicable to materials such as textiles and gauzes and to specially grooved surfaces, where a deposit is able to penetrate the interstices and so be held in position by a clamping or dove-tail effect. It is important to note that shot blasting, the usual method of pre-spray preparation, rarely produces a re-entrant type configuration.

Research, supported by industrial experience, confirms that significantly higher bond strengths are achieved (no exceptions are known) on roughened surfaces than on smooth ones of the same material. In fact, in the laboratory, zero adhesion has resulted with many combinations of metals, when using both conventional spraying and specially developed "single drop" techniques on polished substrates. Although this apparently supports the interlocking hypothesis of bonding it now seems likely that a simple mechanical concept is impossible because of four discoveries (RARDE, 1962–66), they are:

   (i) thermal isolation of the projections on roughened surfaces,
  (ii) the existence of a critical temperature for adhesion,
 (iii) penetration of the oxide film on roughened aluminium by high velocity spray particles,
 (iv) reduction of bond strength at shot blasted surfaces by a film of contaminant.

These findings suggest there is an intrinsic adhesion to rough surfaces, facilitated by roughness and cleanliness which is not caused by mechanical interlocking. This intrinsic adhesion could conceivably be achieved by classical metal/metal or chemical bonds, both of which require a high interfacial temperature and surface cleanliness. As existing theories for such bonding are adequately established, they can be accepted without discussion.

It should not be assumed that substrate roughness plays no important role, just because mechanical interlocking ceases to be the correct explanation for high values of adhesion. On the contrary, it is still necessary but for new and different reasons. For example, metal/metal bonding may take place under the ideal conditions produced when penetration of an oxide layer on the many surface asperites occurs as a result of shear deformation by high velocity spray particles. The significance of particle velocity may be visualized from the fact that tin particles projected at a modest 100 m/s will produce, on impact, pressures of 80 kg/mm^2, which is the mean stress needed for full plastic flow of mild steel (see also Vaistukh and Slepukha, 1971). Another reason is that the specific surface area of a roughened (shot blast) surface is several times greater than that of a smooth polished one. This would indicate that existing considerations of interfacial stress as a two dimensional phenomenon is no longer viable and should be replaced by a three dimensional model.

Experimental work has revealed an increase of bond strength with decrease of thin sheet metal thickness, a finding also supported by Nikiforov and Privezentsev (1969). Heat flow analysis supports this finding by predicting thermal isolation effects on sheet, corresponding to less than approximately three times the flattened spray particle thickness, i.e. typically thinner than about 0·5 mm. In a similar vein, strong bonds are achieved by spraying onto the projections of roughened substrates and onto metal powders. If mechanical interlocking was the predominant factor, thermal isolation effects should not be as important as they undoubtedly are.

Preheating is important not only for its direct contribution in promoting bonding but also because it ensures the removal of residual films of moisture which so strongly inhibit adhesion. All factors which increase the thermal content of interface regions are important. They include latent heat, kinetic energy and exothermic reactions, all of which release substantial amounts of heat.

It is now well established that oxide layers of the appropriate kind can facilitate adhesion by aiding the formation of chemical bonds. The satisfactory spraying of aluminium and alumina onto oxidized iron is almost certainly the result of a spinel bond. If, however, the oxide layer has been produced as a result of rusting by a moisture film, it may be fragile and loosely bonded, in which event strong adhesion is improbable.

There are several methods of avoiding (see also Veretnik *et al.*, 1971) or minimizing adhesion in order to achieve the complete unsullied removal of a deposit from the substrate, as is necessary with certain types of spray forms. Usually the workpiece is precoated with an appropriate "stop-off layer". This may be applied by dipping, painting or spraying and the substances used include salts, oxide, molybdenum disulphide, colloidal graphite or even lime wash. A more sophisticated treatment, applied to copper mandrels, entails the production of a controlled film of copper oxide on the polished substrate.

One of the greatest problems of studying bond strengths is the lack of a really satisfactory quantitative adhesion test. Current methods based on the use of adhesives are associated with a high degree of scatter and complex modes of failure (Catherall and Kortegas, 1972). It is thought that the ultra-centrifuge (Dancy, 1965) may provide the best hope for an absolute test for determining the strength of adhesion in the future but even this may be plagued by stress concentrations at the interface.

VIII. Properties, fitness for purpose, cost effectiveness

The three closely related items in the title of this section are just as funda-mental and important as any other more technical aspect, yet they are rarely mentioned. All too often the "best" materials and the "best" deposition techniques are employed without integrated reference to the properties, life and cost of the sprayed product as a whole. The service requirements of a product ought to be realistically appraised at the design stage (Brit. B.S. 4891 : 1972). The concepts of fitness-for-purpose and cost effectiveness could then guide the choice of materials and processes. This overall approach should minimize the risk of using sprayed deposits which may fail prematurely, or which may possess unnecessarily superior properties during a restricted service life. Similarly, it would indicate whether or not the inspection standards themselves are correct. The spraying standards for, say, a moon rocket and a shovel are unlikely to be the same.

ACKNOWLEDGMENT

This paper is published with the permission of the Controller of Her Britannic Majesty's Stationery Office. British Crown Copyright reserved.

REFERENCES

Anon. (1963). *Machinery* **102**, 2619, 190.
Anon. (1970). *The Engineer* 29 Jan., 18.
Atkins, F. J. (1967). *Inst. Weld. Conf. Paper No. 11*, London.
Ballard, W. E. (1963). "Metal Spraying". Griffin, London.
Bidmead, G. F. and Hutchings, B. E. (1972). *The Prod. Eng.* 337.
Brit. Stan. Inst. BS 4891 (1972). "Quality Assurance."
Brit. Stan. Inst. CP 3012 (1972). "Cleaning and Preparation."
Calabrese, S. J. and Pan, C. H. T. (1971). *Chem. Abs.* **75** (10), 70616C.
Catherall, J. A. and Kortegas, K. E. (1972). *Metal Finishing J.* **18** (205), 40–23.
Dancy, W. H. and Zavarella, A. (1965). *Plating* **52**, 1009.
Donovan, M. (1965). Inst. Weld. Spraying Conf., London.
Elyutin, V. P., Kostikov, V. I. and Shesterin, Yu. A. (1969). *Chem. Abs.* **71** (12), 52800W.
Gross, B; Grycz, B. and Miklóssy (1968). "Plasma Technology". Iliffe, London.
Hasui, A. and Kitahara, S. (1964). *NRIM (Japanese)* **7**, 117.
Jordan, G. R. and King, L. (1965). *Brit. J. Appl. Phys.* **16**, 431.
Kudinov, V. V. (1965). *Weld. Prod.* **12**, 8, 6.
Lemoine, A. and Le Goff, P. (1970). *Chem. Abs.* **72** (24), 126214C.
Levinstein, M. A. (1960). *Met. Finishing J.* 12, 467.
Levinstein, M. A. (1962). 3rd Inter. Metal Spray Conf., Madrid.
Marchandise, H. (1970). Plasmatechnologie Grundlagen und Anwendung, 8. DVS GmbH, Düsseldorf.
Marynowski, C. W., Halden, F. A. and Farley, E. P. (1965). *Electrochemical Tech.* 3, 109–115.
Mash, D. R., Weare, N. E. and Walker, D. L. (1961). *J. Metals*, 7, 473–478.
Moss, A. R. (1963). Inst. Weld. Autumn Meeting, London.
Moss, A. R. and Young, W. J. (1964). *Powder Met.* 7, 261–289.
Moss, A. R. (1971). SCRATA Conf., Harrogate.
Moss, A. R. and Young, W. J. (1971). Inst. Elect. Engs. Conf., Sheffield.
Nachman, M. and Gheroghiu (1969). *Rev. Roum. Phys. Bucarest* 14, 327.
Nikiforov, G. D. and Privezentsev, V. I. (1969). *Chem. Abs.* **70** (20), 90142V.
Okada, M. and Maruo, H. (1968). *Brit. Weld. J.* 8, 371.
Palermo, J. R. (1963). Inst. Weld. Autumn Meeting, London.
RARDE (1962–66). Unpublished researches.

Rosenbery, J. (1966). *Metalworking Prod.*, 6 April, 57.

Steffens, H. D. (1965). Inst. Weld. Spraying Conf., London.

Svirskii, L. D., Krokhin, V. P. and Gordienko, Ya. I. (1971). *Chem. Abs.* 75 (10), 66771G.

Vaistukh, I. M. and Slepukha, V. T. (1971). *Chem. Abs.* 75 (6), 39583 r.

Veretnik, L. D., Evdokimov, K. K., Podol'skii, B. A. and Shapiro, I. S. (1971). *Chem. Abs.* 74 (6), 24785P.

Production of thick film circuits by plasma spraying

R. T. SMYTH

Department of Electrical Engineering, Imperial College, London, England

I. INTRODUCTION

Dr. Loasby has described (Loasby, 1974) the method of producing thick film circuits by screen printing; in the past year we have been examining the feasibility of producing thick film circuits by arc plasma spraying.

A. Advantages

The arc plasma spraying technique appears to offer some advantages over the screen printing method for this application. The firing stage for fusing the coatings is not required and is removed from the production process; this allows low melting point, low cost substrates to be used. Also because the thickness is gradually built up during manufacture, the electrical resistance may be continually monitored, and the process stopped when the desired values are obtained. More accurate control of resistance is thus possible and there may be no need for subsequent trimming. The absence of glass also removes some materials compatibility problems.

The reason why the arc plasma process has been used as opposed to the other thermal spray methods (i.e. flame, arc) is because a wider range of materials can be deposited and a higher quality coating can be produced. The lower feed rate normally associated with arc plasma spraying is not a disadvantage because it allows more accurate control of coating thickness and the small coating thickness required for this application (up to 40 µm) does not involve long spraying times.

B. Basic process

The basic method of producing thick film circuits by plasma spraying is very simple. The substrate is placed in front of the gun, a stencil type mask is laid directly on top of the substrate and the gun then traverses the workpiece and sprays the required material on the mask and substrate (Fig. 1).

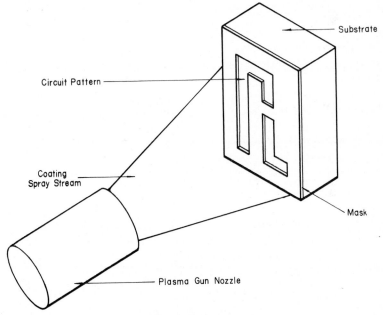

Circuit Pattern

Coating
Spray Stream

Substrate

Mask

Plasma Gun Nozzle

FIG. 1. Basic process.

When the desired thickness is reached, the process is stopped and the mask removed, revealing the coating pattern. This process may be repeated to produce succeeding patterns with different masks and materials until the circuit is complete.

C. *Inherent technical problems*

The technical problems may be divided into four areas, namely:

(i) The production of good quality coatings; the coating requirement is for a thin, smooth-surfaced coating of even thickness with minimal evenly distributed porosity and good adhesion to a smooth substrate.

(ii) The production of patterns; good edge definition is essential to provide the well defined, high density patterns required.

(iii) Finding suitable coating and substrate materials to satisfy the electrical requirements of the thick film circuit.

(iv) Developing a system capable of being used as a production process.

II. INITIAL FEASIBILITY TESTS

A. *Powder size*

Initial spraying tests using a wide range of materials showed that these physical requirements were best obtained by using fine powders. As a result all subsequent work was carried out using powders in the size range 1 to 10 μm.

It is well known that feeding and melting of fine powders is difficult but we have found that by careful selection of type of powder hopper, using short pipe runs, adjustment of arc plasma conditions and, most important of all, vacuum drying the powder, it is possible to feed and subsequently melt fine powder without any difficulty.

B. *Coating thickness*

By using these fine powders we have found it possible to produce coatings as thin as 3 μm, but most test samples have been produced with a thickness of between 5 μm and 20 μm. The possibility of producing even thinner coatings by using still finer powders cannot be ruled out.

C. *Pattern production*

The edge definition produced by the stencil type mask obviously depends partly on the thickness of the mask. For most of our test work we have been using metal shim masks 50 μm thick. Mild steel has been found to be the most suitable material because it not only withstands the spraying process but good mechanical contact between the mask and the substrate can be achieved by using magnets. The life of a mask is determined by the build-up of material on it during spraying and the present life has been found to be about three cycles. The limits of resolution obtainable by this type of mask are a minimum width of continuous conductor path of 150 μm and a minimum separation of parallel conductor paths of 200 μm.

Two new masking techniques, for arc plasma spraying, have recently been developed in our laboratory and show great promise for circuit production, they are:

(i) Photo resist film.
(ii) Screen printed paint.

Both of these methods have the advantage that pattern design is more versatile as the mask may now consist of areas not connected together. These masking techniques have limits of resolution much higher than the spraying process. A minimum width of 80 μm for a conductor path is a limitation due to the coating process at the present time.

D. Substrate materials

Substrate materials must be both electrically insulating and capable of withstanding the spraying process. The added requirement of low cost leads one to consider glass and plastic as being the most suitable materials. When using these low melting point substrate materials, the critical parameters are the gun-to-substrate distance, the traverse rate, and the rate of surface cooling (which is accomplished with inert gas jets).

E. Coating materials

The selection of materials for the electrical components is a problem. The electrical properties of a sprayed material differ greatly from those of the corresponding bulk material. The platelet structure of the coating with its inherent multiple boundaries, contamination and porosity is largely responsible for these different properties. The rapid high temperature heat cycle that each particle undergoes can also cause fundamental changes in the structure of the coating. The possibility of chemical reaction with arc and powder gas constituents further increases the possible difference between the coating and the bulk material. The selection of materials thus has to be determined by practical tests to establish the electrical properties in the sprayed state. Because of the different electrical requirements of the components in building up a circuit, each component requires a coating of a different material:

1. *Conductors*

The production of conductor paths was simply resolved by spraying metals such as aluminium, copper, nickel and tin: all these materials have an electrical resistivity of less than 2×10^{-4} ohm cm (2×10^{-6} ohm m) in the sprayed state.

2. *Resistors*

The production of a range of resistors is more of a problem. If metal/metal oxide mixtures are used then, as the percentage of metal in the metal oxide is increased, the switch from being a good insulator to a good conductor is very rapid: this makes control of the resistivity impractical. For this reason it was decided to try mixed oxide materials, the apparent disadvantage here being that the metal oxides normally have negative temperature coefficients of resistance (t.c.r.). However, after spraying a wide selection of oxides, it was found that two oxides produced positive t.c.r.s. This enabled us to use these oxides mixed with other oxides to produce controlled t.c.r.s. Further

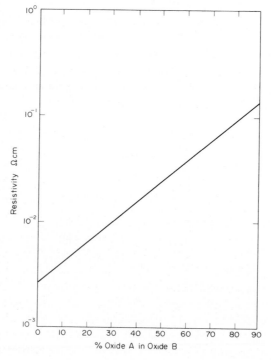

FIG. 2. Resistivity vs. composition for a mixed oxide resistor.

advantages were that the change in resistivity with different oxide percentage was gradual, thus allowing good control of resistivity (Fig. 2) and also the t.c.r. could be made either positive or negative (Fig. 3). The values of resistivity so far produced are in the range 1 to 500 ohms/square, and the t.c.r. can be easily controlled to within ±150 parts per million per degC from 25–125 °C. The limited life tests carried out indicate good long term stability. More work is required to increase the range of resistivities available.

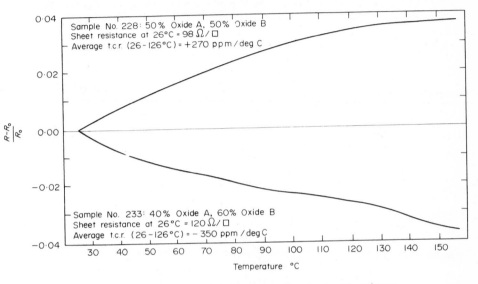

FIG. 3. Temperature variation of resistance for mixed oxide resistors.

From the work carried out to date it is possible to form a few general rules for the manufacture of resistors that may be used as guide lines for future work:

(i) Mixed oxide resistors are more suitable than metal/metal oxide mixtures.

(ii) Component materials of the mixture should be selected so that their resistivities are as close as possible to give good control of resistivity. The t.c.r.s. of the major constituent materials should have opposite signs.

(iii) All materials to be in the finest powder forms sprayable.

3. *Crossovers*

The production of crossovers has also been examined; the requirement here is that the material separating the conductor paths should have low dielectric constant but good insulating properties. The most suitable materials to date are plastics; some ceramics have also been tried but these have been found to have the disadvantage that a thicker coating has been required to prevent electrical breakdown.

4. *Capacitors*

We have recently started work on the production of parallel plate capacitors and it is obvious that they have more stringent coating requirements than the other circuit components.

F. *Production techniques*

The suitability of arc plasma spraying as a production technique is well established but the production of the coatings required for this application needs further consideration. The method used for batch production in the laboratory is shown in Fig. 4. The substrates and masks are mounted on the

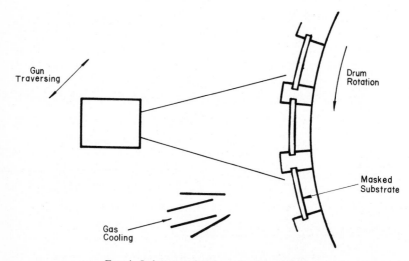

FIG. 4. Laboratory batch production method.

periphery of a drum that is rotated while being traversed by the spraying gun. This ensures that a thin coating is gradually built up and therefore it is possible to monitor the coating whilst it is being sprayed. The cooling gas jet is used to maintain a low surface temperature of the coating so that no temperature effects are produced. Automation of the process would be a relatively simple task.

III. Conclusions

In conclusion it is worth noting that no reason has been found to indicate that the arc plasma spraying technique cannot be developed to fulfil all the requirements of this application and that the techniques established in this study should allow wider use to be made of arc plasma spraying in both the electrical and general engineering industries.

REFERENCES

Loasby, R. G. (1974). "Science and Technology of Surface Coating" (B. N. Chapman and J. C. Anderson, Eds.), p. 242. Academic Press, London and New York.

The basic principles of electric-arc spraying

Daniel R. Marantz

Flame-Spray Industries, Inc., Port Washington, N.Y., U.S.A.

I. Definition and general process description

Electric-arc spraying is a method of thermal spraying of metal, in which two wires of either similar (conventional electric-arc process) or dissimilar (pseudo-alloy process) materials are melted and the fused particles are projected onto a prepared surface, building up an adherent coating. The necessary thermal energy required to sufficiently melt the wires is produced by means of an electric arc developed between the two wires. A carrier gas, usually compressed air, is used to atomize the molten metal in the arc and propel the fine droplets thus produced onto the surface to be coated.

II. Equipment description

A typical electric-arc metallizing system is shown in Fig. 1. The system consists of three major component assemblies which are:

(i) A simple, easily handled lightweight electric-arc gun.
(ii) A d.c. power source with electrical characteristics tailored to maintain stable arc conditions.
(iii) A wire feed and control mechanism providing for simple adaptation of the electric-arc metallizing system to production process automation.

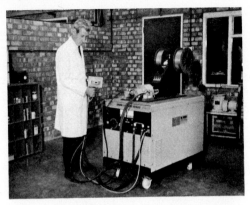

Fig. 1. A typical electric-arc metallizing system.

The interconnections of the components of this typical electric-arc spray system are schematically described in Fig. 2. Functionally, two wires, which serve as consumable electrodes, are drawn from spools, coils or drums through separate insulated wire straighteners by a dual wire drive assembly and are pushed through insulated flexible conduits to the electric-arc gun. As the wires pass through the gun, they make intimate contact with the electrode contact tips (Fig. 3) which are energized by the d.c. power source.

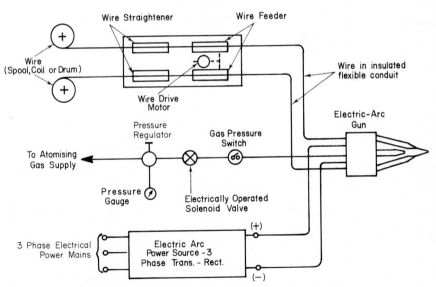

Fig. 2. Schematic diagram of typical system interconnections.

The gun is arranged so that the wires meet at a point with a small included angle between them. An atomizing gas jet (usually air) is located directly in line with the intersecting wires. As the wires contact, an electric arc is drawn which continuously melts the wires as they are fed into the arc. The jet of gas acts to atomize the melted wire and propels the fine atomized particles to the substrate. It has been found in some cases that the use of an inert atomizing gas improves the characteristics of the coating.

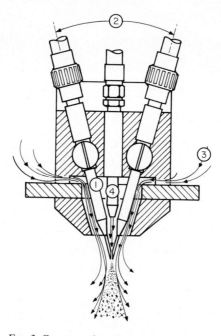

Fig. 3. Cut-away view of an electric-arc gun.
(1) Replaceable electrode tips; (2) wire feed conduits; (3) open area behind arc shield—allows air to be pulled in by suction created by flow of atomizing gas—thus cooling the electrode tips and forming an air sheath around the spray stream; (4) atomizing gas nozzle.

A. The electric-arc gun

The electric-arc gun, shown in Fig. 4, is light in weight (less than 2 kg) and compact, permitting it to be easily hand held or machine mounted. A modified form of this gun, as shown in Fig. 5, incorporates a special attachment to the gun which provides the capability for spraying the bore (internal diameter) of cylinders. A relatively low velocity flow of atomizing gas is deflected just beyond the arc zone, by a high velocity flow introduced normal

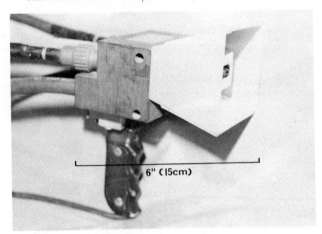

FIG. 4. An electric-arc gun.

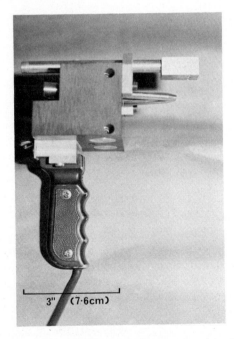

FIG. 5. Electric-arc gun with internal diameter spraying attachment.

312 DANIEL R. MARANTZ

to this low velocity flow, by the attachment which causes the spray pattern to be deflected downward, approximately 90° to the wire feed plane.

Another modified form of an electric arc gun is shown in Fig. 6. As can be noted from this figure, the assembly incorporates the wire feeding mechanism, which pulls the wire from its source to and through the electrode tips to the arc zone. This arrangement thereby maintains only a short spacing between the drive roll and the electrode tip and was developed specifically

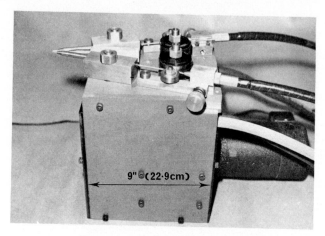

FIG. 6. Front drive electric-arc gun.

for the handling and spraying of very soft materials such as lead-tin solder, which possesses very little column strength. This apparatus is also specifically intended for mechanized production processes. Since its development, applications involving other materials have been found for it and it has now been used for many production spraying applications.

B. The power source

The power source is basically a static type electrical apparatus and has no moving parts with the exception of the motor driven ventilating fan. It contains a ruggedly designed three phase full wave transformer and a silicon bridge rectifier. The transformer is designed to have both static and dynamic electrical characteristics optimized for maximum arc stability for the electric-arc metallizing process. The use of silicon rectifiers provides high efficiency conversion of a.c. to d.c. power and represents the utmost in reliability.

The basic electrical characteristic of an electrical power source may be defined by its static characteristic which can be graphically described by a set of voltage–current characteristic curves (*V–I* curves or volt–ampere curves). Typical sets of *V–I* curves for constant voltage and for constant current type power sources are shown in Fig. 7. Essentially, the degree of slope is the major difference between constant current and constant voltage power sources. Slope is the slant of the *V–I* curve, measured at the output terminals of the power supply under varying load conditions. It is generally referred to as "volts charge per one hundred amperes".

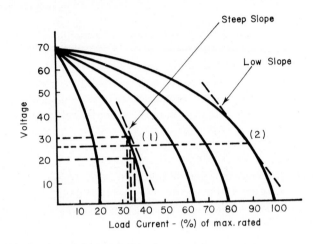

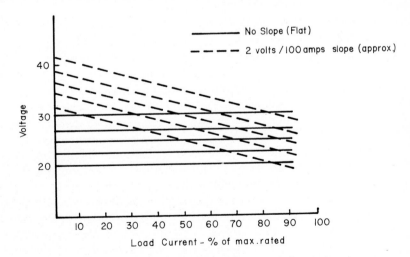

FIG. 7. Typical static electrical characteristics of an electric-arc power source.

It has been suggested that only a power source having a static characteristic of absolutely flat slope is usable for standard two-wire type electric-arc metallizing applications. However, it has been found that in order to utilize a whole range of materials such as molybdenum, steel, nickel, copper, bronzes, aluminium, zinc, etc. over a full range of wire feed rates (starting from the minimum rate consistent with obtaining a constant feed rate up to the maximum consistent with the maximum load capability of the power source) it is most useful, if not essential in some cases, to have varying degrees of controlled slope. The value for slope characteristic which has been found useful in general is 2–3 volts per 100 amperes. This level is most useful, since during the period of arc ignition, a considerably higher voltage is present across the electrodes in order to assure a more reliable initiation of the arc. As soon as the arc is struck, current is drawn from the power source and the voltage drops to a desired lower level optimally selected for stable arc conditions consistent with the minimum voltage required to produce minimum burn-off or chemical change in the material being sprayed. Specifically, this static slope characteristic has been found most useful in the applications involving materials such as copper, aluminium and in general any material which has a relatively high specific electrical conductivity. Therefore, it should be considered that a power source having a slope characteristic of 2 or 3 volts per 100 amperes possesses the facility for automatically regulating arc voltages to be consistent with the conditions of spraying, i.e. arc ignition or running.

Furthermore, it has been found that in applications of electric-arc metallizing with low melting temperature type electrode materials such as zinc, lead, tin, etc. and where it may be found desirable under certain circumstances to utilize low to medium wire feed rates, difficulty has been encountered in maintaining stable arc conditions. The arc condition which has been noted for this type of application with the use of a power supply having minimum slope, is such that there is a continuous making and breaking of the arc, yielding a bursting effect. This bursting effect is cyclic at a rate of approximately 2 to 3 bursts per second for the minimum wire feed rates. Upon visual and analytical study of the arc under these conditions, it was concluded that due to the high intensity of radiation of thermal energy in the region near the arc and due to the low melting temperature of material and its relatively slow motion at low wire feed rates, the electrodes tend to melt back from the arc, causing the arc length to become extended beyond a point which could be sustained by the arc voltage applied to the electrodes. It would be well to note that arc length is a function of arc voltage. At the point where the arc length extended beyond the stable point, the arc extinguishes and re-ignites itself as the electrodes proceed to contact again. A power source having steep static slope characteristics of the constant current type was found to completely eliminate this condition. As can be seen in Fig. 7, an operating point on the static V–I curve is selected such as point (1); the arc voltage can vary over a relatively broad range with negligible effect on the arc current. Under this condition the power supply voltage was found to cyclically vary to follow the cyclic variations of arc length, thereby sustaining stable arc conditions. Essentially then it should be considered that a power supply having steep static slope characteristics is suitable for improving the application capability of electric-arc metallizing systems in the area of low and medium spray rate of low melting point type electrode materials. Since other operating points can be selected such as point (2) in Fig. 7, with considerably less slope than point (1), it is therefore possible to utilize constant current type power supplies for general electric-arc metallizing applications as well as for specific problem solving such as just described.

C. Wire feed and control

The dual wire drive and control assembly provides uniform reliable control of the feeding of the two wire electrodes. The dual wire feeder itself is powered by a high torque, infinitely variable speed d.c. shunt type electric-gear-head motor with a dual shaft output, thus assuring uniform positive feed for the two wires and providing the facility for reliable repeatable feed rates. The controls and motor provide high speed acceleration and dynamic braking for instant responsive "on-off" control of the wire feed. Additionally, control is provided so that when the wire feed is stopped, the wires are withdrawn inside the electrode contact tips of the gun, precluding the possibility of the wires shorting on start-ups. The actual driving of the wires is accomplished by two separate drive roll and pressure roll sub-assemblies, both of which are completely insulated from each other. A V-groove type of drive roll is used, not a knurled type, to avoid disruption of the smooth wire surface. Burred wire surfaces can cause erratic electrical contacting of the wire and also possibly cause wire feeding problems.

Included as part of the wire drive assembly is a pair of wire straighteners used to remove the curvature in the wire, thereby allowing for more uniform feeding, reducing wear in the system and permitting more precise positioning of the two wires with respect to each other in the region of the arc, thereby increasing the stability of the arc condition.

III. CONDITIONS IN THE ARC (SOURCE STAGE)

An electric arc is produced by the passage of an electric current through an ionized gas. Initially, in the electric-arc metallizing process, the ionized gas is created as the two wires, which are electrically energized, advance to an intersecting point and touch at a low contact pressure and a small point-like contact surface. Due to the high density of the electric current, extreme heat is generated at the contact surface, fusing those portions of the metal wires and ionizing the surrounding gas, thereby creating a localized plasma. The plasma now established between the two wires provides a reasonably low resistance path for the flow of an electric current. The high current density flowing through the plasma provides the necessary sustaining power to maintain the ionized state. Within the arc, ionized atoms, which have lost electrons, are left with a positive charge. These positive gas ions flow from the anode to the cathode. At the same time, there is an electron flow from the cathode to the anode. The power expended in the arc, expressed in electrical units, is the product of the current passing through the arc plasma and the voltage drop across it.

The cathode is intensely heated by the impacting of the positive gas ions. This intense heating of the cathode causes the releasing of electrically charged particles, i.e. electrons, from its heated surface by thermionic emission. As the electrons flow from the cathode surface through the plasma, a considerable part of their energy is given away to the plasma, thereby causing the anode to be cooler than the cathode. In addition, since there is a high velocity flow of atomizing gas blasting on the two electrodes and the arc from only one side, as shown in Fig. 8, a large heat gradient is established such that the electrodes, in the region closest to the flow of atomizing gas are cooler than the regions furthest away. Due to these heat gradients, the melting of the electrodes occurs in a manner shown in Fig. 8. It should be noted that the gap between the two electrodes widens in the direction of the flow of the atomizing gas. In addition, the anode is seen to be extended more in the

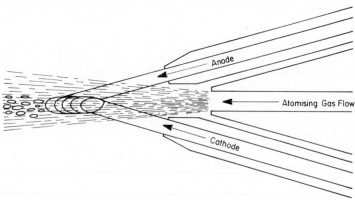

FIG. 8. Fusing conditions in the arc.

direction of the workpiece than is the cathode, since it is cooler than the cathode and thereby melts at a slower rate. Due to the cooler nature of the anode, the particles formed from the anode are larger than those formed by the cathode. Anode particles are in the form of droplets, while at the cathode there is a finer "spray" type, droplet formation.

It is evident that arc conditions must be maintained relatively stable in order to obtain uniform or consistent coating results. Arc stability is dependent upon several variables, namely wire feed rate, atomizing gas pressure and flow rate, arc voltage, arc voltage waveform, electrode size, electrode material properties and geometrical proportions in the arc zone.

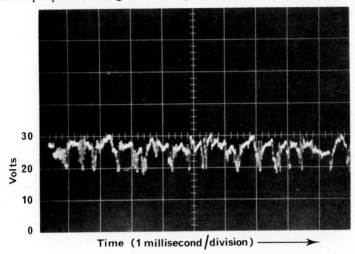

FIG. 9. Typical oscilloscope trace of arc voltage.

Firstly, it should be noted that the included angle between the two electrodes in the arc zone vastly affects the nature of the fusing of the electrodes. At small electrode angles, less than 35°, both electrodes generally melt down uniformly, maintaining very stable arc conditions. At angles exceeding 60°, erratic arc behaviour occurs, due to an increased sensitivity to adjustments of relative electrode positions and their relative position to the atomizing gas flow.

As the metal particles are formed from the anode and cathode and pass through the plasma, they cause perturbances in the plasma. These perturbances have been measured electronically, using cathode ray oscilloscope recordings. Shown in Fig. 9 is a typical oscilloscope recording showing the effect of these perturbances, and it can be seen that there are both low and high frequency components. The disturbances of the plasma considered to be the cause of the lower frequency component, are due to the large droplets from the anode passing through the arc plasma. Since the cathode is intensely heated, "spray" type droplets are formed, which pass through the arc plasma, causing the high frequency component to be modulated on the lower frequency component of the arc voltage.

An analysis of the effect which various operating conditions have on particle size, using this oscilloscope technique, provided the following conclusions:

(i) Conditions at the cathode are not affected greatly by variations in operating parameters. The cathode consistently produces particle sizes which are extremely fine compared with those produced by the anode and the particle size distribution produced by the cathode remains comparatively consistent.

(ii) Varying atomizing gas pressure over the range of 2·8–7·05 kgf/cm^2 (270–691 kPa) causes the anode droplets to be smaller with increase in atomizing gas pressure. See Fig. 10(a).

(iii) By the use of finer wire size, finer anode particles are produced. This was determined by the use of wire diameters of 1·57 mm, 2·28 mm and 3·08 mm. See Fig. 10(b).

(iv) Varying the arc voltage from the optimum level, consistent with maintaining steady arc conditions up to 10 volts above this level, it was found that the anode droplet size increases (Fig. 10(c)) and the rate of producing these larger droplets decreases.

(v) There is little or no effect on droplet size due to spraying at various rates, i.e. increasing the kilograms per hour of metal being melted (see Fig. 10(d)).

Due to the high current density in the arc plasma, extremely high temperatures are achieved. It has been reported that spectrographic measurements indicate that temperatures of 6100 ± 200 K are attained in an arc between iron electrodes with a current of 280 amperes. Under the operating condition prevalent in electric-arc spraying, the kinetics of high temperature reactions must be taken into consideration for any particular material in the arc zone. Drastic processes of diffusion and reaction take place between the sprayed metal particle and the surrounding gases which have a great influence, for example, upon particle composition, spreading and wettability, surface tension, solubility of metallic elements, behaviour of transformation and oxidation. The nature of these high temperature processes is very complex and therefore must be arrived at by working backwards from a study of the composition and nature of the sprayed deposit for any given material.

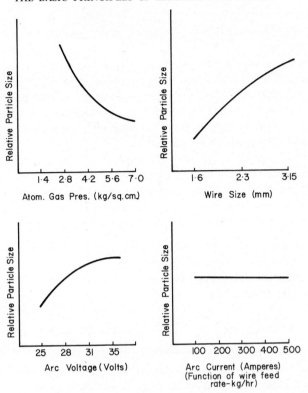

FIG. 10. Typical parameter interdependences: (*a*) atomizing gas pressure vs. particle size; (*b*) wire vs. particle size; (*c*) arc voltage vs. particle size; (*d*) feed rate vs. particle size.

IV. CONDITIONS DURING TRANSFER (TRANSPORT STAGE)

The metal particles leave the arc in a superheated state and have a large specific area. The particles are accelerated and propelled by the atomizing gas and exhibit a relative velocity to the gas flow during the acceleration period. This relative velocity is changing continuously until the particles have reached the velocity of the gas. Since the atomizing gas is usually air, which generally contains oxygen, nitrogen, hydrogen, carbon dioxide, water vapour and possibly other compounds, reactions between the superheated particles and the atomizing gas occur. The gas solubility in each metal, in general, is dependent on temperature as well as on the partial pressures of the gases. Since the temperature and partial pressure vary independently with distance from the arc, their effects must be considered together. With regard to the arc spraying of alloys, each component of a given alloy exhibits different affinities to the various constituents of the gas. Even trace elements in metals, barely detectible by analysis, may significantly influence the reaction behaviour.

With regard to oxidation reactions, various spray particles are formed, clearly distinguishable in a sprayed structure. The different conditions required for both the diffusion of oxygen and mechanical mixing exist, thereby leading to these different formations. Considerable work has been done by Dr. H. D. Steffens regarding the reaction effects on steel alloys, and is well covered in a number of his papers.

The particles reaching the substrate appear in various forms due to the strong turbulence of the gas stream and the different positions the particles had in it. Particles found in the coating are generally fairly uniform in composition; however, large variations can exist between different particles within the coating. Various forms of particles found in arc spray coatings are:

  (i) Relatively pure metal particles.
 (ii) Metal particles with oxidized surfaces.
(iii) Pure metal oxide particles.
(iv) Mixtures of the above.

V. FORMATION OF THE COATING ON THE SUBSTRATE

In considering the phenomena occurring after the arc-spray particles impact on the substrate, the proper conditions on the surface of the substrate should be described. The surface must be clean from oil, moisture, oxides or, in general, any contaminants. In addition the surface should be roughened. This roughening is generally accomplished by grit blasting with iron grit or by rough threading. As the molten spray particle arrives on the roughened surface, it will spread radially from the point of impact, becoming parallel to the surface. As the particle spreads across the surface, its direction will be altered due to the superficial roughness. At the peaks of the roughness, a structural component of the surface is added which is normal to the surface plane, providing the surface with branches which offer good possibilities to connect with subsequent particles, thus helping to hold the sprayed layers together. Depending upon the viscosity, surface tension, arrival velocity and mass of the sprayed particle, different surface shapes and degrees of roughness are formed for the same roughness contours of the substrate.

VI. ADHESION OF THE COATING

Depending upon the energy brought into the boundary layer by the sprayed particles as well as the structural surface conditions and the energy state of the substrate material, the type of metal combination and the surrounding gas atmosphere, different mechanisms for adhesion of the metal sprayed layers may dominate.

One mechanism for adhesion of the sprayed coating to the substrate is due to mechanical interlocking of the coating with undercuts in the surface and also by flowing around roughness peaks. This type of adhesion force is small compared with the total bond strengths generally obtainable with the arc-spray process.

Another mechanism for the bonding of the coating to the base material could be Van der Waals' forces since the sprayed particles have a sufficiently high impact velocity. It is therefore supposed that contact between both materials approaches lattice dimensions. Wetting of the substrate occurs and the surface layers of foreign material such as absorbed gas or water are either absorbed into the sprayed material or displaced by the impacting of the particles. This mechanism of adhesion can contribute greatly to the strength of the bond. However, since a high level of thermal energy does exist in the particle and since adhesion due to Van der Waals' forces takes place non-thermally, other mechanisms involving the influence of an activation energy also must contribute to the nature of the bond.

A metallurgical bond between the sprayed particles and the substrate can originate from diffusion as well as reaction due to thermal energy activation. Based on the fact that absorbed gas and water layers on the substrate surface are displaced or absorbed by the sprayed particles, two-dimensional reaction at the active centres of the surface can take place. The heated portions of the elastically distorted surface zones will recrystallize upon being contacted by the sprayed particle. The depth of such recrystallization zones varies but being generally a few micrometres. In addition, a diffusion may take place, carrying atoms of the base material into the deposit, and atoms of the sprayed particle into the substrate surface. This diffusion process takes place at much lower temperature levels than one might expect, due to the high concentration of lattice defects existing in both the sprayed particle and the surface of the substrate.

In addition to this diffusion, in the case of certain material combinations, exothermic reactions may take place, based on the formation of intermediate phases at the same time as diffusion occurs. Under these circumstances, the velocity of the metallurgical change will be dominated by the reaction process.

BIBLIOGRAPHY

Steffens, H. D. (1966). *British Welding Journal* **13**, 10, 579–605.
Matting, A. and Steffens, H. D. (1967). *Progress in Applied Materials Research* **7**, 91–133.

The basic principles of electrical harmonic spraying

STEVEN B. SAMPLE,* RAGHUPATHY BOLLINI,† DONALD A. DECKER,
AND JOSEPH W. BOARMAN

*School of Electrical Engineering, Purdue University, West Lafayette,
Indiana, U.S.A.*

I. INTRODUCTION

This paper reviews the development of a method for producing collimated beams of small, monodisperse, uniformly charged liquid particles. The method utilizes the harmonic electrical spraying of liquids, wherein the natural (or induced) periodic oscillations of an electrically stressed meniscus at a capillary tip cause a stream of uniformly charged particles to be emitted from the meniscus. The emitted particles are extremely uniform in size and charge, and collinear in trajectory; thus, beams of such particles can be readily focused, accelerated, and deflected by purely electrical means. It is expected that such beams will facilitate the selective coating and abrading of solid surfaces within microscopic tolerances without the need for masks or baffles (Weinberg, 1968). The harmonic spraying process is also a convenient

* Steven B. Sample is currently on leave at the Illinois Board of Higher Education, 500 Reisch Bldg, 119 South Fifth Street, Springfield, Illinois 62701.
† Raghupathy Bollini is currently with SIU, Edwardsville, Illinois.

method for dispersing a liquid in the form of a monodisperse aerosol having a predictable particle diameter.

The diameter of the particles generated by the harmonic spraying process can at present be varied from 50 to 500 μm with a corresponding variation in charge-to-mass ratio from 10^{-2} to 10^{-4} coulombs per kilogram. However, by using smaller capillaries it is expected that particles can be produced having diameters on the order of 10 μm and charge-to-mass ratios on the order of 0·1 coulombs per kilogram.

A number of investigators over the past century have studied various methods for producing charged liquid droplets (Sinclair and LaMer, 1949; Vonnegut and Neubaur, 1952; Magarvey and Taylor, 1956; Mason et al., 1963; Schneider and Hendricks, 1964). (Actually the methods described in the references are mostly for producing uncharged drops which could then be charged by contact charging or by induction.) Particular attention has been given to a method that involves mechanically vibrating a jet of liquid issuing from an electrically charged capillary. The vibrations induce spatially periodic perturbations on the jet, which in turn cause the jet to break up into a stream of closely spaced charged droplets (Magarvey and Blackford, 1962; Mason and Brownscombe, 1964; Schneider and Hendricks, 1964). Sweet (1965) gives an excellent review of the development of this technique, along with a description of how the vibrating jet method can be used for making low-cost oscillographs. The vibrating jet method is also used today in some commercial computer printers.

While highly developed, the vibrating jet method has a number of limitations. First, the particles are inherently closely spaced, and thus the charge-to-mass ratio of the particles must be kept low in order to maintain collinearity. Second, there appear to be severe technical difficulties in producing particles smaller than 50 μm with the vibrating jet method. It is hoped that the harmonic spraying process will be able to overcome some or all of these limitations.

There is a large body of literature dealing with the general area of electrical spraying of liquids from capillaries (e.g., Zeleny, 1914; Vonnegut and Neubauer, 1952; Drozin, 1955; Hendricks, 1962; Cohen, 1963; Hendricks et al., 1964; Pfeifer and Hendricks, 1967). In most of these works the spraying is described as being very random and irregular, resulting in drops of varying size and charge that are emitted from the capillary tip over a wide range of angles. However, some observations by earlier workers (Vonnegut and Neubauer, 1952; Winston, 1962; Hendricks et al., 1964; Cohen, private communication) suggested that, when the spraying configuration is highly symmetric, and under certain conditions of liquid pressure and applied voltage, the spraying process can be somewhat regular and periodic.

The present paper describes our efforts to isolate these naturally periodic (i.e., harmonic) spraying modes (Sample and Bollini, 1972). We have found that rigorously periodic modes do indeed exist, and that such modes give rise to particles that are extremely uniform in size and charge, and collinear in trajectory. Data (with water as the working fluid) are given which show the dependence of these naturally harmonic spraying modes on reservoir pressure, needle diameter, and applied voltage. Data are also presented which show the existence of induced harmonic spraying modes, wherein the frequency of droplet emission can be controlled over a fairly wide bandwidth by the addition of a small a.c. signal to the d.c. spraying potential.

While the original experiments with harmonic spraying were confined to using water in air, our current investigations show that this process can also be used to produce beams of molten metal particles in vacuum (Bollini *et al.*, 1970; Sample *et al.*, 1971). The apparatus for harmonically spraying molten metals is described, and some of the problems attendant to achieving stable spraying modes are discussed. In addition, some preliminary data for molten tin are given.

Finally, a number of potential applications of harmonic spraying are discussed. None of these has actually been achieved in practice; however, given the nature of this Advanced Study Institute, it seemed important to suggest ways in which the harmonic spraying process might be related to the broader interests of those persons working directly in the area of surface coatings.

II. APPARATUS FOR SPRAYING WATER

A schematic of the experimental apparatus for spraying water in air is shown in Fig. 1. A stainless steel hypodermic needle, with the tip ground and polished flat, fits into a plastic syringe, which is in turn connected by a tube to a reservoir of liquid. Variations in liquid pressure head (measured from the needle tip to the top of the liquid in the reservoir) are obtained by adjusting the height of the reservoir. A plate with a small hole in the middle is centred directly under the tip of the needle and about 2 or 3 cm below it. A small cup, located underneath the plate, is used to collect the drops. The needle is connected to one side of a high voltage a.c.–d.c. supply. The grounded side of the supply is connected directly to the plate and is also connected through an electrometer to the collector cup.

The spraying process is observed through a small microscope (not shown in the figure) and is recorded with a conventional 16 mm movie camera focused through a long focal length lens. The high frequency (~400 Hz) periodic spraying process is effectively slowed to about 1 Hz by setting the frequency of the stroboscope close to the spraying frequency. This technique permits the direct measurement of the spraying frequency and in addition allows slow motion movies to be taken at 16 frames/second without any synchronization between the camera and the stroboscope.

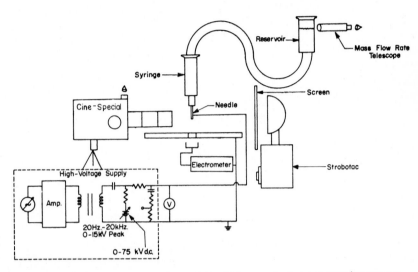

FIG. 1. Experimental apparatus for studying the harmonic electrical spraying process.

The mass flow rate is measured by observing the rate of fall of the reservoir level through a small telescope. The mass of the emitted particles is then computed by dividing the mass flow rate by the frequency of droplet emission (as measured with the stroboscope). Similarly, the charge on each drop is obtained by dividing the electrometer current by the frequency of emission.

III. EXPERIMENTAL RESULTS FOR WATER

All the experiments reported here were performed in an open air environment using distilled water as the working fluid. Since the relaxation time of distilled water ($\sim 3 \times 10^{-6}$ s) is extremely short compared to the period of the spraying process ($\sim 10^{-3}$ s), the working fluid can be considered a perfect conductor (Hendricks, 1962).

Two basic modes can be observed when a d.c. voltage is applied to the needle shown in Fig. 1. The first of these, which occurs at lower potentials, is best described as the dripping mode (Raghupathy and Sample, 1970). At higher potentials the low frequency dripping mode gives rise to the higher frequency harmonic spraying mode.

A. Dripping mode

In order to start the dripping mode, the pressure head is adjusted such that the working fluid drips at the rate of one or two drops per second. In this case, the liquid forms a pendant drop, which hangs from the outer circumference of the needle if the liquid wets the needle or from the inner circumference if the liquid does not wet the needle. The drop will grow until its weight overcomes the net vertical component of the surface tension force. At this point the liquid nearest the needle forms a neck, which eventually ruptures, thereby allowing the main part of the drop to fall from the needle. When a d.c. voltage is applied to the needle, the force on the drop due to the interaction of the induced charge and the electric field is in the same direction as the gravitational force. Therefore, the emission frequency increases with d.c. voltage. The variations in emission frequency and drop diameter with applied voltage are plotted in Raghupathy and Sample (1970). This mode has wide potential application in biological and erosion studies as an inexpensive means for producing large, collinear drops of known size.

B. Harmonic spraying

As the voltage is increased beyond a certain critical point, a sudden transition from the dripping mode to the high frequency spraying mode is observed. The transition is marked by a sharp increase in the frequency of droplet

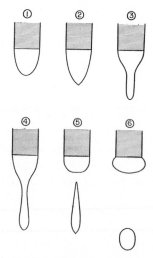

FIG. 2. Six sequential stages in the formation of a drop at the needle tip during harmonic spraying.

emission and a sharp decrease in the mass flow rate. Moreover in the harmonic spraying mode, the residual meniscus which remains attached to the needle after droplet emission is usually larger than the emitted drop, whereas in the dripping mode this residual meniscus is so small it appears as though the drops are pulled off directly from the needle tip.

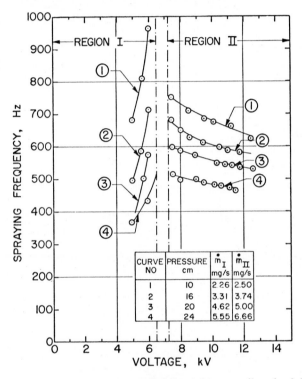

FIG. 3. Harmonic spraying frequency vs. applied d.c. voltage; needle o.d.—0·41 mm, i.d.—0·20 mm. The symbol $\dot{m}$ stands for mass flow rate.

The sequence of events taking place during a single harmonic spraying cycle is shown in Fig. 2. The meniscus at the start of the cycle beings to move down as shown in frame 1 and is distorted into a more or less conical shape as shown in frame 2. A neck then starts to form (frames 3 and 4), which elongates and eventually forms a point of separation from the main meniscus, (frame 5). As the forward part of the liquid separates from the meniscus, it quickly coalesces into a drop and carries with it part of the surface charge that is induced on the meniscus by the high voltage. The main meniscus is now shielded from the electric field by the newly formed charged drop, and thus, as the drop is accelerated towards the plate, the meniscus relaxes back toward the needle. The process described above repeats itself in a regular, periodic manner. The emitted particles are thus uniform in size and charge and collinear in trajectory.

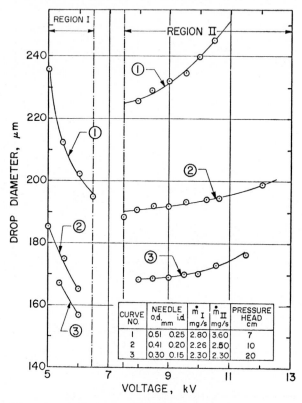

FIG. 4. Emitted drop diameter vs. applied d.c. voltage, for three different needles and pressure heads.

Figure 3 shows the behaviour of the spraying frequency as a function of voltage using the same needle with four different pressure heads, and with water as the working fluid. Figure 4 shows the variations in drop diameter as a function of voltage for three different needles and pressure heads; for purposes of comparison, curve 2 in Fig. 4 was obtained under essentially the same conditions as curve 1 in Fig. 3. It can be seen from the figures that, for a given pressure head and needle size, a band of voltages exists over which no harmonic spraying occurs.

It should be pointed out that the transition from dripping mode to harmonic spraying mode occurs in region I. For voltages between region I and II the spraying is very random and irregular (this voltage region between I and II therefore should not be confused with the region where dripping is

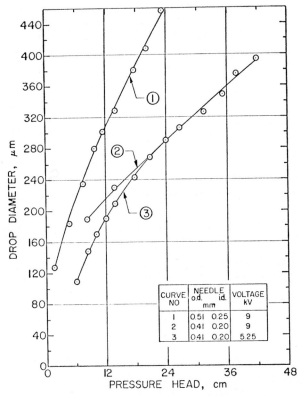

FIG. 5. Emitted drop diameter vs. pressure head, for two different needles and d.c. voltages.

observed). In region I the spraying frequency increases rapidly with increasing voltage, while in region II the frequency decreases slowly with increasing voltage. Figures 3 and 4 do not include data much above 12 kV because random spraying modes appear above this voltage, accompanied (and possibly caused) by strong corona from the needle tip.

Figure 5 shows the dependence of drop diameter on pressure head. It is evident from this figure that the particle diameter can be readily controlled over a wide range by varying the pressure head. In particular, in the case of curve 1, the drop diameters vary over almost a 4 to 1 range, which represents a 64 to 1 range in drop masses. The frequency range for curve 1 is from 750 Hz for the 120 μm drops to 270 Hz for the 460 μm drops.

C. Induced harmonic spraying

It is possible to synchronize an a.c. voltage with the harmonic spraying process, and once synchronization is established, it is possible to vary the frequency of harmonic spraying by simply varying the a.c. frequency.

Synchronization is achieved by setting the frequency of the signal generator to the natural spraying frequency as measured by the strobe, and the gain of the amplifier is adjusted so that the peak value of the a.c. is about 20% of the d.c. potential. Minor adjustments in a.c. amplitude and frequency are then made until the meniscus appears sharply defined and stationary.

It was observed that once the spraying process becomes locked to the a.c. voltage, it is possible to vary the harmonic spraying frequency by varying the a.c. frequency. It was noted that there are upper and lower limits in frequency, between which the droplet emission frequency tracks the a.c. frequency. However, when the a.c. frequency is outside this range, the spraying process reverts to its natural frequency.

The tracking bandwidth (i.e., the range of frequencies over which the spraying frequency will track the a.c. voltage frequency) as a function of pressure is shown in Fig. 6 for a typical needle diameter and voltage. The dotted curve inside the tracking region represents the natural spraying frequency (or natural characteristic frequency of the meniscus) as a function of pressure.

Analysis of the experimental results is given in Sample and Bollini (1972).

IV. Apparatus for spraying molten metals

A diagram of the apparatus used for spraying molten metals in vacuum is shown in Fig. 7. The vacuum chamber consists of a 3 foot (91 cm) section of 6 inch (15 cm) i.d. glass pipe. The stainless steel top plate with its feedthroughs and the inner workings of the vacuum chamber (i.e., furnace, reservoir, collection cup, etc.) are a complete unit independent of the glass pipe and any of its supporting structure. This is accomplished by suspending a stainless steel rod from the top plate and to this rod any necessary components can be attached. Thus, the entire subunit can be easily removed and placed in a specially built rack. With this arrangement, all necessary alignments can be done readily and completely outside the glass pipe.

The furnace has a unique construction which permits easy cleaning of the heater wires, supporting structure for the reservoir and the high voltage lead.

The heater wires are strung vertically between insulators attached to a stainless steel cylindrical framework. A circular disc with a hole in the centre to guide the needle is suspended by means of ceramic insulators from the above framework. The height of the disc is adjustable which in turn enables the

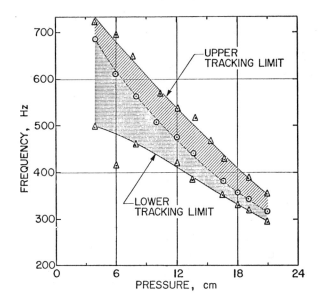

FIG. 6. Tracking bandwidth of the harmonic spraying process vs. pressure head. Applied d.c. voltage—8 kV; a.c. voltage—about 900 V r.m.s.; needle o.d.—0·51 mm, i.d.—0·25 mm.

distance between the needle tip and ground plane to be varied. The high voltage lead is connected to the disc. The stainless steel framework slides inside a nickel plated 3·5 inch (8·9 cm) diameter brass pipe. Brass was used because it was the most easily obtainable material in the size and shape needed. However, since zinc (one of the components of brass) has a high vapour pressure, brass was not suitable to use in a high temperature, low pressure environment. Thus, it was necessary to plate the brass with nickel which is suitable in this environment. Also the plating provided a highly reflective surface which improved the efficiency of the furnace. A small slit in the front of the brass pipe permits observations of the needle.

The reservoir is made of Pyrex tubing with a glass-to-metal seal (stainless steel to Pyrex) at one end. A stainless steel adaptor is welded to the seal to which is attached the needle assembly. The use of stainless steel for the adaptor and the seal permits the usage of acids for cleaning purposes. The top of the reservoir is connected to a valve system for adjusting the pressure on the molten metal for obtaining proper dripping.

Centred directly beneath the furnace is a circular grounded metal plate which serves the same purpose as the ground plate in the apparatus used for spraying water. A heated metal collector cup is placed underneath the ground plate for collecting the drops. The reasons for heating the cup are given in the next section.

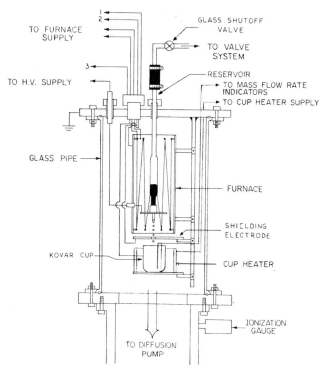

FIG. 7. Apparatus for spraying molten metals in vacuum.

The frequency of droplet emission can be measured by means of a strobe as described in the water experiments or by grounding the cup through a resistor and displaying the voltage pulses which appear across the resistance by means of an oscilloscope. This latter method is used at present since it provides more reliable results.

The charge on the drops is determined, as in the water experiments, by dividing the collector cup current (measured with an electrometer) by the droplet emission frequency.

Despite the careful cleaning process used for the tin, a small amount of slag formed on the surface of the molten tin in the reservoir. As a consequence it was not possible to measure the mass flow rate by the technique used for water. The present method for determining the mass flow rate is to measure the time required for a predetermined volume of tin to fall into the collector cup. The apparatus used consists of a set of graduated length metallic probes, a set of light bulbs, and a power supply. The probes are suspended inside the collector cup with the longest probe acting as a return to the power supply. Each of the other probes is connected to its respective light bulb which in turn is connected to the other side of the supply. As the level of molten tin in the collector cup rises, the molten tin will make contact with each of the probes in sequence thus completing the circuit for the respective indicator lamps. The change in the volume of metal needed to turn on successive indicator lamps is determined outside the vacuum system using mercury. Thus, after calibrating the probes with mercury, it is only necessary to measure the time lapse between turning on successive light bulbs to calculate the mass flow rate.

V. EXPERIMENTAL RESULTS FOR TIN

A. Experimental procedure

During the course of the present attempts to spray molten metals harmonically, a number of experimental problems arose. In order to solve these problems a careful experimental procedure was developed and is described in this section.

Since the reservoir is made of Pyrex, at present only metals which melt below 400 °C can be sprayed. At first Wood's metal (Eutectic alloy, melting point 80 °C) was tried, but it was difficult to obtain clean samples of this alloy or develop a suitable cleaning procedure in the lab. As a result it consistently clogged the needles. Reagent grade tin (melting point 230 °C) was then tried. Since a consistent periodic dripping is required for reliable data, a meticulous cleaning process had to be developed even for reagent grade tin before the molten metal would flow consistently through the small needles.

In order to initiate and maintain dripping, nitrogen had to be introduced on top of the molten tin in the reservoir. The nitrogen pressure required was conducive to breakdown once the high voltage was applied to the needle. The breakdown problem was overcome by introducing a glass on-off valve (shown in the figure as shutoff valve). The valve was closed after a consistent dripping was established and before applying high voltage. Once the valve is closed, the path to ground is interrupted thus alleviating the breakdown problem. In order to maintain a constant pressure on top of the molten tin after the valve is closed, a large cross-sectional reservoir is being used.

Originally experiments were carried out using stainless steel needles mounted in chrome plated brass hubs. However, these needles clogged easily since the chrome had a tendency to flake off and contaminate the molten tin. At present, a system (i.e., hub, needle and an adaptor to hold the needle) made entirely of stainless steel is giving satisfactory results.

The drops collected in the cup solidified very quickly and as a result they piled on top of each other and grew up from the bottom of the cup in the form of stalagmites. The stalagmites went through the hole in the ground plate and reached the needle which in turn shorted the high voltage. Therefore, a heater had to be placed around the cup in order to keep the metal collected in the cup molten. Moreover, the cup heater is now essential for the present method of mass flow rate measurements.

B. Experimental observations and results for tin

The harmonic electrical spraying process for tin is somewhat different from that of water and is therefore described in this section. Since tin does not wet stainless steel the meniscus which forms at the tip of the needle does not extend to the outer diameter of the needle as in the case of water. Therefore, the drop size is controlled by the i.d. of the needle instead of the o.d. as with water.

The dripping rate increases with voltage and at a critical voltage the transition from the dripping mode to the harmonic spraying mode takes place. The transition is marked by an abrupt increase in frequency, but the transition spraying frequency was observed to be in the 60 Hz range and therefore the change in frequency is not as great as in the case of water. No accurate mass flow rate data has been recorded to date which shows a decrease in mass flow rate with the transition.

The range of voltages over which the harmonic electrical spraying of tin is observed is very narrow, normally 9 kV to 11 kV. Above these voltages very erratic spraying and very high currents are observed. Since these high currents are not observed if the fluid flow is stopped and the voltage held constant, it is believed that the high current is due to ion emission (see for example Mahoney et al., 1969). Unlike the water experiments, two separate spraying bands as a function of voltage were not observed for molten tin.

As in the case of water, the natural spraying process for tin could be synchronized with an a.c. voltage. The drop emission frequency tracked the a.c. frequency up to twice the natural spraying frequency, i.e. from a natural frequency of 60 Hz up to about 120 Hz. In fact, the induced spraying mode was observed to track the a.c. signal exactly for a period of over 20 minutes. However, at present, due to the limitations on the amplifier and signal generator used to drive the broadband transformer for the a.c. supply, the tracking of droplet emission below the natural spraying frequency has not been observed.

Since the dripping takes place from the inside diameter of the needle, the drop diameter in the harmonic spraying mode is controlled by the i.d. of the needle. The diameters of the drops generated to date using an 0.75 inch (1.9 cm) long, 27 gauge needle (o.d. of 410 µm and i.d. of 200 µm) varied from 260 µm to 320 µm.

The charge-to-mass ratio data obtained to date indicate that the charge-to-mass ratio for molten tin drops is close to one half the Rayleigh limit, i.e. for 300 µm diameter drops the charge-to-mass ratio was about 5×10^{-4} coulombs per kilogram.

No significant changes in the spraying frequency with change in pressure have been observed.

At present, efforts are being made to spray molten lead and bismuth. If these efforts are successful more insight will be gained into the harmonic electrical spraying process for molten metals.

VI. POTENTIAL APPLICATIONS

The broad range of potential applications of harmonic spraying, stems from the fact that, in principle, any material which melts below the softening point of the spraying capillary, and which exhibits a conductivity greater than 10^{-8} mhos per metre, can be harmonically sprayed. In particular, it should be possible to spray mixtures, solutions, and colloidal suspensions. Thus, a material which does not melt easily might be sprayed by first suspending it in a liquid vehicle, while a material with a very low conductivity might be sprayed after it has been lightly doped with an electrolyte.

Figure 8 illustrates one possible configuration for depositing harmonically sprayed particles in precise patterns without the use of masks. It should be pointed out that there is essentially no physical interaction between adjacent drops in the beam; therefore, provided the deflection amplifiers are sufficiently fast, the beam can be swept instantaneously from one point to another point some distance away without leaving any trace deposition in between.

In the area of engineering applications, some of the potential uses for harmonically sprayed particle beams are:

(1) Integrated circuits. Electrically focused and deflected beams could be used for metallization and deposition of dopants without the need for

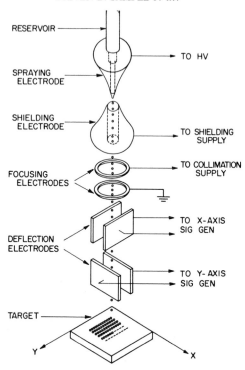

FIG. 8. Possible configuration for depositing harmonically sprayed particles in precise patterns without masks.

masks, thus saving several steps in the fabrication process. This application might be particularly useful for fabricating and modifying experimental circuits, and for very low volume production runs.

(2) Hybrid circuits—deposition of thick and thin film resistors and interconnecting metallization. This application would compete with conventional printing techniques only in those cases where extremely close tolerances were required.

(3) High-impact film deposition. In this application, the sprayed particles could be electrically accelerated to high velocities, thus generating high local temperatures and pressures at the instant of impact. This technique, which amounts to explosion bonding on a microscopic scale (without the bulk deformations that normally accompany macroscopic explosion bonds), might facilitate film deposition on those substrate materials such as aluminium that resist film deposition by more conventional techniques.

(4) Printing. Potentially useful where more precision is required than can be obtained using the vibrating jet method (Lindblad and Schneider, 1965; Kamphoefner, 1972). The process might also prove useful for generating precision printing plates, and for high density information storage.

(5) Micromachining and polishing. In this application, the grinding or polishing agent could be suspended in a liquid vehicle. A beam of particles sprayed from this mixture could be focused on a particular spot of the substrate, so that each particle struck the substrate obliquely, and then bounced off again.

In addition to these engineering applications, the harmonic spraying process has a number of potential uses as a tool for scientific research. Among the applications in this area are:

(1) Preparation of radioactive films. Solutions of radioactive salts could be harmonically sprayed onto thin backing foils. The harmonic spraying process should give much more uniform films than the random spraying process currently in use (Lauer and Verdingh, 1963).

(2) Erosion studies. Harmonically sprayed particles could be accelerated to the desired velocity, and allowed to impinge on the substrate material at the desired angle of incidence. This technique might prove useful for synthesizing the effects of hypersonic rain erosion (Engel, 1961) or micrometeorite bombardment (Shelton et al., 1960) under controlled conditions.

(3) Drop coalescence studies. Particles from two separate harmonically sprayed beams could be caused to collide, either in the open or on various surfaces, thus facilitating the study of the coalescence of similar and dissimilar materials under various conditions (Schneider et al., 1965).

VII. Conclusions

The harmonic electrical spraying of liquids from capillaries is a convenient and relatively inexpensive method for generating highly collimated beams of monodisperse, uniformly charged liquid particles. The addition of an a.c. potential to the process improves the collinearity and uniformity of the particles and can also be used to control the frequency of particle emission.

Experiments to date have used capillaries with diameters on the order of a few hundred micrometres and have produced droplets with diameters in the 50 to 500 μm range having charge-to-mass ratios ranging from 10^{-3} to 10^{-2} coulombs per kilogram. However, there is every reason to believe that smaller capillaries can be used, thereby producing correspondingly smaller particles with higher charge-to-mass ratios. It has been found that droplets produced by harmonic spraying exhibit charge-to-mass ratios that are essentially equal to one-half the Rayleigh limit.

So far, only water and molten tin have been harmonically sprayed, the former in an air environment and the latter in vacuum. However, it appears that practically any conducting liquid is amenable to harmonic spraying. It is this versatility of working fluid, along with the fact that the particle beam can be electrically focused, deflected and accelerated, that makes harmonic spraying a potentially useful method for the precision deposition of coatings.

ACKNOWLEDGEMENTS

The authors wish to thank Mr. Robert Aram for his help in building the apparatus. This work was supported in part by grants from the U.S. National Science Foundation (GK-3394, GK-24868) and in part by the U.S. Army, Navy and Air Force through the Joint Services Electronics Program. The authors wish to thank Miss Charlotte Sexton for her help in preparing the manuscript.

REFERENCES

Bollini, R., Decker, D. A. and Sample, S. B. (1970). *Proc. Natl. Elect. Conf.* **26**, 162.
Burgoyne, J. H. and Cohen, L. (1953). *J. Colloid Sci.* **8**, 364.
Cohen, E. (1963). Rept. ARL-63-88, Space Technology Laboratories, Redondo Beach, California, U.S.A.
Drozin, V. G. (1955). *J. Colloid Sci.* **10**, 158.
Hendricks, C. D. (1962). *J. Colloid Sci.* **17**, 249.
Hendricks, C. D., Carson, R. S., Hogan, J. J. and Schneider, J. M. (1964). *AIAA J.* **2**, 733.
Engel, Olive G. (1961). "Symposium on Erosion and Cavitation". ASTM Special Technical Publication No. 307.
Kamphoefner, F. J. (1972). *IEEE Trans. Electron Devices* **19**, 584.
Lauer, K. F. and Verdingh, V. (1963). *Nucl. Instr. and Meth.* **21**, 161.
Lindblad, N. R. and Schneider, J. M. (1965). *J. Sci. Instr.* **42**, 635.
Magarvey, R. H. and Taylor, B. W. (1956). *Rev. Sci. Instr.* **27**, 944.
Magarvey, R. H. and Blackford, B. L. (1962). *J. Geophys. Res.* **67**, 1421.
Mason, B. J., Jayaratne, P. W. and Woods, J. D. (1963). *J. Sci. Instr.* **40**, 247.
Mason, B. J. and Brownscombe, J. L. (1964). *J. Sci. Instr.* **41**, 258.
Pfeifer, R. J. and Hendricks, C. D. (1967). *Phys. Fluids* **10**, 2149.
Raghupathy, B. and Sample, S. B. (1970). *Rev. Sci. Instr.* **41**, 645.
Rayleigh, J. W. S. (1882). *Phil. Mag.* **14**, 184.
Sample, S. B., Raghupathy, B. and Hendricks, C. D. (1970). *Int. J. Engrg Sci.* **8**, 97.

Sample, S. B., Bollini, R. and Decker, D. A. (1971). In "Record 11th Symposium on Electron, Ion and Laser Beam Technology (R. F. M. Thonley, ed.). San Francisco Press, U.S.A.

Sample, S. B. and Bollini, R. (1973). *J. Colloid Sci.* (in Press).

Schneider, J. M. and Hendricks, C. D. (1964). *Rev. Sci. Instr.* **35**, 1349.

Schneider, J. M., Lindblad, N. R. and Hendricks, C. D. (1956). *J. Colloid Sci.* **20**, 610.

Shelton, J., Hendricks, C. D. and Weurker, R. F. (1960). *J. Appl. Phys.* **31**, 1243.

Sinclair, D. and LaMer, V. K. (1949). *Chem. Rev.* **44**, 245.

Sweet, R. G. (1965). *Rev. Sci. Instr.* **36**, 131.

Vonnegut, B. and Neubauer, R. L. (1952). *J. Colloid Sci.* **7**, 616.

Weinberg, F. J. (1968). *Proc. Roy. Soc.* **A307**, 195.

Winston, C. R. (1962). U.S. Patent 3 060 429.

Zeleny, J. (1914). *Phys. Rev.* **3**, 69.

The basic principles of evaporation

M. MARTINI

Simtec Industries Ltd., Montreal, Quebec, Canada

I. INTRODUCTION

Evaporation *in vacuo* is a well established technique for depositing thin films of metals, alloys, semiconductors, metal oxides and other oxides on various substrates (mainly insulators, plastic materials and semiconductors).

As we endeavour here to describe in a paper of necessarily limited size a subject extensively treated in textbooks (Holland, 1961), we have chosen to give the details of the evaporation of a thin metallic film on a semiconductor (e.g. *n*-type silicon) in order to achieve either a strongly rectifying or an ohmic contact.

In particular this paper describes: (1) The preparation of the source (i.e. the metal to be evaporated and the heater) and especially of the substrate (the slice of semiconductor where the contact is to be made). (2) The evaporation process, with details of the vacuum which must be achieved, the temperature of the source, the rate of deposition and the gauging of the thickness of the evaporated film. (3) The electrical characteristics of the resulting device with some explanations on the physical mechanisms which lead to these characteristics.

II. Preparation of the Source and the Substrate

A. The source

The preparation of the source, which can be (typically) Au for a strongly rectifying contact on *n*-type silicon, is quite simple because the only requirement is a high degree of purity of the material to be evaporated and normal precautions in keeping both this material and the heater as clean as possible.

The geometry of the source varies widely according to the different requirements. For instance, a "point source" is easily obtained by wrapping a wire of the metal to be evaporated around a V-shaped filament. The filament is resistively heated by passing a large current through it. On melting, the evaporant "wets" the filament (Fig. 1).

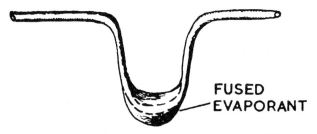

FIG. 1. Point source configuration (from Holland, 1961).

On the other hand, when it is necessary to deposit uniformly a film over the internal surface of a spherical holder, then the vapour source emission corresponds to that of a small plane area ("Knudsen source", see Fig. 2*a*, *b*, *c*).

B. The substrate

The preparation of the substrate (*n*-type silicon in our example) is a much more complicated matter, where witchcraft and local traditions play an important role: "The successful thin film vacuum worker must be almost as much an alchemist as a physicist" (Tolansky, Foreword to Holland, 1961).

A typical procedure for obtaining a good rectifying contact by evaporation of gold on *n*-type silicon is the following (Coche and Siffert, 1968): a slice of *n*-type silicon is cut and subsequently lapped and polished by means of pastes with grains of decreasing size, the finest ones having linear dimensions of approximately 1 μm. Care is taken so that the polished sample has a thickness of approximately 60 μm over the final desired value. The dimensions of the slice and the resistivity of the starting material depend on the particular use of the device. After carefully degreasing, possibly in an ultrasonic bath, the etching process is carried out; this is a most important step since a very clean surface is needed. In general the etching is performed with a mixture of nitric, hydrofluoric and acetic acids (3:1:1). The nitric acid acts as an oxidizer, the hydrofluoric acid partially dissolves the oxide formed, and the acetic acid,

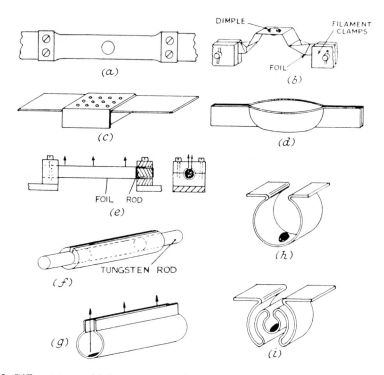

FIG. 2. Different types of foil vapour source heaters. Heaters (a), (b) and (c) give Knudsen surface emitting sources (from Holland, 1961).

by slowing down the reaction rate, allows easier control of the process. The etch (performed normally at room temperature) is quenched with double distilled deionized water.

The intermediate step between the etching and the gold deposition is by no means standardized, some authors recommend proceeding directly to the gold deposition while according to others the sample should be kept for some-time in air or other gases or oxidized in boiling water or by other means.

III. THE EVAPORATION PROCESS

A. General considerations

Under conditions of high vacuum (10^{-5}–10^{-6} torr; $1\cdot3$–$0\cdot13$ mPa), metals begin to evaporate rapidly when their temperature is raised to the point where their vapour pressure is higher than approximately 10^{-2} torr ($1\cdot3$ Pa).

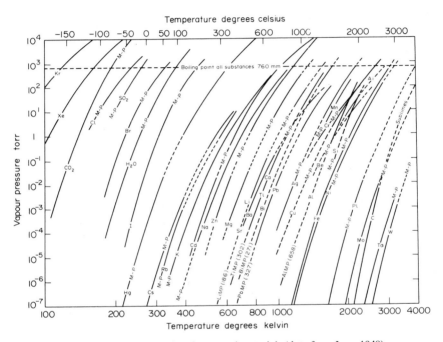

FIG. 3. Vapour pressure data for several materials (data from Law, 1948).

Figure 3 shows the vapour pressure of several metals as a function of temperature. During the evaporation process, in equilibrium conditions, molecules are continuously passing from the condensed phase to the vapour phase, but at the same time, another stream of molecules is condensing out of the vapour and the two processes balance each other (Kennard, 1938).

The maximum possible rate of evaporation would occur if one could remove the vapour as fast as it is formed: however the phenomenon is not observable. Therefore the only way to calculate the evaporation rate is to utilize the fact that, in equilibrium, it is equal to the condensation rate, and the latter can be calculated from the kinetic theory of gases. The equation of the condensation rate R (in g/cm^2 s),

$$R = 5\cdot85 \times 10^{-5}\alpha p \sqrt{\frac{M}{T}}$$

(where M is a mol of the substrate to be evaporated, T the absolute temperature and p the pressure of the saturated vapour), gives the upper limit of R because α (called the "condensation coefficient") is an unknown constant slightly lower than one, taking into account the fact that only a fraction of the molecules impinging on the surface of the substrate do condense.

B. The evacuation

The above considerations apply to evaporation *in vacuo*. In order to better define the term, we can again use the kinetic theory of gases for calculating

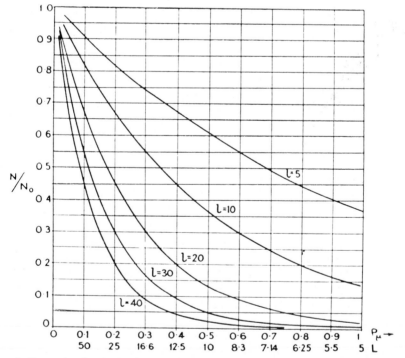

FIG. 4. Curves showing the fraction N/N_0 of the emitted molecules which traverse a distance l without suffering a collision as a function of the system's vapour pressure (in units of 10^{-3} torr ($0\cdot13$ Pa)) and of the mean free path L (in cm). The vapour atoms are assumed to have a mean free path of 50 cm in the residual gas at 10^{-4} torr (13 mPa) gas pressure (from Holland, 1961).

the fraction of the molecules emitted from the source which traverse a distance l without suffering a collision. An idea of the situation is given in Fig. 4. From these data one could infer that a vacuum of the order of 10^{-4} torr (13 mPa) is adequate for most evaporations.

FIG. 5. A modern coating unit. The data on this unit are given in the Appendix on p. 348.

However, it is a well known rule that in most cases (including the particular one that we are following here) vacua of the order of 10^{-5}–10^{-6} torr (1·3–0·13 mPa) are necessary, and moreover that these values should be reached in as short a time as possible, i.e. of the order of a few minutes. The reason for these requirements on the vacuum is essentially the necessity of avoiding contamination of the substrate by residual gas in the vacuum chamber. Such contamination can arise even if the distance between the source and the substrate is sufficiently small and the gas pressure is sufficiently low so that collisions between molecules in transit can be neglected.

C. The deposition

Once a suitable vacuum has been obtained, one can proceed to heat the source for the deposition. Often a glass bell jar permits one to control the process visually (Fig. 5). The temperature necessary for evaporating Au [taken as the temperature at which the vapour pressure is equal to 10^{-2} torr (1·3 Pa)] is approximately 1700 K. In an arrangement such as the one shown in Fig. 1, most of the power (of the order of 100–1000 watts) is employed in heating the filament rather than in supplying the latent heat of evaporation to the metal. A more efficient system of heating the source is, for instance, by electron bombardment. With this method the electron beam, usually accelerated at a voltage of 1 kV, produces the highest temperature in the vapour emitting surface and not in the evaporant support material.

Before concluding these notes on the deposition process, it is important to mention how the thickness of the deposited layer can be measured. The most accurate method of checking the thickness is by using an instrument based on the variation of the frequency of a crystal quartz oscillator. Instruments of this type now available feature a digital readout in Ångstroms and an accuracy of 20 Å (2 nm).

IV. THE SCHOTTKY BARRIER DEVICE

In this paper we have briefly described the fabrication of a rectifying device, i.e. a device whose V–I characteristic is of the type shown in Fig. 6. This characteristic is by no means unique to the evaporation of gold on n-type silicon. Actually, "the fact that given an arbitrary metal deposited upon an arbitrary semiconductor the result has a probability of well over 90% of being a rectifier . . . has represented to many a development engineer one of the prime examples of the innate perversity of nature" (Mead, 1966).

The early experimenters were also puzzled by the fact that the high resistance was measured with the negative bias applied to the metal, which was known to be much richer in free electrons than the semiconductor. A very basic explanation of the physical mechanism at the basis of the metal-semiconductor rectifier (Schottky barrier) is the following (Mead, 1969). In a metal most of the electrons are below the Fermi level. The distance (in the

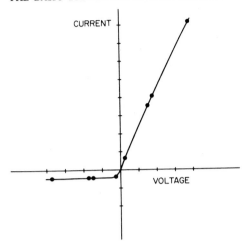

FIG. 6. Typical V–I characteristics of a rectifying device.

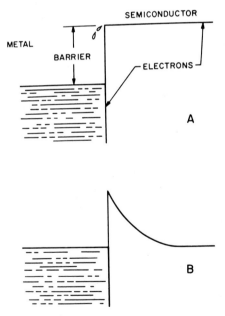

FIG. 7. Simple model of metal–semiconductor interface. The electrons "dribble" from the higher energy state at the bottom of the conduction band of the semiconductor (A) until a potential barrier which stops the process is created (B) (from Mead, 1969).

space of energy) between the Fermi level and the energy of a free electron is the work function of the metal (Fig. 7). In a semiconductor (*n*-type), however, the electrons are crowded at the bottom of the conduction band, whose distance from the energy of a free electron is smaller than in the metal. Therefore, when a metal and an *n*-type semiconductor are brought in close contact, the electrons "dribble" into the metal (Fig. 7*a*) until the built-in space charge stops the process (Fig. 7*b*). This simple model explains qualitatively the characteristic shown in Fig. 6.

It should be noted that this mechanism does not apply to those semiconductors (e.g. GaAs) in which an important role is played by the surface charge induced in surface states due to the termination of the crystal lattice.

A detailed discussion of the physics of the semiconductor-metal interface is beyond the scope of this paper where only an example of the physical mechanism which determines the behaviour of such an interface is needed.

V. APPENDIX: PRINCIPAL CHARACTERISTICS OF THE COATING UNIT SHOWN IN FIG. 3

Pumpdown time:	9 minutes to 8×10^{-7} torr, 25 minutes to 2×10^{-7} torr.
Ultimate pressure:	$3 \cdot 5 \times 10^{-8}$ torr.
Chamber:	Pyrex glass 18 inches (46 cm) inside diameter, 30 inches (76 cm) high.
Baseplate:	Stainless steel, 20 inches (51 cm) diameter.
Evaporation circuit:	2 KVA, 100% duty cycle (3·5 KVA, 20% duty cycle) at 5, 10, 20, or 40 volts. Includes variable transformer, on–off switch, and ammeter.
Power requirement:	115 volt, 57 amp, 2 wire (plus equipment ground) single phase, or 230 volt, 36 amp, 3 wire (plus equipment ground) single phase.
Water requirement:	20 gallons per hour (25 cm^3 S^{-1}) at $20° \pm 5$ °C inlet temperature.

(1 torr = 133·3 Pa.)

ACKNOWLEDGEMENTS

The author is grateful to the Defence Research Board of Canada for partially supporting this work (Project DIR E–205), and to Dr. Holland of Edwards High Vacuum Ltd. for having kindly given permission to reproduce three figures of his textbook.

REFERENCES

Coche, A. and Siffert, P. (1968). In "Semiconductor Detectors" (G. Bertolini and A. Coche, eds.), pp. 144–147. North Holland, Amsterdam.

Holland, L. (1961). "Vacuum Deposition of Thin Films". John Wiley, New York.

Kennard, E. H. (1938). "Kinetic Theory of Gases". McGraw-Hill, New York.

Law, R. R. (1948). *Rev. Sci. Instr.* **19**, 920.

Mead, C. A. (1966). *Solid State Electronics* **9**, 1023.

Mead, C. A. (1969). In "Ohmic Contacts to Semiconductors" (B. Schwartz, ed.). The Electrochemical Society, New York.

Decoration by vacuum metallization of plastics and metals

S. KUT

E. Wood Ltd., Talbot Works, Stanstead Abbotts, Ware, Hertfordshire, England

I. Introduction

Vacuum metallization is the deposition under high vacuum of metals and non-metals on to prepared surfaces of plastics, metals, glass, paper, textiles and other substrates to produce finishes for both decorative and functional applications.

The decoration of plastics by vacuum metallization and painting started some thirty years ago, and was for many years confined to the novelty field, the performance requirements of which were relatively low, appearance being the prime requirement.

In the late 1950's, the automotive industry in the U.S.A. took an increasing interest in vacuum metallization followed by the European industry in the 1960's. Rigorous specifications are now met by vacuum metallized components.

A. First surface vacuum metallization

The decoration of the upper side of the plastic—that is the surface being viewed. Opaque or clear plastics may be used. The reflective coating is a very thin film of brilliant aluminium. Figure 1 illustrates schematically the system.

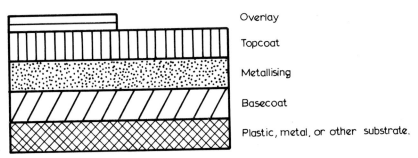

FIG. 1. First surface vacuum metallization.

(i) *Basecoat, first surface*
This is a coating applied directly to the substrate, such as plastic, glass or metal. The vacuum metallized layer is deposited on this basecoat.

(ii) *Topcoat, first surface*
This is a protective coating applied over the metallizing film. The topcoat may be clear (transparent) or tinted to give a variety of attractive transparent shades.

(iii) *Overlay*

In addition to the topcoat, a further decorative coating may be applied. This is in particular the practice for automotive components, such as instrument panels. An overlay is normally a pigmented opaque colour—gloss, eggshell or flat.

B. *Second surface vacuum metallization*

Decoration of the reverse of the plastic—a transparent plastic (such as polystyrene or an acrylic) has to be employed, and the surface in service is the plastic itself. Typical examples are badges, appliance escutcheons, horn buttons, medallions, nameplates and trade marks. Figure 2 illustrates schematically the second surface vacuum metallized system.

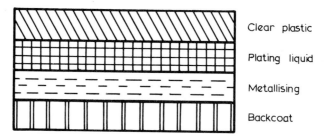

Clear plastic

Plating liquid

Metallising

Backcoat

FIG. 2. Second surface vacuum metallization.

(i) *Plating liquid, second surface*

This describes a coating applied direct to the plastic prior to vacuum metallizing, on the reverse side of the plastic. This coating can be clear or tinted. If tinted gold for example, an aluminium deposited film will confer brilliance, and the decorative effect will be gold. The vacuum metallized film is then deposited on this plating liquid.

(ii) *Backcoating*

A backcoating is applied to the metallized film. This is generally an aluminium or pigmented opaque paint acting as the outer protective coating of the system, and to lend opaqueness to the decorated part as normally viewed, that is, through the plastic.

II. ADVANTAGES

The advantages of vacuum metallized plastic mouldings may be summarized as: light weight—for example ABS (acrylonitrile–butadiene–styrene) is one-sixth of the weight of die castings; greater processing speed than for metal stampings or die castings; lower cost than chrome plating metal; functional properties maintained; greater freedom in design—injection moulding allows shapes, contours and textures not possible in die casting; faster and less expensive mould construction; extensive variations in decoration using vacuum metallizing in conjunction with painting, wood graining or hot stamping.

III. CHARACTERISTICS OF THE METALLIZING COATINGS

A. *First surface metallization—basecoat*

The essential functions of the basecoat are:
(1) To seal off the plastic substrate to prevent outgassing and to eliminate occluded gases on the plastic surface, which might adversely affect adhesion of the metallized film.
(2) Simultaneously, to form a smooth, even film, filling minor imperfections; the coating must therefore have adequate build.

The thin, highly reflective metallized film highlights any imperfection in the substrate or in the coating—such as "orange peel", due to inadequate flow. It is essential that the solvents of the basecoat do not visibly attack or etch the plastic thereby showing up as a serious decorative defect after metal-lizing. At the same time, the basecoat must have good adhesion to the sub-strate, be receptive to the metallized film and to the topcoat.

Curing temperatures must be reasonably lower than the warpage tem-perature of the plastic part.

After curing or drying, the basecoat must not emit any residual solvent or other organic vapours which would interfere with the pumping cycle and metallizing. Certain plasticizers in the basecoat can cause outgassing, and also lead to plasticizer migration difficulties with the substrate.

B. *First surface metallization—topcoat*

Vaporized aluminium coatings are very thin, and are not abrasion resist-ant; a "topcoat" is therefore applied for protection. A metallized finish is in effect an ultra-thin film of aluminium, sandwiched between two layers of coatings. Clearly, the nature of the latter control the properties of the finished component.

The topcoat must not attack the basecoat nor reduce the general brilliance, and must attain the specified decorative and service requirements, such as resistance to abrasion, humidity, perspiration, ultra violet radiation and heat.

Some defects due to factors other than the topcoat are sometimes not apparent until the topcoat has been applied. Depending on service requirements, topcoats are air drying or baking types.

IV. VACUUM METALLIZING—LACQUERING

Essential stages are:

(1) Application of basecoat by spraying, dipping or flowcoating. After allowing 15–30 minutes for solvent flash-off, follows the process of baking to cure the coating. In view of the low temperature, the period of baking is generally considerably longer than in the normal industrial baking of paints for metals, which can be heated to far higher temperatures. Baking time is also related to oven efficiency.

(2) After baking, the basecoated parts are placed in the vacuum metallizing chamber and the metal, generally aluminium, deposited.

(3) The parts are removed from the chamber, and the topcoat applied, solvents allowed to flash-off, and baked or air dried. As desired, an overlay may be spray applied to the topcoat, either without prebaking, or partially or fully baking out the topcoat, or stoving out fully after application of the overlay, as is most suitable to the particular plant conditions.

V. BAKING TEMPERATURES FOR PLASTICS

Typical temperatures at which some plastics are baked after lacquering without distorting the moulding are:

Polystyrene	140–150 °F (60–65 °C)
Acrylic	150–175 °F (65–80 °C)
ABS	175–195 °F (80–90 °C)
Thermosetting, as phenolics and urea	230–300 °F (110–150 °C)
Polycarbonate, polypropylene, poly-sulphone, PPO	265–300 °F (130–150 °C)

The maximum temperatures depend on the particular grade of plastic used and the individual moulding.

VI. COLOURED TOPCOATS

A clear topcoat gives a bright silver finish. When a coloured effect is desired, the transparent topcoat may be dyed to give many shades, such as gold, copper and brass (Kut, 1967).

VII. SECOND SURFACE—THREE DIMENSIONAL METALLIZING

This can be a complicated process with a multicolour finish, vacuum metallizing frequently in conjunction with painting and hot stamping.

This process is effective only by the use of clear plastic. Acrylics are one of the most commonly used. In light-transmission qualities acrylic is superior to standard plate glass. Acrylics also have excellent light stability, withstanding weather, and are resistant to sunlight in tropical conditions.

Polystyrene is an alternative choice, particularly for appliance components.

VIII. COMMERCIAL APPLICATIONS

A. Metallized automotive components

Typical components vacuum metallized in the automotive industry are: armrest bases; locking rod knobs; steering column bezels; radio hole covers; brand name medallions; transmission selector panels; instrument cluster panels; instrument cluster bezels; instrument cluster escutcheons; seat belt safety switch housings; radio bezels; clock bezels; dome light bezels; air conditioner controls; tail-light reflectors.

B. Non-automotive outlets

In the majority of cases these are metallized mainly for appearance—to increase sales appeal. Some typical examples are: bottle closure—such as for whisky, perfume and cosmetic lotions; lipstick holders, face powder compacts; control panels and nameplates for domestic appliances; novelty items; funeral furniture and fittings; costume jewellery; badges; furniture fittings; television and radio components, such as control buttons; switches and knobs for electrical equipment; toys, plastic and metal; headlights and lamp housings—plastic and metal; bottles—particularly for cosmetics; bicycle light housings; Christmas and display ornaments; clock bezels.

IX. ADHESION OF LACQUERS TO PLASTICS

In order to meet these broad criteria, one of the most critical requirements is to attain good adhesion, with no embrittlement of the lacquer film. Adhesion is generally determined by the standard procedure of cross-hatching; good adhesion is indicated by no removal of any of the lacquer. Embrittlement is generally already observed on cutting the film before taping and in evaluating new products, it is advisable to re-check the adhesion after some days.

X. Mould strains

A not uncommon defect on lacquering plastics is "crazing"—cracking of the moulding. This appears to be related to the effect of the solvents on releasing mould strains. Correct solvent formulation, annealing, and possibly moulding conditions, will correct this defect. In general, with thermoplastics, strain-free mouldings are required to avoid distortion during baking, particularly at higher temperatures.

XI. Non-plastics

Vacuum metallizing lacquers applied to metals, such as zinc die castings or aluminium, and to glass must have good adhesion and meet the performance requirements. They must also not "move" on curing the topcoat or if exposed to heat during service. Such movement leads to distortion of the thin aluminium film—showing up as irridescence (a "rainbow effect") which mars the appearance and is unacceptable.

XII. Mould release compounds

It is highly desirable for satisfactory metallizing that no mould release agents are used. Any such contamination can lead to subsequent problems—surface defects, such as "cissing", and poor adhesion (Kut, 1967).

XIII. Annealing

Injection mouldings, such as acrylics, are generally annealed by heat treatment before decorating, thereby avoiding stress-relieving crazing due to solvents.

Size of the part, and the specific plastic used, will determine the time and temperature, generally between 140–160 °F (60–72 °C), for annealing.

XIV. Etching of "difficult" plastics

Certain "difficult" plastics, such as polyolefines, may require more drastic pretreatment to attain satisfactory adhesion of the coating. For example, polypropylene or polyethylene are pretreated with flame, an acid or a vapour etch, or other priming procedures. Fuller details have previously been described (Kut, 1968, 1969).

A. Polypropylene

Polypropylene, in contrast to other "difficult" plastics, no longer requires such drastic pretreatment, with the availability of a special primer applied direct to the plastic. This is left to air dry for five minutes and is then followed by the appropriate lacquer.

XV. Methods of application

A. Masking

First surface metallized mouldings are today frequently of sophisticated design, in particular, automotive instrument panels. The general procedure is to overall vacuum metallize the mouldings—applying basecoat and top-coat, giving a bright chrome-like finish. For enhanced decorative appearance, and to reduce glare, overlays are then applied by the use of masks over certain selected areas, e.g. satin coatings to give a dull aluminized effect, or coloured paints leaving other areas brightly metallized.

Similarly, second surface decoration is often highly decorative and complex involving selective metallizing over an area, with possibly four, five or more differently coloured paints over the remaining surface, possibly in conjunction with hot stamping.

In order to apply these various adjoining decorative effects, very accurate masking is essential. The best results are obtained with electroformed masks —made for each decorative step—and the excellent decorated parts being produced demonstrate the accuracy and versatility of these masks. In designing the moulding, the need for subsequent masking should be borne in mind, by allowing for "steps" to hold the masks (Kut, 1968, 1969).

B. Application techniques

(i) Spray application of lacquers

Spray application is the most frequently used, and versatile, procedure for the application of lacquers to plastics. Manual spraying with suction or pressure feeding, and manual handling is still widely used, particularly for smaller runs. But even for smaller production, semi-automation is possible, operating the spray gun or masks by foot pedals.

Automatic spraying for vacuum coating lacquers on plastics was initiated in the U.S.A., and is widely used there because of the large runs possible (Kut, 1968, 1969).

(ii) Dipping

Hand or automatically controlled dipping is employed to a very limited extent only for automotive work, and dipping has been mostly displaced by flowcoating.

(iii) *Flowcoating*

Flowcoating essentially involves exposing the components in an enclosed area of space to streams or jets of the lacquer. The parts, mounted on a spider, are rotated. After thorough wetting with the lacquer, rotation is briefly continued in the vapour phase. The spider is then removed, mounted on a rotating rack, and rotation slowly continued for 20–30 minutes, to ensure even flow out of the lacquer. After baking, the topcoat may be applied in a similar manner (Kut, 1967).

(iv) *Topcoating*

This proceeds as for basecoating, followed by oven curing.

XVI. Ovens

Elevated temperatures are employed to oxidize and/or polymerize lacquers to form resistant films. Ovens which are solvent-laden, with uneven temperature areas, do not give good results and are the cause of many possible defects. Ovens should be well ventilated, whether batch or conveyor type, and the essential requirements have been discussed elsewhere (Kut, 1967).

XVII. Radiation curing

A major automobile manufacturer has the first fully operational plant for the radiation curing of coatings on plastics. Typical components coated are plastic radiator grilles and instrument panels.

Coatings on plastics, as for instrument panels, are cured in a matter of seconds. A very large throughput is, however, required in view of the capital expenditure involved for the equipment.

Amongst advantages of the process are speed of curing, in one or two seconds; no temperature sensitivity problems with plastics yet achieving well cured, resistant coatings without heating; minimal pollution with the "solvent free" coatings employed; high rate of throughput; and far less floor space than is required for thermal polymerization.

XVIII. Safety considerations—glare

Increasing attention has been paid to dulling, overpainting or dispensing with vacuum metallized areas on instrument panels, except, for example, on edges, due to safety factors.

Safety requirements being imposed by the U.S. Government specify that certain items of equipment installed inside passenger cars must have a matt finish to avoid dangerous glare. It has been shown that bright trim of the right shape can reduce this unpleasant light reflection. Whilst glare varies with intensity, it also varies with image size and, using a comparator, it has been possible to study the variation of discomfort glare when intensity is cut at constant size compared with variation when the image is reduced at constant intensity (Kut, 1968, 1969).

XIX. Exterior metallizing

Some non-automotive exterior parts have been vacuum metallized to a limited extent for some years, and there is continuous development to attain a high order of weathering resistance.

Electroplating is by its very nature competitive in terms of performance, but lacquers for exterior vacuum metallizing are likely to find their place.

XX. Current position in the automotive field

Overall bright vacuum metallized large components such as instrument panels were, after their initial novelty, not aesthetically fully acceptable. This, combined with the need to reduce the glare, led to increasing proportions of such metallized instrument clusters being overpainted, or coated with a satinized lacquer.

As a result, large components were vacuum metallized overall—using base and topcoat—and then selectively painted, or in part incorporating wood-graining. The areas left as vacuum metallized tended to be edges and instrument inserts.

The natural result has been, in some instances, not to vacuum metallize overall such components, but paint them only, and to highlight bright areas by hot stamping. There is, therefore, a distinct trend away from vacuum metallizing some automotive components. Future variations in designs, exhibiting larger areas of attractively dulled vacuum metallization, could reverse this trend.

XXI. Vacuum metallizing problems

Problems and defects in production can occur which are quite specific to this process. Typical defects have previously been outlined (Kut, 1967) but with greater experience, improved equipment and lacquers, competitive, economic production is today readily achieved with a minimum of rejects.

XXII. Performance requirements

Detailed specifications to which vacuum metallized plastics must conform are laid down by automotive and other manufacturers. Typical requirements —not all in any one specification—are as follows: good, flaw-free decorative appearance; excellent adhesion of the vacuum metallized system to the plastic or other substrate—for first and second surface decorated mouldings; simulated weathering cycles; hot–cold cycle; humidity resistance; salt spray resistance; abrasion resistance; resistance to cleaners, wax, detergents; odour test; fat resistance; resistance to perspiration; environmental spotting tests—to resist spotting with water, soap, acid or other agents; product resistance—for example, closures to resist bottle contents.

The above is an indication of the high standards of performance that can now be met with decorated plastic components. Fuller details have been published in an earlier paper by the author (Kut, 1967).

XXIII. Plastics currently vacuum metallized

Details of plastics currently vacuum metallized have previously been published (Kut, 1967).

REFERENCES

Kut, S. (1967). *Trans. J. Plastics Institute* August, 621.
Kut, S. (1968). *Product Finishing* **21**.
Kut, S. (1969). *Product Finishing* **22**.

Reactive evaporation

R. F. BUNSHAH

University of California, Los Angeles, California, U.S.A.

I. INTRODUCTION

This paper is concerned with the deposition of compounds. To date, much of this work is related to the production of thin films for the optical and semi-conductor industry.

Direct evaporation of compounds from a single vapour source has been used for the deposition of compounds (Holland, 1966). However, this presents two problems. First, partial dissociation of the compound can occur, thus resulting in a deposit of less than full stoichiometry. Thus, direct evaporation of Al_2O_3 results in an oxygen deficient film (Hoffman and Liebowitz, 1971) which can only be restored to full stoichiometry by evaporation in the presence of a partial pressure of oxygen or possibly a post-deposition annealing in oxygen. Second, for high melting point compounds such as Al_2O_3, TiC, etc., a high power density source has to be used to obtain appreciable evaporation rates which often causes an operational problem with the source for long time runs under steady conditions.

Another aspect to be considered is whether low deposition rates (<0.1 μm per minute) or high deposition rates (>0.1 μm per minute) are desired. The latter is often an economic consideration for applications involving thick coatings, self-standing shapes, semi-continuous production, etc. On the other hand, low deposition rates are very adequate for thin film production. The distinction between low and high deposition rates is arbitrary and depends on the application and its economics.

To circumvent the problems associated with direct evaporation of compounds, the development and exploitation of the process of reactive evaporation started (Brinsmaid *et al.*, 1957; Auwarter, 1960). In this process, metal atoms are evaporated from an evaporation source with a partial pressure of a reactive gas present in the chamber. The reaction between the metal atoms and the gas atoms takes on the substrate, in the vapour phase or on the evaporant source surface, to form a compound. For example:

$$2Al(vapour) + \tfrac{3}{2}O_2(gas) \rightleftharpoons Al_2O_3(solid\ deposit)$$

Since it is much easier to evaporate a metal than a compound, reactive evaporation is a preferable process over direct evaporation.

Historically, Soddy (1907) found that calcium vapour reacted with all gases except the inert gases and Langmuir (1913) studied the formation of tungsten nitride by vapour phase reaction between tungsten and nitrogen.

II. Theoretical considerations

The reaction between metal vapour and the reacting gas (e.g., Al and O_2 to form Al_2O_3) or between two elemental vapours (e.g., Zn and S to form ZnS) can take place at one or more of three possible sites—on the substrate, in the vapour phase or on the evaporant surface. The last mentioned site, i.e. the evaporant surface, may not be too important since the compound may well dissociate at the high temperatures present at this site. In all cases, there are three essential factors: (1) An adequate supply of reactants; (2) Collision between the reactant species; (3) Reaction between the colliding reactant species.

The probability of collision between the reactant species depends on the mean free path which is inversely proportional to the partial pressure (Dushman and Lafferty, 1962) and the source–substrate distance. If the mean free path is longer than the source–substrate distance, no collisions between the reacting species will take place in the vapour phase and the reaction will occur only on the substrate. This is typically the case for low partial pressures and hence low deposition rates. For higher deposition rates, one needs a greater supply of the reacting species, i.e. a higher partial pressure and hence a smaller mean free path leading to collisions in the vapour phase as well as on the substrate. Vapour phase reactions are essential for high deposition rates.

The collision between the reacting species does not insure that a reaction will occur between them. The factors influencing the yield of a reaction are much more subtle and undoubtedly include the chemical nature of the reacting species, the stability of the reacting gas or compound species, the free energy of formation of the compound, the dissociation pressure and dissociation temperature of the compound, the temperature of the substrate, condensation behaviour of the reactant species on the substrate which in turn is sometimes influenced by the temperature of the substrate and the partial pressure of the reactants, removal of the reaction products from the reaction interface, etc.

The probability of reaction on collision between the reactants can be enhanced by activating one of the reactants. For example a gas can be activated by a microwave or electrical discharge (Auwarter, 1960; Wank and Winslow, 1968; Kosiki and Khang, 1969; Bunshah and Raghuram, 1972). Again, for high deposition rates, the yield of the reaction must be high and hence it is sometimes mandatory to activate the reacting gas. Such a process is therefore called activated reactive evaporation (Bunshah and Raghuram, 1972).

III. Results and Discussion

A few examples will illustrate some of the concepts discussed above.
Consider the following two reactions for compound formation:

$$2Y + \tfrac{3}{2}O_2 \rightarrow Y_2O_3; \qquad \Delta F = -429{,}000 \text{ cal/mol (102 kJ/mol) at 298 K} \quad (1)$$

$$Ti + \tfrac{1}{2}C_2H_2 \rightarrow TiC + \tfrac{1}{2}H_2;$$
$$\Delta F = -76{,}500 \text{ cal/mol (18·3 kJ/mol) at 298 K} \quad (2)$$

The free energy of formation, ΔF, is more favourable for reaction (1) than for reaction (2). Reaction (2) also has a more complex reactive gas species than reaction (1), i.e. C_2H_2 vs. O_2. Reaction (2) can also be driven in the reverse direction by increase of H_2 partial pressure in the reaction zone, which is why the removal of reaction products from the reaction interface is important. This does not apply to reaction (1).

For all these reasons, one might expect the kinetics or reaction efficiency for the formation of Y_2O_3 by reactive evaporation to be more favourable than that for TiC.

In our experiments (Bunshah and Raghuram, 1972), the synthesis and high rate deposition of Y_2O_3 by reactive evaporation was readily achieved whereas evaporation of Ti in the presence of a reactive gas such as C_2H_2 failed to produce TiC. However, activation of the reaction by ionization of C_2H_2 (Bunshah, 1971) permitted the deposition of TiC at high deposition rates (3 to 12 μm per minute). In fact, the stoichiometry of the deposit can be varied by changing the ratio of the evaporation rate of Ti to the partial pressure of C_2H_2.

In the formation of CdS by reaction between Cd and S evaporated from two sources (deKlerk and Kelly, 1965), it was found that sulphur would deposit on the substrate only at temperatures below 50 °C whereas cadmium would deposit only at temperatures above 200 °C. The former is due to re-evaporation of the sulphur from the substrate as its temperature increases and the latter is due to a critical beam density requirement (Holland, 1966). At temperatures between 50 and 200 °C it was found that neither Cd nor S would deposit from the vapour unless the other was present to combine with, forming CdS on the substrate surface.

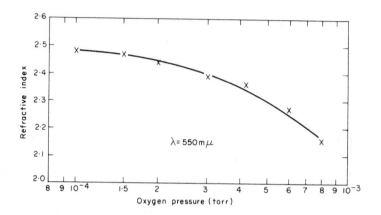

FIG. 1. The dependence of refractive index on partial pressure of oxygen (Ritter, 1966).

Another example is the formation of TiO_2 by reaction between TiO and O_2 (Ritter, 1966). The oxygen concentration should be high enough to allow a useful reaction rate; however, the oxygen concentration is limited by the pressure of the gas. If, in the gas phase, the evaporated molecules suffer too many collisions with the oxygen molecules, the refractive index and the hardness of the films decrease. Figure 1 shows the dependence of the refractive index of TiO_2 films, prepared by evaporation of TiO in an oxygen atmosphere of 10^{-4}–10^{-3} torr (13·3–133 mPa), on the oxygen pressure. The source-to-substrate distance is 40 cm and the substrate temperature is 300 °C. If the evaporation is carried out at 1×10^{-4} torr (13·3 mPa), complete oxidation cannot occur in the gas phase, because the chance for the necessary gas phase collisions is not high enough. On the other hand, there are about 3×10^{16} oxygen molecules impinging on 1 cm² of the substrate in 1 sec, which means, in the case of medium evaporation rates, an appreciable excess of oxygen for one condensing atom or molecule (namely, a factor of 30 in the example given above, of TiO evaporation). This makes complete reaction possible, as shown by the fact that the refractive index has the highest value of 2·48 at 1×10^{-4} torr (13·3 mPa) and falls with increasing oxygen pressure.

A literature search revealed that thin films of the following were produced at low deposition rates (<0.2 μm per minute) by reactive evaporation:

—oxides of Ca, Ti, Si, Al, Zn, Sn, Zr, Cu (Holland, 1966; Hoffman and Leibowitz, 1971; Brinsmaid et al., 1957; Auwarter, 1960; Soddy, 1907; Langmuir, 1913; Dushman and Lafferty, 1962)

—nitrides of W, Nb, Ta, Al (Hoffman and Leibowitz, 1971; Rairden, 1968; Rairden, 1969)

—β-silicon carbide (Learn and Haq, 1970)

—sulphides of Cd and Zn (deKlerk and Kelly, 1965)

Examples of the activated reactive evaporation process using an electrical or microwave discharge to activate the vapour species and thus increase the reaction yield are listed below. In all these cases, the activation was carried out OUTSIDE the reaction zone and produced low deposition rates (<0.2 μm per minute). They are:

—oxides of Si, Zn, Al, Ti, Zn, Sn (Brinsmaid et al., 1957)

—nitrides of Al and Ga (Wank and Winslow, 1968; Kosicki and Khang, 1969)

At the other end of the spectrum, Bunshah and Raghuram (1972) report the high rate deposition (3 to 12 μm per minute) of Y_2O_3, TiN, TiC, VC, ZrC, HfC, NbC and TaC. In all cases, the metal was evaporated at a high rate in the presence of a partial pressure in the range of $2-8 \times 10^{-4}$ torr (27–106 mPa) of O_2, N_2 or C_2H_2. In all cases except Y_2O_3, the high deposition rate was obtained only by activation of the gas species in the reaction zone, as described in the following section and shown in Fig. 2. Y_2O_3 could be deposited with or without the activation.

The adhesion of the film to the substrate is good. This observation is based on the fact that none of the authors reported any adhesion problems. This is not surprising since the substrate is usually thoroughly cleaned and outgassed prior to deposition. Furthermore, the process of vapour deposition produces a perfect mechanical locking between the deposit and the surface grooves and the high substrate temperature produces a diffusion zone in many cases. This is shown in Fig. 3 for a TiC deposit on stainless steel (Bunshah and Raghuram, 1972).

A very important advantage of this process is that it permits the use of lower substrate temperatures than would be the case with competing processes such as chemical vapour deposition. The substrate temperature in turn governs the grain size and other properties of the deposit. Thus it is possible to produce very fine grained deposits as shown in Fig. 3 by keeping the

substrate temperature low. The low substrate temperature also permits one
a greater latitude in selection of substrate material.

IV. Experimental procedure

The experimental arrangements for reactive evaporation are very similar to
those for normal evaporation processes. The process is carried out in a

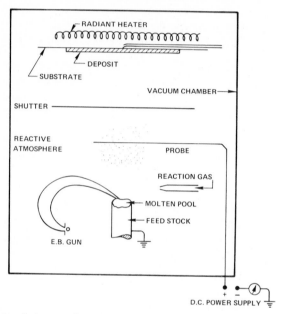

Fig. 2. Schematic of the experimental arrangement for the activated reactive evaporation
process.

vacuum chamber using one or more resistance, induction or electron beam
heated sources. The substrate is located at some distance above the source
and its temperature is controlled by a heater located behind it. Rate monitors
are often present. A source of gas with a variable leak valve provides the
reacting gas when needed. The gas can be activated by a microwave discharge
or an electrical discharge either in the reaction zone, i.e. the space between
the source and the substrate, or outside the reaction zone. The former is
essential for high deposition rates, as otherwise the ionized gas may recom-
bine to neutral molecules before it has had a chance to react. Figure 2 shows
such an experimental arrangement for the deposition of carbides (Bunshah
and Raghuram, 1972; Bunshah, 1971) at high deposition rates. The 1 inch

(2·5 cm) diameter evaporation source is electron beam heated to give an evaporation rate of 0·66 g/min (11 mg/s) and the gas pressure is 5×10^{-4} torr (67 mPa). A probe biased to 100–200 V attracts the electrons from the pool which ionize the gas thus causing a high reaction yield and a high deposition rate of 3 to 12 μm per minute at source to substrate distances of 24 to 15 cm respectively.

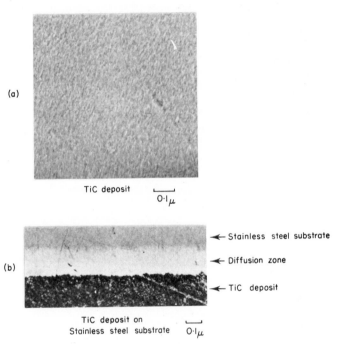

(a)

TiC deposit   0·1 μ

(b)

← Stainless steel substrate

← Diffusion zone

← TiC deposit

TiC deposit on
Stainless steel substrate   0·1 μ

FIG. 3. (a) Cross-section of TiC deposit; (b) cross-section showing diffusion zone between substrate and deposit.

V. CONCLUSIONS

The reactive evaporation process is a well developed process for the deposition of compounds at low and high deposition rates. It permits one to use low substrate temperatures. In fact the substrate temperature can be independently varied if one adjusts the experimental conditions so that the reaction occurs in the vapour phase. Adhesion between the substrate and deposit is good. It is predicted that this process will be very useful in the synthesis of new materials of desirable optical, electrical and magnetic properties as well as for improved abrasion and corrosion resistant coatings.

REFERENCES

Auwarter, M. (1960). U.S. Patent 2 920 002.

Brinsmaid, S., Keenan, W. J., Koch, G. E., and Parsons, W. F. (1957). U.S. Patent 2 784 115.

Bunshah, R. F. Patent Application.

Bunshah, R. F. and Raghuram, A. C. (1972). *J. Vac. Sci. Tech.* **9**, 1389.

deKlerk, J. and Kelly, E. F. (1965). *Rev. Sci. Inst.* **36**, 506.

Dushman, S. and Lafferty, J. (1962). "Scientific Foundations of Vacuum Techniques". John Wiley and Sons, New York.

Hoffman, D. and Liebowitz, D. (1971). *J. Vac. Sci. Tech.* **8**, 107.

Holland, L. (1966). "Vacuum Deposition of Thin Films". Chapman & Hall, London.

Kosicki, B. B. and Khang, D. (1969). *J. Vac. Sci. Tech.* **6**, 592.

Langmuir, I. (1913). *J. Am. Chem. Soc.* **35**, 931.

Learn, A. J. and Haq, K. E. (1970). *Appl. Phys. Letters* **17**, 26.

Rairden, J. R. (1968). *Electrochem. Tech.* **6**, 269.

Rairden, J. R. (1969) "Thin Film Dielectrics" (F. Vratny, ed.), p. 279. Electrochemical Society, New York.

Ritter, E. (1966). *J. Vac. Sci. Tech.* **3**, 225.

Soddy, F. (1907). *Proc. Roy. Soc. Lond.* **78**, 429.

Wank, M. T. and Winslow, D. K. (1968). *Appl. Phys. Letters* **13**, 286.

The basic principles of sputter deposition

L. HOLLAND

Physics Dept., Brunel University, Uxbridge, Middx,, England

I. INTRODUCTION

When the gas pressure in an envelope is reduced to a fraction of a torr (1 torr = 133 Pa) and a cold cathode discharge is excited between metal electrodes inserted in the envelope, the cathode electrode begins to emit atoms and atomic clusters, i.e. "sputter". Cathode material is ejected by the bombardment of positive ions and energetic particles in the electrical discharge and the liberated substance is condensed on the envelope walls or in a coating system on adjacent receivers. A metal deposit can form rapidly when the gas pressure is sufficiently reduced for the applied voltage exciting the glow discharge to be 500 V or above, i.e. the glow discharge is in the *abnormal regime*. This "critical" voltage is an apparent effect because the onset of sputtering can be detected (in the absence of collisions between gas molecules and target atoms) at ion energies down to a few electronvolts ($1\,\text{eV} = 1\cdot 6 \times 10^{-19}$ J). When the gas pressure is sufficiently high the liberated particles, although initially energetic, can lose energy by collision with gas molecules and cool to the gas temperature. They then diffuse in the gas with the probability of being returned by collisions to the cathode.

Sputtering occurs in glow discharge lamps operated in the 1–10 torr ($0\cdot 13$–$1\cdot 3$ kPa) pressure region but the resistance offered by the gas to particle flow reduces the deposition rate on the envelope to a negligible quantity. This led to the erroneous belief that sputtering did not occur in the *normal regime* of a glow discharge. One can, however, observe a sputtered film on neon lamps operated in this regime at domestic mains voltage after extended use.

Film deposition from cathode sputtering is a phenomenon which accompanies the passage of a high voltage discharge between electrodes in a gas at about 10^{-1} torr (13 Pa) or less, with film growth rates of a few atoms per second. Consequently as soon as studies of electrical discharges in low pressure gases were commenced, sputtering was observed and turned to practical use for film formation. The ease with which many metals can be sputtered to form thin films made early (and some contemporary) workers neglect to apply existing knowledge of particle emission and transport to the design of deposition systems. The situation was not simplified by the controversy, which existed until fairly recently, on the cause or causes of sputtering. Development of our understanding of the phenomenon (and our ability to design sputtering systems to obtain desired emission and condensation conditions) is related to the technological ability to provide ion beams and vacuum systems. With these targets can be sputtered at sufficiently low pressures that the residual gas molecules influence neither the target emission nor the film growth by adsorption or transit collisions.

At 10^{-6} torr (133 µPa) the impingement rate of air molecules per square metre is about equal to that of positive ions with unit electron charge and a current density of 1 A/m^2. If the gas molecules were chemisorbed on a target on first impact and the sputtering yield of the surface gas was one molecule per impacting ion, no sputtering of the base target would occur. Generally, sticking coefficients for active gases are less than one, but reduction in the target yield occurs sharply for active gas pressures $\geqslant 10^{-4}$ torr (13 mPa). When the gas pressure in the target vessel is sufficiently raised for the sputtered particles to have a mean free path less than their transit distance [generally about 10^{-2} torr ($1\cdot 3$ Pa) for target/receiver distances of 50 mm], then gaseous diffusion as well as adsorption occurs.

We are concerned here with the application of ion impact sputtering to thin film deposition, but ion/target impact produces several other phenomena which have practical application and brief mention of some will help to enhance comprehension of target and receiver effects which can occur during thin film growth. These other uses are: structural and profile etching of target surfaces; target analysis by mass spectrometry of secondary positive and negative target ions; and the gettering and ion pumping of gases by sputtered deposits.

A wide range of substances can be sputtered to form coatings of metals, alloys, semiconductors, metal oxides, some sulphides, carbides, and complex mixtures of inorganic substances such as cermets. Techniques used in their preparation are as follows:

(i) When sputtering targets made of alloys and mixtures, there is an induction period during which the concentrations of the surface components adjust themselves so that the components are emitted in proportion to their bulk content. Their reformation as a film depends, for each component, on: the transport rate; loss, if volatile, by pumping; condensation coefficient and mobility at the substrate temperature. Compounds may dissociate or sputter in the form of atomic clusters and molecules.

(ii) Films of mixed substances can be prepared by sputtering simultaneously from several targets on to a single receiver or multilayers prepared by sputtering the targets in sequence.

(iii) Active gas may be added to the deposition atmosphere to react with a growing metal film to form a compound or restore a gas component partially lost from a target by pumping—the process being termed *reactive sputtering.*

(iv) Generally organic compounds release atoms and fragmented species when ion bombarded, but some, e.g. PTFE (polytetrafluoroethylene), form condensates of low molecular weight material released from the polymer.

Many types of deposition systems have been devised for preparing the foregoing range of thin film materials on different kinds and shapes of substrate. Also, attention has been given to the sputtering of dielectrics and the prevention of film contamination by impurity gases. However, it would clearly be beyond the scope of this paper to review all these topics, therefore an attempt has been made to analyse the conditions existing in typical sputtering systems so that the dependence of film structure and properties on the mode of deposition can be better appreciated. It will be shown that the conditions of film growth can vary considerably with the design of the sputtering system. We shall commence by considering how target material is released and then discuss briefly some generalized forms of sputtering system.

II. Sputtering

A. Chemical and physical sputtering

Target material can be released by reacting with active components, in an ionized gas, which form volatile compounds e.g. metal hydrides or halides. Such compounds may be unstable and dissociate outside of the active discharge zone to deposit target material. The process has been termed *chemical sputtering*. However, the principal source of target emission by energetic ion impact, even when such active gas is present, is *physical sputtering* which arises from energy released in the target.

B. Target effects

Bombardment of a solid by energetic ions can result in:

  (i) Sputtering of energetic neutral target species with small amounts of secondary positive ions ($\sim 1\%$) and negative ions ($\sim 10^{-3}\%$),
  (ii) structural damage,
  (iii) ion implantation,
  (iv) back-scattering of energetic ions and neutral atoms, and
  (v) the emission of secondary electrons and electromagnetic radiation.

Most of these effects are of importance when sputtering is used for film deposition as the sputtering yield (atoms/ion) may change as the target surface is damaged and implanted with gas. Also the back-scattered and emitted species may reach the substrate with the sputtered particles and influence the film growth.

C. Energy exchange

An incident ion imparting energy to a solid can eject atoms out of the structure in two ways: it can, if the conditions are appropriate, knock an atom directly out of the solid or it can set in motion a collision sequence amongst the atoms, which leads to atom ejection. The mechanism which is dominant depends on: the mass, energy, and related cross section and incidence angle of the ion, and the mass and structural arrangement of the target atoms. If the target is a single crystal, the impacting ions may set up collision cascades in crystallographic directions of greatest atomic density. Wehner (1955, 1956) was the first to observe the formation of sputtered deposits showing patterns which arose from nearest neighbour collisions. Depending on the cross sections of and degree of separation between target atoms, a primary collision at an angle to an atom row can be *focused* along the row. However, the impinging ions can damage the lattice and form interstitials which, before complete randomization of emission, can modify the emission to make it appear of different crystal origin (Anderson and Wehner, 1960). Thompson (1959), using a gold foil target with preferred orientation of the crystallites, found that protons in the MeV range, with sufficient energy to sputter Au atoms from the far side of the foil, ejected the atoms preferentially in the close packed directions.

Localized collision sequences can occur in a target which have been descriptively termed "thermal spikes". These are regions of decaying collision sequence which arrive at the surface with weak energy and only release target atoms if their sublimation energy is low or the *bulk* temperature has been raised near to the sublimation value.

D. Ion energy, mass, and cross section

The number of collisions an ion of energy E_I makes in a given target depth depends on its collision cross section σ so that the sputtering yield S (atoms/ion) is approximately given by

$$S = K\sigma \frac{m_I m_T}{(m_I + m_T)^2} E_I$$

where m_I and m_T are the masses of the ions and target atoms respectively and K is a proportionality constant. There is a limited target depth from which recoil collisions can reach the surface with sufficient energy to eject atoms and this depth is about 10 nm. When the ions have enough energy to penetrate the target past the critical recoil zone, the sputtering rate attains a maximum value with rise in ion energy. The collision cross section of an ion decreases as its energy rises and fewer collisions occur in the critical depth but more energy is transferred on each impact and S tends to a maximum value.

The greater the ion mass the larger will be its cross section and more collisions will occur in a critical recoil depth, and if m_I approaches m_T in value more energy will be transferred on each collision. Thus when using Ar^+ ions with a mass of 40 a.m.u. (atomic mass units; 1 a.m.u. $= 1.7 \times 10^{-27}$ kg), S typically reaches a maximum for energies above 30 keV but with Hg^+ ions their great mass (200 a.m.u. or 3.32×10^{-25} kg) helps to limit their penetration to the recoil zone effective for sputtering and S reaches a maximum only in the 100 keV region. For sputtering efficiency in thin film deposition systems, one should use heavy inert gas ions, e.g. argon, with energies in the 1 to 10 keV range raising if necessary the ion current density to increase the sputtering rate.

E. Collision depth and ion incidence

As stated above, target atoms are released either by a direct collision between an impinging ion and a surface atom or by penetration of the ion in the solid, producing internal collisions with recoils which reach the surface. Release of material will be enhanced if the ion incidence angle to the target surface normal is raised, because more energy will be transferred to the surface layer within which cascade collisions can reach the surface with sufficient energy for sputtering. The depth t of this layer is usually about 10 nm and the sputtering yield S (atoms/ion) will tend to be proportional to $t/\cos \theta$,

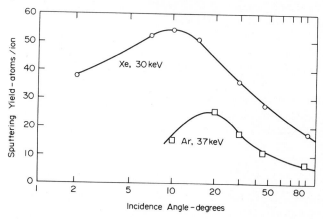

FIG. 1. Sputtering yield of copper as a function of Xe^+ and Ar^+ ion glancing incidence angles $(90°-\theta)$ showing effect of ion mass. After Cheney and Pitkin (1965).

where θ is the ion incidence angle (to the surface normal). However, as ions are reflected at high incidence by surface electric fields and can be neutralized by extracting electrons from the surface without energy exchange, S reaches a maximum at $\theta \simeq 60$–$70°$.

As the energy and mass of an ion are raised, both the value of S and that of θ at which S_{max} occurs are increased because the more energetic ions are not so easily reflected and the surface penetration is less for heavy ions.

Cheyney and Pitkin (1965) have reported the yield for copper bombarded by 37 keV Ar^+ and 30 keV Xe^+ as a function of the glancing incidence angle 90–θ as shown in Fig. 1. Bach (1968, 1970) has given values of S (molecules per ion) as a function of θ for a number of glasses and dielectrics bombarded by argon ions at 5·6 keV; silica and glasses are polymeric and must emit atoms or polymer fragments.

F. Angle of preferred emission

As the ion incidence angle θ to the target normal is increased, the emission may show a maximum yield in a forward direction partly related to the incidence and passage of the ions. A useful study of the phenomenon has been made by Weijsenfeld (1966). Betz *et al.* (1970) used different gas ions

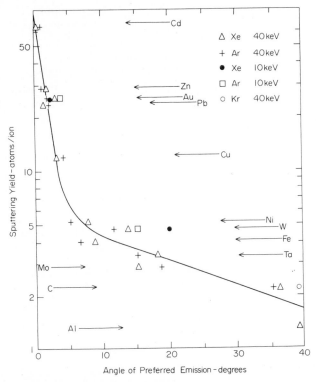

FIG. 2. Angle of preferred emission for ion incidence of 60° to the target normal measured by Betz *et al.* (1970) and plotted by them as a function of the total sputtering yield S reported by Almén and Bruce (1961).

incident at $\theta = 60$ ° on several metals and measured the angle of preferred emission (to the surface normal). Their results are plotted in Fig. 2 against the total yield values S measured by Almén and Bruce (1961) for 45 keV Kr^{84} at normal incidence bombarding the same metals. They found that:

(i) the lower the sputtering yield the more forward emission was preferred, and

(ii) this was independent of the mass of the Ar^+ and Xe^+ ions used and of their energy in the 10 to 40 keV range.

Thus as a simple generalization one can state that thermally volatile materials when sputtered tend to emit atoms with negligible relation to the ion incidence angle. This is because collision cascades with a large number of interacting target atoms can still liberate sufficient energy at the surface to release atoms. (Many early workers studying the sputtering of volatile metals such as phosphorus, cadmium, zinc etc., actually observed thermal evaporation due to bombardment raising the target temperature.) Obviously the greater the scope of the collision cascade, the less dependence momentum exchange at the surface has on the direction of the incident ion. Energy liberation under these conditions tends to approach locally the random atom agitation otherwise produced in bulk by thermal means.

G. Energy spread from sputtered species

As sputtering arises from either direct ion/target atom collisions or target collision sequences, the released particles can have energies well in excess of

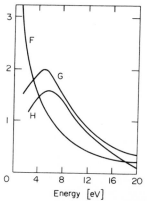

FIG. 3. Energy spectra of neutral sputtered copper collated by Macdonald *et al.* (1970). F— Farmery and Thompson (1968); G—Stuart and Wehner (1964); H—Macdonald *et al.* (1970), 10 keV Ar^+.

thermal values for comparable evaporation rates. The energy of incident ions when accelerated by applied voltages of 1 kV or more corresponds to a temperature of several million degrees centigrade (as kinetic energy = $\frac{3}{2}kT$, then 1 eV is equivalent to a temperature of 7740 K).

Many measurements have been made of the dependence of the energy spectra of sputtered particles on the energy of incident ions. Shown in Fig. 3

is the spectrum obtained by Macdonald *et al.* (1970) for 10 keV Ar$^+$ ions ejecting neutral material from polycrystalline Cu. Macdonald *et al.* have also plotted the results obtained by other workers for neutral sputtered copper, and the emitted particles have a mean energy of ~10 eV. Similar results are obtained for the energy spread of secondary positive ions ejected from a target by ion impact.

Atoms released from thermal spike regions have an energy of typically ~1 eV, i.e. above that of the bulk target temperature but lower than that from primary encounters and focused collisions.

H. Atom and cluster emission

The ejected material may be in atomic or cluster form depending on the nature of the target and the ion/target interaction. Woodyard and Burleigh Cooper (1964) sputtered polycrystalline copper with 0–100 eV ions extracted from a low pressure magnetically confined argon discharge. The sputtering source resembled a "triode" system (discussed below) as used for film deposition. The neutral sputtered atoms entered, via an orifice, the electron impact (Nier type) ion source of a mass spectrometer.

For 100 eV Ar$^+$ ions, the Cu_2:Cu atomic ratio was 0·05:1 with Cu appearing with a target voltage of -19 V and Cu_2 with -50 V; Cu_3 was not detected. However, Stirling and Woodward (1970, 1971) give evidence that appreciable emission of clusters can occur when sputtering metals in plasmas, but the writer believes that these contradictory results arose because impurity gas in the plasma reacted with the target, forming compounds which sputtered with a high yield of molecules.

I. Summary of effects

From the foregoing discussion of the sputtering mechanism several things have emerged of importance when film deposition is being considered:

  (i) sputtered particles have higher energies than those of thermally evaporated substances and their mass distribution for a given material may differ from that for vacuum evaporation;

  (ii) the target particle emission may tend to follow a cosine law or be preferred depending on the ion incidence angle and energy and the target sublimation energy and atomic structure;

  (iii) the sputtering yield is enhanced by increasing the ion incidence angle; and

  (iv) a neutralized ion may rebound from a target with appreciable energy because of incomplete energy accommodation.

When thin films are prepared by vacuum evaporation we must, for a given substrate and growth temperature, consider the influence on film structure and composition of the condensation rate, vapour incidence and gas sorption. However, in sputtering these factors and several more related to the emission mechanism are operative during film growth.

III. Basic types of sputtering system

A. Diode d.c. and a.c. systems

The simplest form of sputtering system (but by no means the simplest to analyse!) is the cold cathode diode arrangement of which several forms are shown in Fig. 4. Unwanted glow discharges can be prevented from parts of the cathode and lead-in electrode surfaces by placing anode shields so that secondary electrons are trapped before ionizing sufficient gas molecules to sustain the discharge. The electrode spacing to achieve this must be smaller than the Crookes or cathode dark space (c.d.s.) as in Fig. 4. As the receiver must be placed just outside of the c.d.s. in order not to obstruct the discharge passage, the c.d.s. depth d is roughly that of the cathode receiver gap.

Diode systems may be made using alternating h.t. supplies and an a.c. planar type with an electrode spacing less than d for obtaining a discharge is shown in Fig. 4 for coating two opposed substrates. Arrangements have been devised for coating both surfaces of a plane receiver using opposed target electrodes, each with an earth backing shield which obstructs the electrons promoting ionization.

Wire cathodes give an enhanced sputtering yield because a high proportion of the ions arrive obliquely to the wire surface. Also, when sputtering in a low pressure gas, the deposition efficiency for a wire is greater than that of a massive target when the wire diameter ($2r$) is commensurate with or smaller than the mean free path L of a liberated particle in the gas (see Fetz, 1942). Elevated temperatures can be reached with wire cathodes and thermal evaporation may occur. Wire targets are not commonly used because, although efficient in terms of deposition yield, their ratio of thermal flux to deposition rate at the receiver can be higher than that of a water-cooled target for the same deposition rate. Also, it is not easy to provide uniform films with wire cathodes, because the localized target can produce undulations in the film thickness distribution.

Sputtered films may absorb gases during growth either by implantation of energetic atoms from the glow discharge or by chemical reactions. Chemisorption by the deposited coating of impurities in the inlet gas can be reduced by arranging the gas to flow through a sputtering region before reaching the

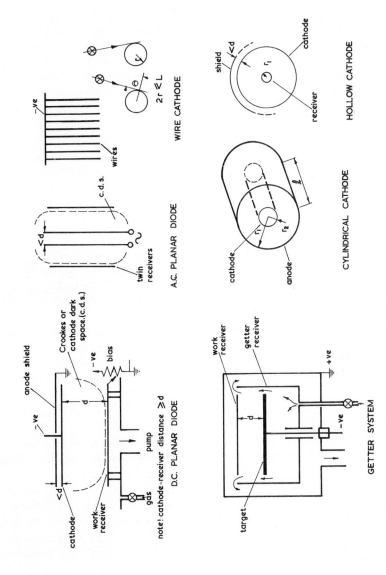

Fig. 4. Cold cathode glow discharge sputtering systems.

deposition zone (Fig. 4); for a discussion of getter sputtering see Berghaus and Burkhardt (1943), Theuerer and Hauser (1965) and Holland and Priestland (1972).

Getter sputtering can be used for depositing metal films with a minimum content of active gas when using conventional high vacuum apparatus and pumps. However without analysis of the gas in the deposition zone it is impossible to know the degree to which precleaning by gettering has been effective. Gas clean-up undoubtedly occurred fortuitously in some early sputtering plants, because internal shields fitted to keep the vessel free from deposits could make the inlet gas flow past a sputtered film which gettered the impurities.

Alternatively one may bombard the receiver with electrons or ions (bias sputtering) to reduce gas sorption during growth, but this can promote other forms of contamination such as ion implantation or degradation by electrons impinging on adsorbed organic molecules, e.g. from pump vapours.

B. Crossed-field systems

One can increase the path of an electron, and thereby reduce the operating pressure of a cold cathode discharge, by making the electron follow a curved path in crossed electrical and magnetic fields. Also the electron can be made to oscillate by directing the accelerated particle into a retarding electrical field. This is the basis of the Penning vacuum ionization gauge which Penning (1936) proposed could also be used for thin film sputtering. The writer (1954) modified, as shown in Fig. 5(c), the well known opposing disc cathode/ring anode form of the Penning system to prevent undesirable bombardment of the receiver by charged particles. The modified electrodes operated at a gas pressure of $\sim 10^{-3}$ torr (0·13 Pa) so that the sputtered species were able to escape freely from the target. Gill and Kay (1965) have described an inverted magnetron form of the Penning discharge (Fig. 5(b)).

The degree of ionization of the residual gas can be enhanced by elongating the electron paths in the low pressure region. A quadrupole magnetic field was used by Kay (1963) and Kay et al. (1967), with a planar diode system to obtain enhanced ionization. Mullaly (1971) has modified this geometry using a concave target and an anode ring, as shown in Fig. 5(a). The cusp magnetic field is produced by passing current in opposing directions through the exciting coils. A typical arrangement using a coil with a magnetic flux density of 200 gauss (16 kA/m), an applied voltage of 0·5 kV and operating at a gas pressure of 5×10^{-3} torr (0·65 Pa) gives a total current of about 5 A for a receiver of 0·15 m diam. Uniform films with rapid growth rates are possible.

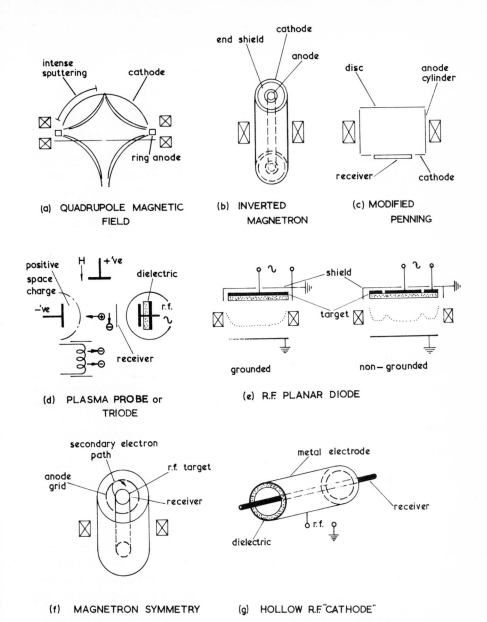

FIG. 5. High vacuum and plasma probe sputtering systems. To simplify drawing, the vacuum vessel, pump and gas inlet connections have been omitted from each system. Usually the magnetic field is provided by an external coil and is between 200–700 gauss. (*a*) Quadrupole Magnetic Field, after Mullaly (1971) see Kay (1963), Kay *et al.* (1967); (*b*) Inverted Magnetron, Gill and Kay (1965); (*c*) Modified Penning, Holland (1954) (see Penning, 1936); (*d*) Plasma Probe, d.c. probe—Wehner (1955); r.f. probe—Anderson *et al.* (1962); (*e*) R.F. Planar Diode, Davidse and Maissel (1965)—Grounded, Holland *et al.* (1968)—Non-grounded; (*f*) R.F. Magnetron Symmetry, Rivlin (1971); (*g*) Hollow R.F. "Cathode", Davidse and Whitaker (1971).

C. Plasma probe or triode

One can produce a plasma by electron impact using a hot cathode emitter as in Fig. 5(*d*) or ionize a gas with an r.f. electric field. Positive ions can be extracted from the plasma to a target probe and this technique was used by Wehner (1955) for studying sputtering by Hg$^+$ ions at reduced gas pressure. The development of r.f. sputtering, as discussed below, first for use with dielectric targets and later as a general deposition tool irrespective of target electrical conductivity, has tended to make plasma probe or triode systems of less interest; particularly as it is difficult with the latter to provide uniformly bombarded targets with dimensions above about 100 mm square.

D. R.F. sputtering

When a body is immersed in a plasma and insulated from ground, it quickly acquires a negative charge from the electrons which bombard its surfaces. The less mobile positive ions flow to the body forming a positive space charge zone under the influence of the field created by the accumulated negative charge on the body. At equilibrium an equal number of electrons and positive ions will flow in a given time to the body to restore those removed from the surface by recombination. The surface of bodies can be cleaned by weak sputtering which results from immersing them in the plasma of a glow discharge (positive column) or that formed by a radio frequency electric field. Usually under these conditions the positive sheath voltage is only about 20–30 V.

A plasma can be excited by an r.f. field giving an ion density comparable to that of a cold cathode discharge but at a tenth of the gas pressure. Also a dielectric body can be more energetically bombarded if instead of immersion in the plasma it is backed by one of the r.f. electrodes used to excite the plasma. A metal target can of course be ion bombarded in a d.c. system, but if used with an r.f. supply capacitively coupled to the target, the ion bombardment can be achieved at a lower pressure. Thus r.f. sputtering has proved useful for sputtering a range of substances regardless of their individual electrical conductivity and systems have been built containing several targets for depositing multilayers of metals, semiconductors and dielectrics in chosen sequences. With a target capacitance of 10^{-7} F/m^2 and a positive ion current density of 10 A/m^2, the target would acquire a surface voltage by ion neutralization at a rate of 10^8 V/sec if the ion current flow was constant. If the electric field changes sign in a time interval of about 10^{-7} sec (e.g. at a frequency of 13·56 MHz) and the peak applied voltage was of the order 1 kV, then a previously acquired negative charge (during a positive field half cycle) would not be greatly diminished. In fact the target surface can retain a negative potential over most of the r.f. cycle and the positive sheath which forms appears similar to that of a cold cathode discharge but the gas pressure is 10 to 100 times lower.

Although r.f. sputtering was observed some 40 years ago, Anderson *et al.* (1962) were the first to propose a system which could be used for deposition of sputtered dielectric materials. They employed a plasma probe system as described above with an r.f. potential applied to the backing electrode of the probe. Obviously one may use the r.f. supply to both generate and extract the positive ions. Davidse and Maissel (1965) achieved this with a planar target, shown in Fig. 5(*e*), grounding the remaining electrode. If one grounds the r.f. electrode, sputtering of earthed fixtures is negligible if there is not an accumulated negative charge to create a field for positive ion flow. Also a large area of earthed metal fixtures will reduce the ion current density J and this is related to the sheath voltage V by $J \propto V^{3/2}$. However, insulating oxide films on grounded fixtures and insulating substrates resting on grounded receivers can accumulate a negative charge under an applied r.f. potential similar to that of the target. For these reasons and to localize the plasma the writer and his co-workers developed the non-grounded system (Holland *et al.*, 1968) shown in Fig. 5(*e*).

A characteristic of r.f. sputtering systems is the intense electron bombardment of the receiver which can arise from secondary electrons released from the target being accelerated across the positive sheath of the target; Holland *et al.* (1968) reported that electron bombardment could be reduced by arranging the magnetic field to bend the electron paths. Rivlin (1971) has described a magnetron symmetry (Fig. 5(*f*)) in which the magnetic field bends the paths of the secondary electrons so that they are captured on a grid. With this system he has been able to deposit films on to plastics which are easily degraded by electron bombardment.

If one applies an r.f. potential to a metal cyclinder containing a dielectric target, also in the shape of a cylinder, a hollow r.f. "cathode" discharge, Fig. 5(*g*), can be created for sputtering on to an axial receiver (Davidse and Whitaker, 1971). Of course one may use a metal target cylinder capacitively coupled to the r.f. supply and obtain metal sputtering at reduced pressures. This type of discharge is the same as that obtained in an r.f. ion source using a glass tubular envelope with external electrodes.

E. Deposition conditions

The foregoing sputtering systems have been classified in accordance with their mode of ion production and extraction, but to appreciate how film structure and composition is influenced by the deposition conditions, one must classify the systems in terms of the nature of the sputtered particle flow and the type and energy of the particles bombarding the target and the receiver.

Generally when the product of the target/receiver distance d and the gas pressure p is less than 10^{-2} mm torr (1·3 mPa m), sputtered particles flow to the receiver without losing appreciable energy in the gas or being deflected by collision. Such energetic particles (~ 10 eV) can be implanted in the substrate surface forming adherent deposits. However, neutralized ions can rebound from the target with appreciable energy and if these and accelerated secondary electrons impinge on the receiver they can influence the structure and purity of the condensing deposit.

F. Ion beam system

Gas molecules impinge on both target and receiver surfaces in all systems and to reduce the rate of formation of contaminant compounds the pressure of active gas (or partial pressure of active gas in an inert atmosphere) must be in the high, or ultra high, vacuum region. For this reason it is desirable to use a partial pressure gauge to determine the vacuum or low pressure conditions. For basic research studies on the sputtering mechanism it is common to use an ion beam system in which positive ions are formed in a cell (e.g. r.f. gas ion source) and after extraction and acceleration made to pass as an energetic beam into a target vessel. Monoenergetic beams of a selected ion mass and charge are normally required. Interest is growing in the use of such systems for film deposition as both the target and receiver can be kept under high vacuum conditions. For coating purposes one can employ a multitarget head, and close control of the ion beam energy and charge distribution is usually not essential. For a further discussion of this subject see Gaydou (1967), Chopra and Randlett (1967), and Weissmantel et al. (1972).

REFERENCES

Anderson, G. S., Mayer, Wm. N. and Wehner, G. K. (1962). *J. Appl. Phys.* **33**, 2991–2992.
Anderson, G. S. and Wehner, G. K. (1960). *J. Appl. Phys.* **31**, 2305–2313.
Almén, O. and Bruce, G. (1961). *Nucl. Inst. Methods* **11**, 257–278.
Bach, H. (1968). *Naturwissenschaften* **55**, 439–440.
Bach, H. (1970). *Non-Crystalline Solids* **3**, 1–32.
Berghaus, B. and Burkhardt, W. (1943). German Pat. 736 130.
Betz, G., Dobrozemsky, R. and Viehböck, F. P. (1970). *Nederlands Tijdschrift V. Vacuümtechniek* **8**, 203–206.

Cheney, K. B. and Pitkin, E. T. (1965). *J. Appl. Phys.* **36**, 3542–3544.
Chopra, K. L. and Randlett, M. R. (1967). *Rev. Sci. Instrum.* **38**, 1147–1151.
Davidse, P. D. and Maissel, L. I. (1965). "Third Inter. Vacuum Congress, Stuttgart", Vol. 2, Part III, 651–655.
Davidse, P. D. and Whitaker, H. L. (1971). Brit. Pat. 1 242 492.
Farmery, B. and Thompson, M. W. (1968). *Phil. Mag.* **18**, 415–424.
Fetz, H. (1942). *Z. Phys.* **119**, 590–595.
Gill, W. D. and Kay, E. (1965). *Rev. Sci. Instrum.* **36**, 277–282.
Gaydou, F. P. (1967). *Vacuum* **17**, 325–27.
Holland, L. (1954). Brit. Pat. 736 512.
Holland, L., Putner, T. I. and Jackson, G. N. (1968). *J. Phys. E.* **1**, 32–34.
Holland, L. and Priestland, C. R. D. (1972). *Vacuum* **22**, 133–149.
Kay, E. (1963). *J. Appl. Phys.* **34**, 760–769.
Kay, E., Campbell and Poenisch, A. P. (1967). U.S. Pat. 3 325 394.
Macdonald, R. J., Ostry, D., Zwangobani, E. and Dennis, E. (1970). *Nederlands Tijdschrift V. Vacuümtechniek* **8**, 207–212.
Mullaly, J. R. (1971). *Res. and Dev.*, February, 40–44.
Penning, F. M. (1936). U.S. Pat. 2 146 025.
Rivlin, J. (1971). Trans. Conf. and School "The Elements, Techniques and Applications of Sputtering", Brighton, pp. 53–57. MRC, London.
Stirling, A. J. and Westwood, W. D. (1970). *J. Appl. Phys.* **41**, 742–748.
Stirling, A. J. and Westwood, W. D. (1971). *J. Phys. D.* **4**, 246–252.
Stuart, R. V. and Wehner, G. K. (1964). *J. Appl. Phys.* **35**, 1819–1824.
Theuerer, H. C. and Hauser, J. J. (1965). *Trans. Metall. Soc. AIME.* **233**, 588–591.
Thompson, M. W. (1959). *Phil. Mag.* **4**, 139–141.
Wehner, G. K. (1955). *J. Appl. Phys.* **26**, 1056–1057.
Wehner, G. K. (1956). *Phys. Rev.* **102**, 690–704.
Weijsenfeld, C. H. (1966). "Yield Energy and Angular Distribution of Sputtered Atoms". Thesis, Rijksuniversiteit Te Utrecht.
Weissmantel, Chr., Fiedler, O., Hecht, G. and Reisse, G. (1972). *Thin Solid Films* **13**, 359–366.
Woodyard, J. R. and Burleigh Cooper, C. (1964). *J. Appl. Phys.* **35**, 1107–1117.

Applications of sputtering—past, present and future

SHELDON WEINIG

Materials Research Corporation, Orangeburg, New York, U.S.A.

I. INTRODUCTION

The most important generalization we can make about the sputtering process is that it can "deposit a thin film of any material (target) onto almost any other material (substrate)".

Other features of sputtering that must be considered when examining the practical applications of this technique are listed below:

(*a*) It is a very *thin* film process (1 to 50,000 Å, or 0·1 nm–5 μm).

(*b*) The films have a high degree of surface adherence (particles are ejected from the target material at energies of 2–10 eV).

(*c*) The film is dense.

(*d*) Compositions of materials that are not necessarily stable or compatible may be deposited.

However, the real advantage of the process is the vast range of materials that may be used to coat or be coated.

II. "COOL" SPUTTERING

I would like to expand on one aspect of sputtering which makes this possible. This is the tubular configuration. It is mainly because of this development that not only metal and ceramic, but paper and fabric may be coated with a film of many different materials.

In the tubular configuration (shown in Fig. 1) the substrate is located out-side of an anode screen and is not in the electrical circuit. The neutral sputtered particles are not affected by the potential on the anode screen and will pass through, hence heating the substrates placed around the chamber. Some electrons will be collected by impact with the screen but most will travel through the screen toward the outer walls of the chamber causing

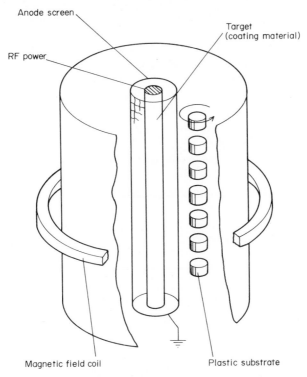

Anode screen

Target
(coating material)

RF power

Magnetic field coil   Plastic substrate

FIG. 1. Schematic diagram of the tubular configuration.

substrate heating. The mesh of the screen has to be large enough to preclude shadowing of the substrates. However, if a magnetic field of approximately 100 gauss (8 kA/m) is imposed parallel to the tubular target, the secondary electrons are deflected in spiral paths around the target. The neutral sputtered atoms are not affected by the magnetic field and continue through the anode screen. In this manner, the effective opening in the screen for the passage of electrons is significantly reduced. This increases the collection efficiency of the screen and shields the substrates from electron bombardment and sub-sequent heating.

What this means in practical terms is that any material that can be formed into a target (sacrificial element) can be sputtered onto almost any surface.

III. Applications

Let us now discuss some significant applications of sputtering; past, present and future. One of the earliest recorded applications of sputtering was in the production of phonographic records in 1928. The phonograph wax master required a conductive film on its surface. Sputtering was convenient to use as it could produce a film directly on the wax, yet faithfully reproduce the surface. The technique proved sufficiently successful so that it was used in 1933 under the trade name of "Sputtered Master".

The advent of microelectronics in the 1950's made serious demands upon the materials scientists to produce high integrity thin films. Although many films were required, the one area which was most exploited was that of tantalum and tantalum alloy circuits. This was given a massive impetus by virtue of its direct application in telephone circuitry. Silicon is primarily used for transistors, diodes and low value resistors of a non-critical nature whereas for precision resistors, capacitors and interconnects among silicon integrated circuits, tantalum thin films are preferred. There are today thousands of applications of sputtering in microelectronics and it is extremely difficult to choose which is the most important.

IV. Beam leads

The beam lead technology probably could not even exist without the availability of sputtering to produce thin films. A device with a beam lead requires only one application of heat and pressure in order to bond multiple leads to a supporting substrate. The conventional technique requires two bonds per lead using wire bonding methods.

One of the features of the beam lead system is the use of platinum, in the form of a compound with silicon. A particular difficulty with platinum silicide formation lies in the stringent cleanliness requirement of the silicon. It is most important that the silicon surface should be as clean as possible and ideally should be atomically clean to facilitate the formation of platinum silicide of the preferred phase over the whole of the contact area. This is virtually impossible with normal cleaning methods, particularly under production conditions, and sputtering provides a unique opportunity for achieving the cleanest possible site for the deposition of platinum. Sputter etching cleans the silicon slice in general and the contact areas in particular. This is followed by deposition of platinum, ensuring the most effective use of available contact areas. In present day equipment the slice is heated and platinum silicide formed without the device being exposed to air.

V. Razor blades

The next important sputtering application was the venerable razor blade in the late 1960's. The primary objective was corrosion resistance. The eternal contest of making a sharper blade was really not the question at issue, but

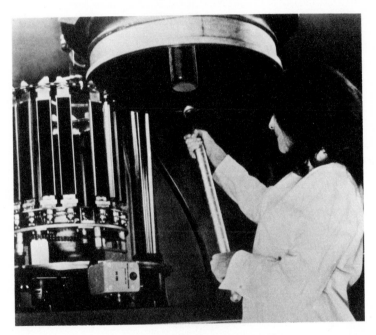

Fig. 2. A tubular target sputtering machine.

rather, how long would the "sharpness" last. The first application was with pure chromium. Although this was reasonably successful, competition quickly forced its replacement with the ordering alloy Cr_3Pt which also permitted control of edge hardness by subsequent thermal treatment. Whether or not any of the manufacturers actually utilized this fascinating feature of the alloy is questionable, but the "sex-appeal" of a platinum alloy far surpassed that of a pure chromium coating. The tubular target machine developed for razor blade coatings is shown in Fig. 2. It is interesting to note that this was the first meaningful use of the tubular configuration. Although the problem was to prevent blade edge thermal deterioration, the configuration also has a high packing density. The tubular target deposited the Cr_3Pt composition using bands of platinum interspersed along the length of the tube.

VI. Decorative coatings

Another major area is that of decorative and functional coatings on plastics. This application is new and much must still be learned. Sputtering on plastic would not be possible without the tubular configuration and its effective maintenance of a cold substrate. Excellent adhesion of the sputtered coatings is obtained on most plastic substrates. This is undoubtedly due to the high energies of the sputtered particles which are several orders of magnitude higher than the energies encountered in the electrochemical or thermal evaporation techniques. However, the exact bonding mechanism between the sputtered particle and the plastic substrate is not completely understood as there are a number of plastics upon which the coating will not adhere. The following table (Table I) lists those plastics that have been successfully

TABLE I

Polypropylene	Polyimide
Polystyrene	Polyvinyl chloride
Polyester	Polyethylene (high density)
Polycarbonate	Polyvinyl chloride/acrylic
Nylon	Glass-reinforced epoxy
ABS	Impact polystyrene

coated with metals. Substrate geometries included injection moulded parts such as bottle closures. Typical coating thicknesses used are in the 1000–2000 Å (0·1–0·2 µm) range.

Some industrial applications of sputtered plastics are transparent mar resistant coatings on polycarbonate lenses, transparent corrosion resistant coatings on silver, and copper coating for subsequent chrome electroplating. It is interesting to note that colour may be varied with different alloys.

The materials used to coat lenses are aluminium oxide and silicon dioxide. Coatings in the 1500 Å (0·15 µm) range provide improved life and scratch resistance for eyeglasses.

Sputtered decorative coatings are being used by several purveyors of cosmetics. The adhesion and corrosion resistance of the sputtered coating is quite effective against the attack of the cosmetics. (It is amazing that our skin does so well against the attack of these beautifiers.)

VII. Future

What does the future hold for sputtering? It seems audacious for me to speak about the future of a technology that has already celebrated its 120th birthday. (Sputtering was invented in 1852.) One would expect that it is very mature and that instead of speaking about its future, we should be concerned about its demise. Its potential future impact on industry will be based upon a greater comprehension of materials and the awareness that many "thick film" applications really only need good "thin films". If we concern ourselves with the future of this technique as related to the availability of equipment, we completely misunderstand its significance. People speak of the limitations of vacuum; this is nonsense; vacuum chambers can be built to almost any size.

However, before we continue to the future of sputtering, let us consider one equipment idea which I propose as a challenge. The practical solution has enormous potential in many areas. A flat plate target has a one to one ratio of target area to substrate area. In the tubular configuration we enhance the substrate area relative to the target area, that is, we achieve a greater area of substrate than target. I would like to examine the reverse situation. Consider a 6 in. (15 cm) diameter flat target where all of the emitted particles converge to a point. If we consider deposition at the point it will be the summation of the entire surface of the target. For example, if the "point" was approximately 0·1 in. (2·5 mm) diameter we would have a deposition rate multiplier of 3600 × . If the target in question sputters at 1000 Å (0·1 μm) per minute, we would have a deposition rate of 0·014 in. (360 μm) per minute on a substrate placed at the "point". It is also conceivable that vacuum would not be a required part of the equipment. The positive exit of the material through the point may well act as a dynamic seal. We could then "write" with this point in material of our choosing. (Do not confuse this approach with a hot plasma approach.) You could write in material through a programmed pattern, or the material could be built up in areas that could not be feasibly placed in vacuum—such as parts of the human body.

What is the practical problem of building the apparatus? Essentially it is the determination of the forces required to converge neutral particles. The usual force fields will not work so we will either have to artificially charge the particles or invent a force that will work on neutral particles. The exact nature, I leave to your creative imagination. We now have an interesting piece of apparatus with exciting possibilities.

We could deposit directly into tooth cavities within the mouth with a material that approaches the composition of a tooth. The high adherence potential of sputtering and the range of colours available would allow absolute matching. Or we could coat teeth with a hard impervious material such as sapphire. This will prevent decay.

We can even consider *in vivo* application of sputtered materials. Consider the frail hip of the human being. This particular joint tends to erode. The present approach is the substitution of a metallic prosthetic ball joint. Unfortunately, this operation, which is surgically glamorous in concept, is extremely weak in execution. The rate of rejection and replacement of these hip joints is very high. Consider the possibility of *in vivo* deposition on the fretted and worn portion with a material that will firstly build up the decayed area and then coat with a material that will add to lubricity and wear resistance.

One can obviously go from here to other materials in the human body.

The real future of this fascinating process is mostly dependent upon the elimination of material restrictions. Consider combinations of organics, ceramics (alumina) and metals in special applications. Only your imagination limits the amazing number of materials that can be deposited. No other coating technique has an arsenal of so many materials available. There is no limitation as to how we might put our creative minds to work finding new and exciting applications for sputtering.

The basic principles of ion plating

R. Carpenter

Allen Clark Research Centre, The Plessey Company Limited,
Caswell, Towcester, Northants, England

I. Introduction

Ion plating is a name given by Mattox (1964) to describe a vacuum deposition process in which partial ionization of the metal vapour is used to increase the adhesion of the film to the substrate. It is distinguished from ion implantation, a well known process in microelectronics, by the much lower ion energies involved and the degree of ionization used.

In order to discuss the basic principles of the technique it is useful to start by considering in very general terms the factors which may affect film adhesion.

II. Adhesion

Adhesion, as commonly observed, results from a summation of the atomic forces acting across the film-substrate interface. In the simplest possible case (Fig. 1(a)) of a single atom adsorbed on a linear array of identical substrate atoms, symmetry indicates that in the position shown the adsorbed atom will be bound equally to its two nearest neighbours. Evidently there are a number of similar positions along the row where the atom could be adsorbed.

In the case of a surface, the array of substrate atoms is two dimensional and the picture is correspondingly more complicated. If the array is irregular, the forces binding the atom may vary over the surface. Irregularity may be introduced locally by a missing atom of a different type of atom incorporated in the array, or on a larger scale by changes in the overall array. Hence, it is only when the substrate surface is a perfect single-crystal plane that the adsorption can be pictured at all easily. Most substrates are not in this category and in general the atoms are ordered in the same fashion over relatively small areas corresponding to the particular crystal plane exposed. In these

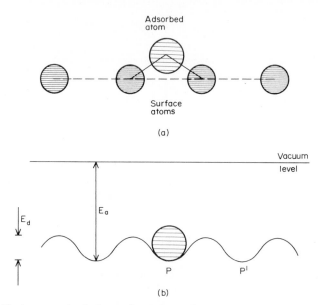

FIG. 1. (a) Single atom adsorbed on a linear array of atoms. (b) Corresponding energy level diagram.

cases the adhesion forces binding the adsorbed atom will vary from one part of the substrate to another. On amorphous substrates, such as glass, there is no long range order at all and it is then not possible to think of fixed positions for the substrate atoms but only in terms of positions which are the most probable (Klug and Alexander, 1962).

So far only the adsorption of a single atom has been considered. With a metal film where there are many atoms, interactions between the film atoms themselves, which give rise to cohesive forces, result in additional constraints being imposed so that the interfacial binding forces may be affected by the film structure. Consequently, the interfacial atomic forces depend both on the types of the atoms and the structures in which they are situated.

Adsorption phenomena are conveniently classified according to the nature of these binding forces into chemisorption and physisorption. The latter is associated with longer range forces, where the atoms are further apart, and is often encountered with gas atoms. It is usually weaker than chemisorption, which is associated with chemical bonds. An account of these effects has been given by Kaminsky (1965). In the discussion which follows, no attempt is made to distinguish between different types of bonds, but features are considered generally.

Instead of considering the physical picture of atoms in the surface, it is often useful to describe the situation by a simple energy diagram for adsorption, such as that shown in Fig. 1(b). Here P and P' represent positions of lowest energy, in other words possible sites for the film atoms, E_a is the adsorption energy for a single atom and E_d the diffusion energy, the activation energy needed for moving an atom across the surface to an adjacent

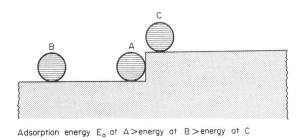

Adsorption energy E_a at A > energy at B > energy at C

FIG. 2. Adsorption of atoms on a stepped surface.

site. E_d is an important parameter in nucleation theory (Lewis, 1971), where it is assumed that atoms condense and then migrate across the surface in a series of hops. The time spent at each site is proportional to $\exp(E_d/kT)$ and nucleation occurs when the number of atoms at a given site exceeds a critical number. This model can be extended to include a distribution of site energies, in which case the sites with a large E_a (or E_d) are filled first of all, and is consistent with the island-type growth observed in thin films.

Returning to adhesion, it is evident that for this to be large, E_a should be large. In other words either the bond strength or the number of bonds per atom should be increased. In practice the bond strength will be fixed by the nature of the atoms. However, it may be possible to increase the number of bonds by modifying the surface structure. One way is illustrated in Fig. 2 which shows diagrammatically the adsorption of single atoms on a stepped surface. The adsorption will be greatest in positions where the atom stands the greatest chance of making most bonds. Clearly this is at the internal corners. Conversely, the least firmly bound atom lies near the external corner. Experimental evidence for the existence of sites with different adsorption energies is the "decoration" of single crystal substrates (Basset, 1958). Figure 3 shows an extraction replica of a very thin gold deposit on rock salt (NaCl) to illustrate the effect. Certain preferred positions in which the atomic clusters have condensed are evident and these lie mainly in well-defined

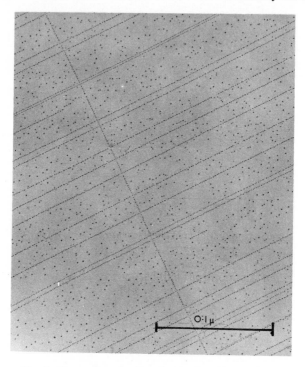

FIG. 3. Decoration of a cleaved rock salt surface with gold.

directions. The surface of the rock salt is not perfectly flat, but the layers of surface atoms terminate in a series of abrupt steps, so that a small section looks like Fig. 2. As the deepest adsorption sites are filled first of all, then this is where the decoration occurs. Decoration effects are apparent only on highly ordered substrates; with commoner substrates, such as glass, no long-range order exists and variations in site energy will be random and therefore indistinguishable.

The adsorption phenomena discussed so far are those on an atomic scale. As indicated earlier, adhesion is a macroscopic effect and is the summation of forces over astronomical numbers of interfacial atoms. Consequently, the adhesion, besides being dependent on individual bond strengths and numbers of bonds per atom, will depend directly on the number of atoms in the interface. This number will be proportional to the effective substrate surface area, which is generally considerably greater than the apparent area. The adhesion can be increased by increasing the effective area.

A method of making absolute adhesion measurements has yet to be found (Heavens, 1950; Butler *et al.*, 1970). However, if a comparison is made between an evaporated and a sputtered film of the same metal deposited on a similar substrate, it is observed that the sputtered film adheres more strongly. The principle upon which the ion plating technique depends is based on the reason for this difference.

III. Evaporated and Sputtered Films

When metal atoms evaporate from a heated source, they do so with energies which are given by a Maxwellian distribution. The mean energy is proportional to the absolute temperature of the source and is about 0·1 eV for 1000 °K. The arrival energy of the atoms at the substrate will be the same as the atoms had on leaving the source, for under normal conditions used for evaporation, the chances of an atom losing energy by collision with a gas atom are small. This energy is lost to the substrate within the period of one or two lattice vibrations ($\sim 10^{-14}$ s).

In contrast, the characteristics of sputtered atoms are markedly different. For example, in the case of diode sputtering, a glow discharge is sustained between an insulated electrode, usually known as the target, and another electrode, which is generally the earthed metal parts of the vacuum system. The target is composed of the material to be sputtered and is operated at a potential of a few kilovolts below earth. Positively charged gas ions, formed in the discharge, are accelerated across the dark space and strike the target. Here, their kinetic energy is dissipated as heat and in the removal of target atoms by sputtering (Holland, 1956). The numbers of atoms sputtered and their spatial and energy distributions depend in a complicated way on the discharge characteristics and target structure (Carter and Colligon, 1968). In a typical glow discharge, the majority of sputtered atoms will have energies between 1 and 10 eV on leaving the target (Stuart *et al.*, 1969). However, these are the energies close to the target; by the time the sputtered atoms reach the substrate some distance away, their energies and energy distribution will have been modified by collisions with gas atoms in the intervening space. The trajectories of the sputtered atoms are also altered in the collisions. This gas scattering becomes more significant at higher pressures and with heavier gas atoms. It occurs in practice because diode sputtering is usually carried out in an argon atmosphere and a relatively high pressure ($\sim 10^{-2}$ torr (1·3 Pa)) is needed to sustain the discharge. For practical cases it is not possible to calculate the effects of scattering; however, it is evident that the mean energy will be reduced and the directional characteristics smoothed out by collisions.

So far only the sputtered atoms have been considered; however, it is of relevance to ion plating to consider what happens at the target. Atoms are sputtered not only from the surface layer, where individual atoms are removed. Various types of collision sequences can also occur between the gas ions and target atoms, which result in structural modifications well below the surface (Venables, 1970). After losing its charge and energy, the gas "ion" may become trapped beneath the surface and not desorbed until later, after the intervening target layers have been sputtered away. The penetration depths of monoenergetic gas ions have been determined experimentally and can be computed approximately (Carter and Colligon, 1968). For argon ions of 10 keV energy the penetration depth is ~ 0.01 µm.

These effects at the target are atomic scale phenomena. It is relevant here to point out that topographical modifications on a larger scale may be introduced by the bombardment. If sputtering rates vary over the surface, as they often do along grain boundaries where different crystal planes are exposed, preferential sputter etching can take place and the surface area may be increased (Kaminsky, 1965). Another effect has been noted by Stewart and Thompson (1969) and by Wehner and Hajicek (1970). These workers observed spikes developing with certain targets as a result of impurities on the surface sputtering more slowly. This effect again can lead to an increase in the target area.

The observed adhesion differences between evaporated and sputtered films can be attributed to the differences in the arrival energies of the atoms. As the difference is between 0·1 eV and a few eV, any increase in average arrival energy above that associated with evaporation would be expected to lead to an improvement in adhesion. This is the objective of the plating technique.

IV. ION PLATING

The process originated by Mattox (1964) basically employs evaporation through an argon glow discharge to increase the average energy of the metal atoms. A typical apparatus is shown in Fig. 4; in this case it is mounted above a 4 in. (10 cm) oil diffusion pump, fitted with a liquid nitrogen cold trap and baffle valve. The directly heated evaporation source can be isolated from the substrate by a moveable shutter. The substrate is mounted against an aluminium cathode plate which is electrically insulated from the rest of the system by the glass wall of the vacuum chamber. Ideally, the edges of the cathode are rounded to eliminate the high fields which would otherwise occur. Adjustment of the pressure within the chamber to operate the glow discharge is made by altering the argon flow with the needle valve. The pump baffle valve is conveniently used as a fine control. The power supply will give

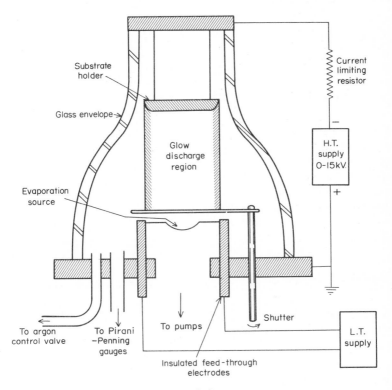

FIG. 4. Ion plating system.

up to 15 kV d.c. on open circuit and is unsmoothed. However, for discharge stability it is essential to use a ballast resistor, or some other means of limiting the current in the cathode circuit.

In a typical deposition, the system is evacuated to a low pressure $\sim 10^{-5}$ torr (1·3 mPa) and is backfilled with argon to $\sim 10^{-2}$ torr (1·3 Pa). A glow discharge is initiated by applying the voltage to the cathode; this sputters the substrate away. The discharge current is about 0·5 mA/cm². The source temperature is slowly raised during this time to outgas the charge. After a few minutes the source temperature is increased again until the charge starts evaporating. The shutter is then opened, so that sputtering and deposition occur at the substrate simultaneously. This is the actual ion plating stage and for a net deposition, the deposition rate onto the substrate must be greater than the rate at which the film is removed by sputtering. This condition is easily met as evaporation is intrinsically a much faster process than sputtering. Ion plating is continued at least until the substrate is covered completely by the film material, which will typically required an average film thickness of ~ 0.01 μm. Finally, the pressure is reduced and the deposition continued by evaporation alone.

The process can be considered to be composed of three stages: pre-plating, ion plating and evaporation. Contamination is removed during pre-plating so that a clean surface is exposed for the subsequent deposition. Penetration of the surface by the argon ions also occurs; this may disrupt the existing structure and lead to the production of deeper sites. Finally, roughening may be produced on a macroscale. As discussed earlier, all these processes are beneficial and promote adhesion of the film. These effects still occur during ion plating, but they will generally be masked by the impinging film atoms. Most of these atoms will be neutral, but as they will have made collisions with the gas molecules in the discharge, their average energy will be somewhat above that associated with the source temperature and consequently their adhesion to the substrate will tend to be greater. In addition a small proportion of the metal atoms, generally about 0·1%, will be ionized and accelerated into the cathode to produce much deeper sites. The production of deep sites tends to blur the interface; a similar effect would be produced by solid state diffusion, but much higher substrate temperatures would be needed. The magnitudes of these effects depend on the ratio of the rate at which metal atoms arrive to the rate at which they are sputtered away and can be controlled by adjustment of the source temperature.

The reason for reducing the pressure is to economize in the use of charge metal; this can be advantageous when the metal is expensive. During plating, gas scattering occurs and the metal atoms are distributed around the vacuum chamber. At a lower pressure the evaporating atoms travel directly from source to substrate, consequently, for the same charge a thicker film can be deposited if the ion plating is curtailed. As the film adhesion is determined by the interfacial atoms, nothing is lost in following this procedure. However, it is sometimes beneficial to utilize gas scattering, particularly when the substrate is an awkward shape, as it is possible to coat surfaces which are not directly exposed to the source.

Figure 5 summarizes the mechanisms occurring during the various stages of the ion plating process.

There are advantages and disadvantages in ion plating. An advantage, shared with other vacuum deposition processes, is that it is intrinsically clean as the film is deposited directly without using a carrier. Providing the vapour pressures are compatible it is easy to deposit alloy films. The technique usually gives a higher adhesion than a directly evaporated film without the necessity of using the much slower sputtering process.

The disadvantage of a vacuum environment is that the process is relatively costly and is only likely to be economic for small parts, where many can be

STAGE I PRE-PLATING

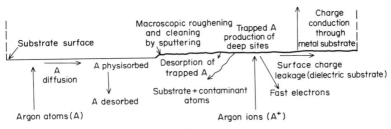

STAGE 2 - ION PLATING   (The following processes occur in addition to those above)

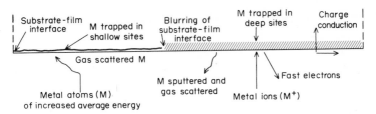

STAGE 3 (OPTIONAL) — EVAPORATION

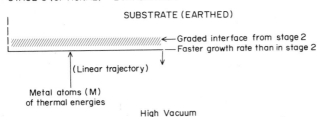

FIG. 5. Processes occurring at the substrate during plating.

plated together, or where the component is costly. In order to deposit the film at a useful rate the vapour pressure of the metal at the source should be at least 10^{-2} torr (1·3 Pa). Hence crucible materials are needed to contain the charge at high temperatures. Not all metals can be evaporated in this way. Some metals are normally deposited by electron bombardment using the metal, either in the form of a rod or wire or contained in a water-cooled hearth, as the anode. Ion plating in conjunction with electron-bombardment evaporation seems to be difficult in view of the high pressure required for the glow discharge and the necessity of avoiding spurious discharges in the gun. Evaporation and, to an even greater degree, ion plating are wasteful in terms of the utilization of source material, as there is no way of restricting the deposition to the substrate alone. Unless considerable elaboration is introduced in the system, the film thickness will be limited by the capacity of the source.

The substrates must be able to withstand the vacuum environment and the thermal gradients which occur during ion plating. It is convenient if they are flat. Ideally they should be conductors as charge build-up on the surface is avoided and the process is more efficient. Even with dielectric substrates the adhesion can be improved (Mattox, 1965), possibly because of charge leakage across the surface. A subsidiary mesh cathode can also be used (Spalvins, 1971). Copper films can be deposited directly on high purity alumina without using an underlayer (Carpenter, 1971). However, better results could probably be obtained with this type of substrate by using a radio-frequency high-voltage supply, as this overcomes surface-charge effects (Anderson *et al.*, 1962).

V. Conclusions

Ion plating can be very useful if poor adhesion occurs with the evaporated film. However, there are many interdependent variables and their relative importance to adhesion has not been evaluated. Consequently, it is necessary to adopt an experimental approach to determine the optimum deposition parameters for a given film-substrate combination. Run-to-run variations may still occur. As it is the first few atomic layers of film atoms on the substrate which determine the adhesion, and as the ionization efficiency of the system described is low, it would be interesting to see if further improvement could be made by using an ion source to deposit the first monolayers and then to plate to the desired thickness by evaporation alone.

Acknowledgements

I should like to thank the Plessey Company Limited for permission to publish this paper and Dr. B. Lewis for many interesting discussions.

REFERENCES

Anderson, G. S., Mayer, W. N. and Wehner, G. K. (1962). *J. Appl. Phys.* **33**, 2991.

Basset, G. A. (1958). *Phil. Mag.* **3**, 72.

Butler, D. W., Stoddart, C. T. H. and Stuart, P. R. (1970). *J. Phys. D.* **3**, 877.

Carpenter, R. (1971). *Component Technology* **4**, 6, 17.

Carter, G. and Colligon, J. S. (1968). "Ion Bombardment of Solids". Heinemann, London.

Heavens, O. S. (1950). *J. Phys. Radium* **11**, 355.

Holland, L. (1956). "The Vacuum Deposition of Thin Films". Wiley and Sons, New York.

Kaminsky, M. (1965). "Atomic and Ionic Impact Phenomena on Metal Surfaces". Springer-Verlag, New York.

Klug, H. P. and Alexander, L. E. (1962). "X-ray Diffraction Procedures". Wiley and Sons, New York.

Lewis, B. (1971). *Thin Solid Films* **7**, 179.

Mattox, D. M. (1964). *Electrochem. Tech.* **2**, 295.

Mattox, D. M. (1965). *J. Am. Ceram. Soc.* **48**, 385.

Spalvins, T. (1971). *Lubrication Engineering* **27**, 40.

Stewart, A. D. G. and Thompson, M. W. (1969). *J. Mat. Sci.* **4**, 56.

Stuart, R. V., Wehner, G. K. and Anderson, G. S. (1969). *J. Appl. Phys.* **40**, 803.

Venables, J. A. (1970). In "Atomic Collision Phenomena in Solids" (D. W. Palmer, M. W. Thompson and P. D. Townsend, eds.), p. 132. North Holland Publishing Co.

Wehner, G. K. and Hajicek, D. J. (1970). *J. Appl. Phys.* **42**, 1145.

The basic principles of ion implantation

M. Martini

Simtec Industries Ltd., Montreal, Quebec, Canada

I. Introduction

Ion implantation, possibly the most recent technique for surface doping of semiconductors, consists essentially of the "introduction of atoms in the surface layer of a solid substrate by bombardment of the solid with ions in the keV to MeV ranges" (Mayer *et al.*, 1970).

Although this technique is currently employed for applying contacts on semiconductors, there is no reason why it should not be used in any application where incorporation of dopants into surface layers is required.

The bombardment is performed in vacuum by means of a small accelerator of the type shown in Fig. 1(*a*). A typical processor is shown in Fig. 1(*b*); its characteristics are given in the Appendix.

The profile of the density of implanted ions is shown (for an amorphous substrate) in Fig. 2, where R_p is the "mean depth of penetration" (called also "projected range") of the implanted ions. This figure also shows for comparison the density profiles obtained by means of evaporation and diffusion, two well established techniques for making contacts on semiconductors. As evaporation is a surface process in which the evaporated atoms do not penetrate the substrate, a meaningful comparison can be made only between diffusion and ion implantation.

Essentially, ion implantation has all the advantages of an externally controlled, non-equilibrium process, whereas in surface doping by diffusion all the key parameters are dominated by unchangeable physical constants (solubility limit, diffusion constants).

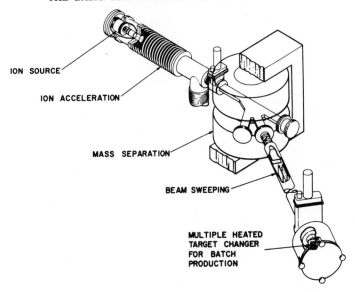

ION SOURCE

ION ACCELERATION

MASS SEPARATION

BEAM SWEEPING

MULTIPLE HEATED
TARGET CHANGER
FOR BATCH
PRODUCTION

FIG. 1. (*a*) Schematic drawing of an ion implantation accelerator (from Mayer *et al.*, 1970).

(*b*) A modern ion implantation processor (courtesy of Ortec Inc.).

Some typical advantages of ion implantation over diffusion are as follows:

(1) Ion implantation can be performed at temperatures at which normal diffusion is totally negligible.
(2) The deposition rate can be varied between wide limits (5×10^9 ions/cm^2 s^1 to 2×10^{11} ions/cm^2 s^1 in a commercial processor) by simply varying the beam current.
(3) The concentration of the dopant is not limited by solubility limits and a much wider variety of dopants can be used.
(4) By varying continuously the energy of the ion beam, and, in consequence, the value of R_p, in principle one should be able to control the dopant profile with a high degree of precision.

Against those advantages it must be remembered that ion implantation is still a much more sophisticated and expensive technology than diffusion.

II. Details on ion implanted contacts

In a single crystal substrate, the density profile of the ion implanted material can be widely different from the one shown in Fig. 2, because of the "channelling" of the implanted atoms. This latter phenomenon occurs when, due to a favourable angle of impact of the ion beam on a single crystal substrate, the implanted ions can channel along a well defined axis of crystalline symmetry (Figs 3 and 4).

However, if one assumes that the substrate is either amorphous or, if it is single crystal, is implanted in a "random" direction, than the channelling effect can be minimized and the profile of the density of the implanted ions can be characterized by means of the previously defined parameter R_p.

Figure 5 shows the mean depth of B, P and Sb ions in Si, Ge and CdTe. It is easily seen that the value of the projected range can be controlled by varying the energy of the beam.

The implanted atoms become electrically active only after heating the implanted samples for a period of time between a few minutes and a few hours at temperatures in the range 300–800 °C. The heating process is performed in order to anneal the damage provoked by the ions, which, as they come to rest, displace many lattice atoms from their positions. At high doses ($\geqslant 10^{15}$ ions/cm^2) an amorphous layer is present at the surface of a single crystal substrate after the implantation because the clusters of damage provoked by every single ion overlap. Implantations can be made on heated substrates in order to anneal the clusters of damage as they are formed.

The objective of an implantation is, clearly, to produce a superficial region of controllable electrical characteristics obtained by the dopant atoms located

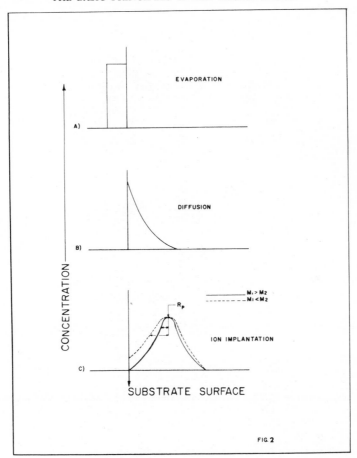

FIG. 2. Schematic profile of the density of: (*a*) evaporated, (*b*) diffused, (*c*) implanted atoms on a solid substrate. The implanted profile refers to the case of an amorphous substrate or of a single crystal substrate aligned in a "random" direction. The dotted and the continuous line for the ion implanted layer refer to the case in which the mass M_1 of the implanted atom is respectively lower and higher than the mass M_2 of the atoms of the implanted substrate.

at well defined positions in the semiconductor lattice. However, it has been found that, in general, there is no simple correlation between the position in the lattice of the dopant atoms and their electrical activity.

III. PERFORMANCE OF ION IMPLANTED CONTACTS

Both good rectifying and ohmic contacts can be obtained by means of ion implantation. For instance (Meyer and Haushan, 1967) in a successful

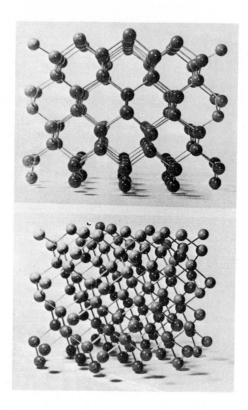

FIG. 3. A diamond type lattice as viewed along the $\langle 100 \rangle$ direction (top) and along a "random" direction at $10°$ from the $\langle 100 \rangle$ (bottom) (from Mayer *et al.*, 1970).

attempt to make a nuclear radiation detector (a device with very stringent electrical characteristics) a rectifying contact was obtained on high resistivity (100–1350 ohm m) *n*-type silicon by implanting boron (10^{13} ions/cm^2, at an energy of 10 keV). An ohmic contact was obtained by implanting phosphorus (10^{15} ions/cm^2, at the same energy as for boron). The resulting diodes, with areas of the order of 80 mm^2 and thickness up to 3 mm, showed reverse currents of 2–3 microamperes at bias voltages of several hundred volts at room temperature.

Other devices such as MOSFET and IMPATT diodes on integrated chips are now routinely made by ion implantation.

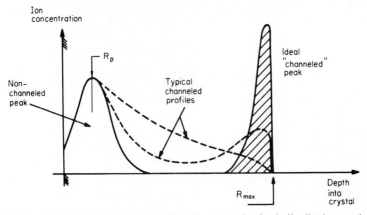

FIG. 4. When channelling of the implanted ion is present, the depth distribution can be quite different from the distribution in an amorphous substrate (from Mayer *et al.*, 1970).

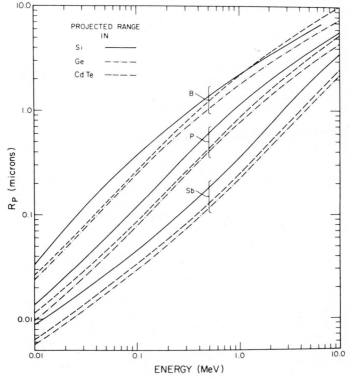

FIG. 5. Projected range R_p vs. energy for non-channelled B, P and Sb ions in Si, Ge and CdTe (from Mayer *et al.*, 1970).

APPENDIX

Principal characteristics of the ion implantation processor shown in Fig. 1(a).

Specifications

Dose Rate (over a 3 in. (7·6 cm) diam. area) (boron): Maximum 2×10^{11} particles/cm^2 s. Minimum 5×10^9 particles/cm^2 s.

Dose Reproducibility: $\pm 1\%$.

Dose Differential Uniformity (over a 3 in. (7·6 cm) wafer): $\pm 1\%$.

Ion Beam Current at Wafer: 3 µA B$^+$, from 5 to 150 keV.

Diameter of Beam Focus at Substrate: 1 cm, from 10 to 150 keV.

Maximum Variation in Angle of Incidence of Beam on Substrate: 2·5 °.

Vacuum: 2 systems, with ion source under constant vacuum and sample chamber cycled; operating pressure during implant 8×10^{-6} torr (1·0 mPa).

Processing Chamber Details: Eighteen 3 in. (7·6 cm) wafers or thirty-six 2 in. (5·1 cm) wafers per batch.

—Each wafer individually implanted.

—Batch trays load directly into chamber.

—Low volume for last pumpdown.

—Optional electron flood for surface neutralization during implant.

ACKNOWLEDGEMENT

This work has been partially supported by the Defence Research Board of Canada (Project DIR–205).

REFERENCES

Mayer, J. W., Eriksson, L. and Davies, J. A. (1970). "Ion Implantation in Semiconductors". Academic Press, New York and London.
Meyer, O. and Haushan, G. H. (1967). *Nucl. Instr. and Meth.* **56**, 177.

Mechanical aspects of solid lubricant and elastomeric coatings

EUGENE F. FINKIN

D.A.B. Industries, Inc., Vincent Avenue, Detroit, Michigan, U.S.A.

I. FRICTION

Where does friction arise, on a surface or within a surface? Most theoretical developments have assumed the former, not realizing that the distinction is very important. Much more likely, friction arises within a thin, near surface region for bulk solids or within the surface film if the solid is film covered. Epifanov has conducted elegant experiments proving this particular point (Epifanov and Aretisyan, 1960; Epifanov and Sanzharovskii, 1962; Akhmatov, 1966). He showed that friction arises from deformation, and adhesion is important in so far as it influences deformation; friction is directly proportional to the real area of shear which in turn is directly proportional to the real area of contact. The shearing action, however, was found not to take place on top of the surface but actually occurs within the surface. For a thin film, it occurs wholly within the film and is prevented from involving the metal beneath. To this phenomenon Epifanov credits the beneficial effects of films. This amounts to saying that friction has surface volume (or thickness) aspects. This concept has been the subject of several lines of development by the author (Finkin 1969, 1970a, 1971).

The friction of bodies having solid surface films is a function of the film thickness. Figure 1, taken from the work of Bowden and Tabor (1954), illustrates the general nature of the phenomenon. Friction, as a function of film thickness, can be separated into two general regimes; those of thin films and of ultrathin films. In the thin film regime friction decreases with decreasing film thickness, whereas in the ultrathin film regime friction increases with decreasing film thickness. These will be dealt with separately.

The analysis of friction requires the analysis of the problem of solid mechanics, of the real area of contact of an indenter pushing against a film-covered half space, and this requires a statement of the mode of film and substrate deformation; i.e. elastic, plastic, viscoelastic, etc. One can postulate

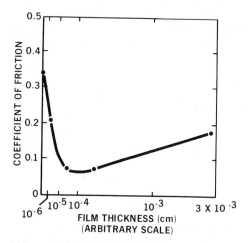

FIG. 1. Effect of film thickness on frictional properties of thin films of indium deposited on tool steel (from Bowden and Tabor, 1954).

a great many theoretical expressions, the differences arising from the assumed modes of deformation. Consequently, if the friction force is directly proportional to the real area of contact, no one equation can uniquely relate film thickness to friction for all materials situations. But the difficulties do not end by the determination, for a given materials situation, of the proper mode of deformation as the mode of deformation itself is a function of film thickness.

The best one can hope to do, therefore, is to formulate special theories for certain types of deformational situations which are known to occur. The present work will consider the problems of an elastic film which has a much lower modulus of elasticity than its substrate.

A. Thin elastic film

The effect of increase in friction with increase in film thickness, exemplified in Fig. 1, and the related load dependence effect, have been the source of much misunderstanding and confusion. Many investigators have failed to realize that these effects exist at all. The consequence was the seeming disagreement of experimental friction results taken under different conditions of load and film thickness. Attempts had been made to explain these seeming discrepancies in terms of materials properties variations without realizing that these seeming discrepancies are the inherent consequences of the applied mechanics of the layered system.

Appendices 1 and 2 from Finkin (1970a) derive the relationships among area of contact, nominal contact stress, film thickness h, and normal load P for films which are much less stiff than their substrates where the deformation primarily occurs in the film. Coefficient of friction, f, obeys the relations

$$f \propto \sqrt{\frac{h}{P}} \quad \text{for spherical contact} \tag{1}$$

$$f \propto \left(\frac{h}{P^2}\right)^{1/3} \quad \text{for cylindrical contact} \tag{2}$$

For a flat on a film-covered flat, no film thickness-induced area of contact effects are anticipated (Finkin, 1970b).

Bowden and Young (1951) examined the friction of diamond sliding against diamond. The diamond surfaces were covered with a graphite film formed by the transformation of diamond to graphite in vacuum at high temperature. The thickness of this film does not seem to have been measured. The contact and sliding of the diamond crystals took place between a natural octahedron apex on one and an artificial dodecahedral flat on the other. Bowden and Young's experimental results for sliding in pure oxygen or vapour-free atmosphere are plotted in Fig. 2 with the theoretical curve from equation (1). The theoretical curve appears to coincide identically with an imaginary median curve drawn through the data points.

A just published study (Finkin, 1973) examines the effect of deformation of the substrate and indentor as well as the film.

Friction coefficient is a systems property, and its treatment from the narrow perspective of consideration as just a materials property, has hampered the development of its understanding.

B. Ultrathin elastic film

Finkin (1971) discusses the ultrathin film case from the point of view that friction arises within the surfaces rather than on them. What is of importance to a film-covered surface is not the shear strain at the surface, but the mean shear strain within the layer. If a surface is covered with a thin layer (i.e., oxide, reaction product film, metal plating, etc.) and another surface is pressed against it which strongly adheres such that the dimensions of contact are much larger than the thickness of the film, the movement of one surface relative to the other corresponds to the applied mechanics problem of a layer bonded to two semi-planes. At low film thicknesses, a film can undergo a substantial apparent stiffening for applied mechanics reasons alone, that is a couple-stress effect.

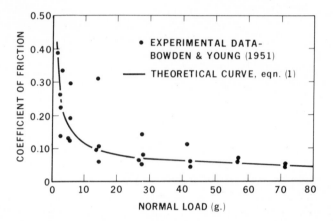

FIG. 2. Friction coefficient of diamond sliding on diamond whose faces have undergone transformation to graphite in the presence of pure oxygen or vapour-free atmosphere (data from Bowden and Young, 1951).

An analysis based on this concept (Finkin, 1971) resulted in the following equation for the film thickness dependence of friction coefficient:

$$f = \frac{A}{1 - (2l/h)\tanh(h/2l)} \tag{3}$$

where A is a proportionality factor, h is the film thickness, and l is the couple-stress constant (an elastic property of the film).

The theory upon which this equation is based is only valid for no slip between the film and the rider; therefore, equation (3) is mathematically valid only for static friction. However, it seems to have empirical applicability beyond that.

Data exists in the literature to test equation (3). Most of these investigators inferred the thickness from the optical interference colour. Kubaschewski and Hopkins (1967) point out that this method is only approximate because three of its assumptions may not be fulfilled. These include:

(a) that the phase change produced by reflection is zero,
(b) that the colour is determined by the wave length suffering maximum interference,
(c) that the refractive index of the oxide film does not vary with the wave length.

To allow for a consistent error in the film thickness determination by the various authors, equation (3) may be modified to read:

$$f = \frac{A}{1 - [2l/(h-e)] \tanh [(h-e)/2l]} \tag{4}$$

where e is the consistent error in film thickness determination and h is the measured film thickness.

1. Campbell's results

A very careful experimental study of the ultrathin film thickness effect has been carried out by Campbell (1939), who measured the static coefficient of friction of sulphided copper surfaces as a function of sulphide film thickness. He used a horizontal plate and slider apparatus, the slider consisting of three 1 in. (2·5 cm) diameter balls held rigidly together by the triangular plates; the slider weighed 297 grams. Extreme care was taken to insure the cleanliness of the system and the atmosphere was carefully controlled.

Campbell's experimental results are given in Table I; the friction coefficients are the means of ten readings. The thicknesses of the films were not

TABLE I
Static friction coefficient as a function of film thickness
for sulphided copper surfaces
(from Campbell, 1939, 1966, and Campbell and Thomas, 1939)

Run No.	Colour of film	Equivalent air film thickness (Å)	New estimate of film thickness on each surface; based on measurements of oxide films (Å)	Static friction coefficient
1	No Film	——	40	1·17
1	Slightly Darkened	1000	120	1·07
1	Rose I	1200	300	0·91
2	Blue I	1600	400	0·88
2	Yellow II	2800	800	0·78
2	Blue II	4300	1450	0·73
3	Blue Green II	4900	1650	0·73
3	Light Blue II	4600	1500	0·78

measured directly but rather were inferred from the tarnish colours, firstly
from their equivalent air thickness colour and more recently (Campbell,
private communication) from later experiments (Campbell and Thomas,
1939) on the tarnish colours of copper oxide films. Campbell (private communication) asserts that the film thickness–tarnish colour relationships of
copper oxide and copper sulphide films are identical and, consequently, one

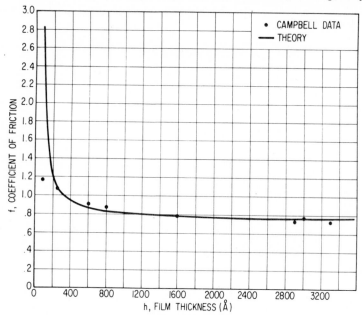

Fig. 3. Theoretical relation and experimental values of static friction coefficient as a function
of film thickness for sulphide films on copper (data from Campbell, 1939, and Campbell and
Thomas, 1939).

may be used for the other. Campbell's improved film thickness estimate is
also given in Table I.

Figure 3 contains a plot of the theoretical expression for friction coefficient,
f, as given by equation (3) and Campbell's experimental data; the thickness
of sulphide film on both surfaces has been summed. In equation (3), A was
taken to be 0·755 and a best fit results for $l = 37$ Å (3·7 nm). The results of
the computations for this data are given in Table II.

2. Greenhill's results

Greenhill (1948), using the Bowden and Leben pin-on-disc apparatus,
studied the dynamic friction of sulphided copper. A summary of the computed results is given in Table II.

TABLE II
Summary of experimental results

Investigator	Materials	Type of friction	Number of Data Points	A	l (Å)	e (Å)	Δ^2	Δ, Standard deviation
Campbell	Copper Rider Copper Sulphide Film Copper Substrate	Static	8	0·755	37	0	0·0013	0·036
Greenhill	Copper Rider Copper Sulphide Film Copper Substrate	Dynamic	8	0·50	250 79	0 730	0·0284 0·0134	0·168 0·116
Levine and Peterson	Steel Rider Iron Sulphide Film Steel Substrate	Static	7	0·35	720 245	0 2450	0·0088 0·0051	0·094 0·071

The data and theory seem to agree well enough. It is encouraging to find that the best fit couple stress constant is within a factor of about two from that of Campbell's experiment. As Greenhill's data is for dynamic friction coefficient rather than static friction coefficient, it appears that the theory applies to the dynamic situation, even though its use cannot be rigorously justified.

3. Levine and Peterson's data

Levine and Peterson (1951) studied the static friction coefficient of steel coated with iron sulphide films of various thicknesses. The rider consisted of three SAE 52100 steel balls of hardness Rockwell C-60, and the SAE 1020 steel plate specimen was of hardness Rockwell A-150. The ball specimens were clamped to the rider at each of the vertices of an equilateral triangle. Relative humidity was maintained in the range between 7 to 10 per cent. Table II contains the computed value of l.

The rather large value of l for steel systems may imply that couple-stress effects are a common feature of the boundary lubrication of steel systems. The consideration of such effects may help to explain much of the experimental scatter and confusion surrounding steel's boundary lubrication.

4. Rabinowicz's data

Rabinowicz (1967b), using a pin-on-disc apparatus, investigated the lubrication of metal surfaces by oxide films and came to the conclusion that the thickness of the film had to exceed a critical thickness before effective friction reduction occurred. This critical thickness was found to be approximately 10 nm.

One may choose to define, in the manner of Rabinowicz, a critical oxide film thickness, h_{crit}, as that film thickness for which the ratio of the coefficient of friction f to that for a thick film, f_{min}, lies in the range

$$1 \cdot 5 \leqslant \frac{f}{f_{min}} \leqslant 2 \tag{5}$$

Substituting equation (5) into equation (3) yields a critical film thickness in the range,

$$4l \leqslant h_{crit} \leqslant 6l \tag{6}$$

From the results of analysing Campbell's data, Table II, it was found that l for copper sulphide is 37 Å (3·7 nm). One could reasonably assume that other metallic oxides have couple-stress constants of the same order. In view of this assumption, the value of h_{crit} should be in the order of

$$h_{crit} = 10^2 \text{ Å} = 10 \text{ nm} \tag{7}$$

Rabinowicz's experimental hypothesis, that metallic oxides need to be in the order of 10 nm thick to properly lubricate a surface, is therefore possibly explained.

The question arises of whether the rise in friction at small values of film thickness is due to the presently suggested mechanism or whether it is due to film penetration and substrate contact, the latter being the intuitive explanation and, in fact, that offered by Bowden and Tabor for their experimental results with an indium film and more recently put forward by Dayson (1971). Levine and Peterson's data showed the friction coefficient to be constant over an order of magnitude range in load (i.e., 350 to 4500 gm) for all film thicknesses studied. If film penetration was significant, one should have expected a variation of friction coefficient with load, as a greater degree of film penetration would have occurred at higher load, in view of the great range of load. This effect did not occur. It, therefore, appears most likely that the explanation for the increase in friction coefficient with decreasing film thickness lies in an apparent change in the film's mechanical properties with thickness at small values of thickness. This in turn supports the theoretical model offered in the earlier part of this paper.

The good agreement between the theory and such a large body of diverse experimental results, indicates a relatively general phenomenon. Though the theory was derived for a static model, it seems to model dynamic friction well enough.

The large values of the couple-stress constants, l, found in analysing the experimental data, imply that significant effects can be expected under many normal lubrication situations. The consideration of such effects may do much to explain the vast experimental scatter now accompanying boundary lubrication resulting from liquid media containing additives and impurities and accompanying dry friction in various gaseous environments which lead to film formation on surfaces.

II. Wear of bonded solid lubricant films

The wear of bonded solid film lubricants was shown in Finkin (1970a) to occur in two very different regimes: high stress and low stress. It is well known (e.g. Hopkins and Campbell, 1968) that under high stress conditions, an optimum film thickness of 0·0002–0·0004 in. (5–10 μm) gives the longest life. The test equipment used in film development corresponds to the high stress situation, so materials development has been aimed at meeting this operating condition.

Low contact stress [say for the moment, 10,000 lbf/in.^{2} (70 MPa) and below] usually occurs in practice with conforming geometries such as those found in sliders, latches, hinges and close tolerance journals in bearings. High contact stresses normally occur in gears, cams, bearing cages and metal-working applications. Most solid lubrication experimentation utilizes configurations which press one or more flat blocks against a rotating ring or squeeze a rotating pin in vee blocks, resulting in very high contact stresses even though the results of this research are primarily intended for low stress application. This inconsistency has not been generally appreciated.

In Finkin (1970a), a careful analysis was made of the findings of seven published independent investigations carried out with a great variety of test machines and solid lubricant films under low stress conditions. It was concluded that the wear of bonded solid lubricant films, at least under conditions of low contact stress, obeys an equation of the form

$$V = KPL \tag{8}$$

where V is the total wear volume, P is the normal load, L is the total sliding distance and K is a wear coefficient representing the sliding conditions, materials properties, and environment. This is the same familiar equation that commonly represents the adhesive or abrasive wear of materials.

The wear equation for low contact stress situations, equation (8), predicts that solid lubricant film life increases linearly with film thickness. If the upper limit on thickness at which this equation is obeyed is significantly above the range 0·0002–0·0004 in. (5–10 μm), film lives can be readily obtained which are much higher than one would predict based on high contact stress experimental results. We will now examine the limits on the applicability of equation (8) as a function of film thickness, using experimental results obtained under low contact stress conditions. This will imply that the use of high contact stress testers has underestimated maximum utilizable film thickness by an order of magnitude (where clearance does not present a problem).

Rabinowicz's data (1967a) for epoxy bonded MoS_2 obeyed equation (8) for the film thickness range 3×10^{-5}–6×10^{-3} cm (12×10^{-6}–$2·4 \times 10^{-3}$ in.). Deviation from the equation occurred for greater film thickness.

The data of Johnson and Sliney (1959) and Sliney and Johnson (1957), for SiO_2 bonded PbO, obeyed equation (8) for their entire range of film thicknesses, which was 0·0005–0·004 in. (12–100 μm).

The experiments of Hagan and Williams (1964) showed wear proportional to film thickness for glass bonded MoS_2, graphite and silver for the thickness range 0·0002–0·0010 in. (5–25 μm).

The data of Hopkins and Campbell (1968) for polyimide bonded MoS_2 and Sb_2O_3 obey equation (8) for their entire range of film thickness, which was 0·0002–0·0032 in. (5–80 μm). Lipp's data (Lipp, 1968) for polyimide bonded MoS_2 at 450 °F (232 °C) obey equation (8) for his entire range of film thicknesses which was 0·0015–0·0068 in. (37–170 μm).

These experimental results show that solid film lubricants can be used to thicknesses substantially greater than is commonly believed, and consequently, solid film lubricant lives are achievable which are substantially greater than those presently obtained with small film thicknesses.

As equation (8) models wear under low contact stress, but not under high contact stress, a major question is what is the proper dividing line between these two stress regimes. Clearly it must depend on both the properties of the sliding system and materials properties. The temperature may be expected to be a major parameter as it significantly influences materials properties. For common solid lubricant films, an examination of the contact stresses seems to indicate that a dividing line of at least 10,000 $lbf/in.^2$ (70 MPa) may be temporarily taken until more experimental information becomes available.

REFERENCES

Akhmatov, A. S. (1966). "Molecular Physics of Boundary Lubrication", p. 328. Israel Program for Scientific Translations.

Bowden, F. P. and Tabor, D. (1954). "The Friction and Lubrication of Solids". Oxford University Press, Oxford.

Bowden, F. P. and Young, J. E. (1951). *Proc. Roy. Soc. Lond.* 444.

Campbell, W. E. (1939). *Trans. ASME* **61**, 633.

Campbell, W. E. and Thomas, V. B. (1939). *Trans. Electrochemical Soc.* **76**, 303.

Dayson, C. (1971). *ASLE Trans.* **14**, 2, 105.

Epifanov, G. I. and Aretisyan, I. S. (1960). Proc. Third All-Union Conference on Friction and Wear **2**. U.S.S.R. Academy of Sciences Press, Moscow. English Translation FTD-61-449, also ASTIA AD-283860, available from Office of Technical Services, U.S. Department of Commerce.

Epifanov, G. I. and Sanzharovskii, A. T. (1962). "Friction and Wear in Machinery", Vol. 15, ASME, 229.

Finkin, E. F. (1969). *Trans. ASME, J. of Lub. Tech.* **91F**, 551.

Finkin, E. F. (1970a). *Trans. ASME, J. of Lub. Tech.* **92F**, 274.

Finkin, E. F. (1970b). *Wear* **5**, 291.

Finkin, E. F. (1971). *Wear* **18**, 3, 231.

Finkin, E. F. (1973) *Trans. ASME, J. of Lub. Tech.* **95F**, 328.

Greenhill, E. G. (1948). *J. Inst. of Petroleum* **34**, 659.

Hagan, M. A. and Williams, F. J. (1964). "Proc. of USAF-SwRI Aerospace Bearing Conference" (P. M. Ku, ed.), p. 91.

Hopkins, V. and Campbell, M. (1968). ASLE Paper No. 68AM-7E-4.

Johnson, R. L. and Sliney, H. E. (1959). *J. Lub. Engrg.* 487.

Kubaschewski, O. and Hopkins, B. E. (1967). "Oxidation of Metals and Alloys", 2nd. edn., p. 182. Butterworths, London.

Levine, E. C. and Peterson, M. B. (1951). NACA TN-2460.

Lipp, L. C. (1968). *Lub. Engrg.* **24**, 154.

Rabinowicz, E. (1967a). *ASLE Trans.* **10**, 1.

Rabinowicz, E. (1967b). *ASLE Trans.* **10**, 400.

Sliney, H. E. and Johnson, R. L. (1957). NACA Research Memorandum No. RME57B15.

Examination of surface coatings by microwave radiometry

M. A. K. Hamid and J. D. Hunter

Antenna Laboratory, Department of Electrical Engineering,
University of Manitoba, Winnipeg, Canada

I. Introduction

In modern technological processes, where greater emphasis is being placed on increasing processing speeds, novel electrical techniques for continuous and contactless monitoring and control of non-electrical quantities are in increasing demand. The parameters of particular interest are high measurement accuracy, low time constant, high sensor reliability under the environmental conditions of a modern plant (vibrations, dust, water and oil vapours, etc.), low investment and maintenance costs, and safety from leakage or radiation hazards.

The technique of microwave radiometry proposed in this paper involves the use of an extremely sensitive radio receiver and a reference noise source. This receiver is provided with a highly directional antenna which is usually a parabolic reflector. The technique consists of directing the antenna at the object or region of space under observation and measuring the natural emission as received by the antenna, after comparison with a reference noise source and appropriate amplification in a superheterodyne receiver. Experience has shown that this detection system offers many basic advantages such as:

(i) High detection capability. The detected signal is determined by (*a*) target emissivity relative to background, geometrical dimensions and orientation relative to background and antenna, (*b*) antenna gain, beamwidth and bandwidth, and (*c*) receiver gain, bandwidth, time constant, noise figure and sensitivity.

(ii) Simple, passive, continuous, contactless, pollutionless, non-destructive, low power and compact (particularly with solid state receivers and active antennas) for mobile use.

(iii) Permits scanning, multifrequency operation, data storage, correlation and display at a central location.

(iv) Has an all-weather capability (particularly at lower microwave frequencies).

Due to these advantages, microwave radiometers have been found very useful for solving many scientific and engineering problems and are potentially capable of solving many others. Some typical applications of microwave radiometry for remote sensing are:

(i) Radio astronomy. Tracking and detection of emission from the sun and radio stars, drift of continents, etc.

(ii) Environmental protection. Measurement of snow and ice thickness; monitoring of flood conditions, hail and rain storms; oil pollution surveillance in lakes, rivers and high seas; monitoring of air and electromagnetic pollution in residential areas.

(iii) Agricultural industry. Monitoring of permafrost, subsurface water, and moisture profiles; optimum frequency for microwave heating and insect control in foods; counting and grading of agricultural products.

(iv) Paper and textile industry. Measurement of moisture content and thickness of web materials.

(v) Aviation industry. Detection of wind and rain storms; detection of airborne aircraft, landing strips and coastal lines.

(vi) Mining industry. Detection of oil, gas and mineral deposits.

Applications in the fields of hydrology, oceanography and sediment classification are discussed in detail elsewhere (Adey, 1970, 1971; Edgerton *et al.*, 1971).

II. Basic concepts

The underlying principle of microwave radiometry is that all objects in the physical universe which are not at a temperature of absolute zero radiate energy in the form of electromagnetic waves. This is not surprising when we realize that matter is composed of electrically charged particles which are constantly undergoing energy changes because of thermal agitation.

In 1860 Kirchhoff demonstrated that a good absorber is also a good radiator. An object which absorbs all radiation incident on its surface is commonly known as a black body and has a reflectivity $r = 0$, absorptivity $a = 1$, emissivity $\varepsilon = a = 1$, while the reverse is a white body. All other bodies in nature are known as grey bodies.

The apparent temperature T_a of an object is the temperature that a black body would have to be at in order to radiate the same amount of power. Since the emissivity of metals is below that of most other grey materials, such as water, oil and paint, there is, in general, strong contrast between metallic substrates and dielectric coatings.

From the classical theory of black body radiation, any perfectly absorbing body emits radiation at a frequency f(Hz) into an isothermal enclosure in accordance with Planck's radiation law:

$$e_{bf}(T) = \frac{2\pi h f^3}{c^3[\exp(hf/kT) - 1]} \text{ watts}/(\text{m}^2 \text{ Hz})$$

where T is the absolute temperature of the body in K, c is the velocity of light (3×10^8 m/s), h is Planck's constant (6.62×10^{-34} J s), and k is Boltzmann's constant (1.38×10^{-23} J/K). It should be noted that most of this power is concentrated near optical frequencies.

Provided T is not near zero, the quantity hf/kT is small at microwave frequencies (0.2–200 GHz) and Planck's law can be accurately approximated by the Rayleigh–Jeans equation:

$$e_{bf}(T) = \frac{2\pi k T f^2}{c^3} \text{ watts}/(\text{m}^2 \text{ Hz}).$$

A system suitable for receiving this power is shown schematically in Fig. 1.

If the black body completely fills the acceptance beam of the radiometer antenna (i.e. large body–narrow beam approximation), the received power is given by (Harris, 1960)

$$P_b = kTB \text{ watts}$$

where the bandwidth B (Hz) is constant for a given radiometer. Thus, at microwave frequencies, the power received by a radiometer is directly proportional to the absolute temperature of the black body.

The power radiated at frequency f (Hz) by a grey body at temperature T can be related to the radiation from a black body at the same temperature by

$$e_{gf}(T) = \varepsilon e_{bf}(T) \text{ watts}/(\text{m}^2 \text{ Hz})$$

where the emissivity ε is a factor less than unity and is, in general, a function of the temperature, frequency and direction. Assuming that ε is constant over bandwidth B at microwave frequencies and over the angles of reception of the radiometer antenna, the power received by the radiometer from a grey body is

$$P_g = kT_aB \text{ watts}$$

where T_a, the apparent temperature, is related to the temperature T of the grey body by

$$T_a = \varepsilon T$$

Thus, the power received by a microwave radiometer from a grey body at a given temperature is directly proportional to the emissivity ε.

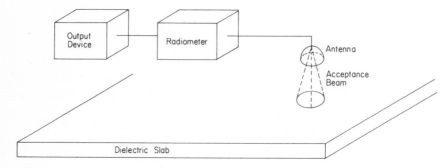

FIG. 1. Radiometric detection from a dielectric slab.

III. RADIATION FROM A DIELECTRIC SLAB

In this and following sections expressions are derived which indicate the relationships between the apparent temperature, thickness, and attenuation constant of a dielectric. In these derivations, no account is taken of the effect of multiples reflections from dielectric–air interfaces.

Consider the dielectric slab of thickness t in Fig. 2. The upper surface of the slab lies in the x–y plane. The radiation from an elemental volume at (x, y, z) is (Sparrow and Cess, 1966)

$$j_{gf}(T) = \frac{\kappa}{\pi} e_{bf}(T) \text{ watts/(m}^3 \text{ Hz unit solid angle)}$$

where the volumetric absorption coefficient κ (m^{-1}) is a function of temperature, frequency and direction. The solid angle ω subtended by unit surface area at $(0, 0, 0)$ to the elemental volume is

$$\omega = \frac{z}{r^3}, \qquad r^2 = x^2 + y^2 + z^2.$$

Hence, the total power density at the unit surface area is

$$J_{gf}(T) = \frac{e_{bf}(T)}{\pi} \int_0^t \int_{-\infty}^{\infty} \int_{-\infty}^{\infty} \frac{\kappa z\, e^{-\alpha r}}{r^3}\, dx\, dy\, dz$$

where α (m^{-1}) is the attenuation constant of the dielectric. For a non-scattering dielectric $\kappa = \alpha$, and it can be shown that

$$J_{gf}(T) = e_{bf}(T)H(\alpha t) \text{ watts/(m}^2 \text{ Hz)}$$

where

$$H(x) = 1 - e^{-x}(1-x) - x^2 E_1(x).$$

$$E_1(x) = \int_x^{\infty} \frac{e^{-t}}{t}\, dt$$

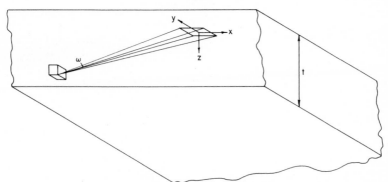

FIG. 2. The geometry of the problem.

The radiation $e_{gft}(T)$ from the dielectric surface can be related to the incident radiation by a transmission coefficient τ ($|\tau| \leqslant 1$), which in general will be a function of α, t, f, T and direction. Thus,

$$e_{gft}(T) = \tau e_{bf}(T)H(\alpha t) \text{ watts/(m}^2 \text{ Hz)}$$

and the emissivity of the surface as a function of thickness t is

$$\varepsilon(t) = \tau H(\alpha t).$$

The power received by a radiometer above the surface is

$$P_g = \kappa\tau TBH(\alpha t) \text{ watts}$$

and T_a is given by

$$T_a = \tau T H(\alpha t) K.$$

The function $H(\alpha t)$ is plotted in Fig. 3. It is apparent that for $\alpha t \gg 1$, $H(\alpha t) \approx 1$, and hence

$$\varepsilon = \varepsilon(t) \approx \tau, \qquad \alpha t \gg 1,$$

where ε is the surface emissivity of the dielectric.

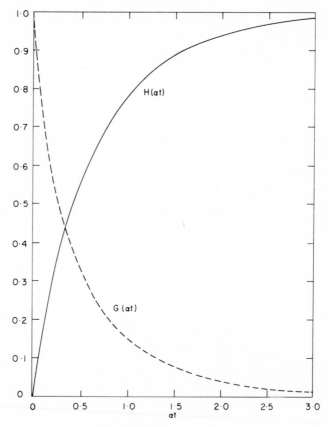

FIG. 3. $H(\alpha t)$ and $G(\alpha t)$ vs. αt.

IV. RADIATION FROM A DIELECTRIC SLAB ON A PERFECTLY CONDUCTING SUBSTRATE

Consider a dielectric slab of thickness t deposited on a perfectly conducting (white) surface. Since any radiation incident upon the conductor is totally reflected, it follows from image theory that the apparent thickness of the dielectric is $2t$, and therefore the apparent temperature is

$$T_a = \tau T H(2\alpha t) \text{ K}.$$

V. ATTENUATION CONSTANT α

From electromagnetic theory, the attenuation constant α is related to the complex permittivity ε^* of the dielectric by

$$\alpha = \frac{2\sqrt{2}\,\pi f \varepsilon''}{c} \left[\varepsilon' + \{(\varepsilon')^2 + (\varepsilon'')^2\}^{1/2}\right]^{-1/2}, \qquad \varepsilon^* = \varepsilon_0(\varepsilon' - j\varepsilon'').$$

Typically, α increases with frequency.

VI. SENSITIVITY CONSIDERATIONS

The sensitivity of a microwave radiometer can be described in terms of the minimum variation in apparent temperature which the radiometer can detect. Figure 3 indicates that the greatest variation of T_a with αt occurs when $\alpha t < 1$. This condition can be used to give a general indication of the region of applicability of radiometry for measurements of αt.

TABLE I

Table of attenuation constants for various materials

Dielectric	Temp. (K)	Freq. (GHz)	α (m^{-1})
Amber (fossil resin)	298	3	8·95
Fiberglas BK174 (laminated)	297	3	3·89
		10	16·03
Glass, phosphate (2% iron oxide)	298	3	0·66
		10	1·97
Neoprene Compound (38% GN)	297	3	4·24
		10	11·0
Paint (dry) (water or oil based)	300	2·45	~2·2
Paper (royal grey)	298	3	5·73
		10	13·58
Plexiglas	300	3	0·58
		10	2·28
Shellac, natural XL (3·5% wax)	301	3	2·71
	343	3	8·45
Water	274·5	3	173
		10	1201
	358	3	25·9
		10	593

Table I gives values of α for various materials at different frequencies and temperatures. In Fig. 4 the variation of T_a with t is plotted for some of these materials when deposited on a perfectly conducting substrate. The slope of these curves, which is a measure of the sensitivity of T_a with thickness, is given by

$$\frac{\partial}{\partial t}\left(\frac{T_a}{\tau T}\right) = 4\alpha G(2\alpha t)$$

where

$$G(x) = e^{-x} - xE_1(x)$$

is plotted in Fig. 3.

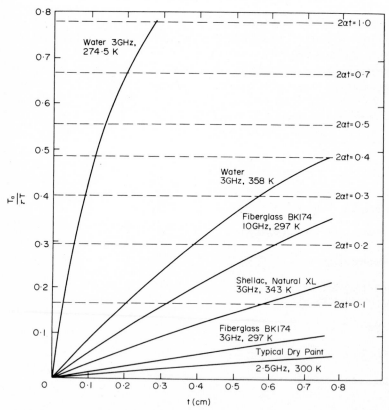

FIG. 4. Variation of apparent temperature with thickness for various dielectrics on a perfectly conducting substrate.

Typically, at 10 GHz and 297 K, Fiberglas BK174 has $\alpha = 16{\cdot}0\,\text{m}^{-1}$, and hence a detectable change of 1 K in T_a is caused by a variation of the order of $0{\cdot}1$ mm when $t = 3$ mm.

Changes in the composition of dielectric coatings can also be detected using radiometry. For example, a layer of water-based paint deposited on a metal surface can exhibit an apparent temperature variation of 100 K while drying, thereby allowing accurate monitoring of the drying process.

ACKNOWLEDGEMENT

This research was supported by the Defence Research Board of Canada, Grant 3801–42.

REFERENCES

Adey, A. W. (1970). "A Survey of Sea-Ice-Thickness Measuring Techniques". Dept. of Communications, CRC Rept. 1214, Ottawa.

Adey, A. W. (1971). AGARD, EM Wave Prop. Panel XVII Am. Symp. on Prop. Limitations in Remote Sensing, Colorado.

Edgerton, A. T., Ruskey, F., Williams, D., Stogryn, A., Poe, G., Meeks, D. and Russell, O. (1971). Aerojet-General Corp. Tech. Rept. No. 9016R-8 (National Technical Inform. Service, U.S. Dept. of Commerce, Accession No. 720388).

Harris, D. B. (1960). *Microwave Journal* 3, April pp. 41–46, May pp. 47–54.

Sparrow, E. M. and Cess, R. D. (1966). "Radiation Heat Transfer". Brooks-Cole, California.

Periodic structure of ultra-thin semiconductor films

Arun P. Kulshreshtha*

Department of Physics, Middle East Technical University, Ankara, Turkey

I. Introduction

In a recent publication Ovsyannikov *et al.* (1971) have theoretically considered the properties of periodic semiconductor structures, consisting of several alternating layers of n- and p-type semiconductor films. The diffusion of electrons from the n- into the p-type regions, and similarly of holes in the opposite direction, gives rise to the depletion of carriers in the space in the vicinity of the p–n junctions. For sufficiently thin layers, the space charge region extends throughout the film such that the average carrier concentration in the structure becomes several orders of magnitude smaller than the ionized impurity. The concentration in the individual layers thus approaches the value of the intrinsic carrier concentration in the semiconductor and the system as a whole becomes homogeneous from the point of view of mobile carriers. In the multilayered structure the concentration of ionized impurity, however, varies periodically along the thickness of the structure, which therefore behaves as a quasi-homogeneous medium. The properties of a structure electronically homogeneous, but with periodically distributed ionized impurities, are expected to be very different from that of a generally known homogeneous semiconductor.

When the thickness of the p- and n-regions are small compared to the diffusion lengths of the minority carriers, and also compared to the depletion layer width, the periodic electrostatic potentials due to alternating p- and n-layers give rise to additional one-dimensional sub-bands in the conduction and valence bands. In so far as the diffusion lengths are large enough, the scattering processes remain of little importance. The sub-band widths and the forbidden gaps between any two sub-bands can be modified by varying the

* Present address: Physics Department, Lancaster University, Lancaster, England.

film thicknesses and the impurity concentrations. This may provide super-conductivity, due to the pairing of electrons in various thin films, and negative differential resistance and controlled oscillations in current-voltage characteristics due to the tunnel transitions between the sub-bands.

If the p- and n-layers are deposited alternately without annealing the film, the junctions could well be approximated to the abrupt type. The depletion region width W in a two-sided step junction is given by Sze (1969):

$$W = \sqrt{\frac{2\varepsilon_s kT}{e^2} \cdot \frac{N_A + N_D}{N_A N_D} \cdot \ln \frac{N_A N_D}{n_i^2}},$$

where N_A and N_D are the acceptor and donor concentrations in the p- and n-type films respectively; n_i is the intrinsic carrier density; ε_s is the permittivity of the semiconductor; e is the electronic charge; k is Boltzmann's constant; and T is the ambient temperature in degrees Kelvin.

In order to obtain ideal periodicity in the structure, both p- and n-films should be equally doped, such that $N_D = N_A = N$ (say), and then upon substituting the value of the intrinsic carrier concentration in terms of the band gap, E_g, and the conduction and the valence bands densities of states, N_c and N_v, respectively, we obtain:

$$W = \sqrt{\frac{8\varepsilon_s}{e^2} \cdot \frac{kT}{N} \cdot \ln \frac{N}{(N_c N_v)^{1/2}} + \frac{4\varepsilon_s E_g}{e^2 N}}.$$

As mentioned before, the sub-bands can be obtained if the depletion region width is sufficiently large compared to the thickness of the individual layer. This evidently puts a limitation on the dopant concentration and the choice of material (through the band gap E_g). Higher temperatures assist in the formation of sub-bands, as may be seen from the above equation.

By substituting appropriate values we find that for an acceptor and donor density of about $10^{21}/m^3$ the depletion layer widths in germanium, silicon and gallium arsenide are 0·7, 1·0 and 1·5 µm respectively at room temperature. The diffusion lengths of carriers in Ge or Si are of the order of 10^4 µm, and are much less ($\simeq 100$ µm) in GaAs. So we should expect the formation of sub-bands in a multilayered structure of any of these semiconductors, provided the individual layers are sufficiently lightly doped and their thickness is of the order of 0·1 µm.

II. EXPERIMENTAL RESULTS

As an initial phase of the programme we chose to investigate the Ge thin film multilayered structure. Germanium was vacuum evaporated at 10^{-5} torr

(1·3 mPa) on a cleaned glass substrate. The heater source was graphite. Davey and Montgomery (1963) have shown that with standard evaporation techniques and boat materials, Ge films exhibit strong p-type behaviour such that the resulting carrier concentration in the film is always of the order of 10^{24}–10^{25} m^{-3} regardless of the electron concentration in the source material, varying from 10^{19} to 10^{25} m^{-3}. The film resistivities in their experiments varied between 0·01 and 0·30 ohm cm.

Our experiments do not agree with the observations made by Davey and Montgomery (1963). We found the resistivities of the deposited films of the order of several kohms and the type of individual films was n- or p- type depending on the source. Once a p–n junction of a very thin layer is formed, however, the meaning of p- or n-type layers loses its sense because of the strong depletion of carriers throughout the film. It is only the periodicity of the ionized impurity, and their densities, that remain of any particular importance.

We are at present investigating the current and capacitance-voltage characteristics in these structures and the detailed results will be published elsewhere.

REFERENCES

Davey, J. E. and Montgomery, M. D. (1963). Trans. 10th National Vacuum Symposium, Amer. Vacuum Soc., Boston.
Ovsyannikov, M. I., Romanov, Yu. A., Shabanov, V. N. and Loginova, R. G. (1971). *Sov. Phys. Semiconductors* **4**, 1919.
Sze, S. M. (1969). "Physics of Semiconductor Devices". Wiley-Interscience, New York.

Glossary

Terms are defined in the context of the chapter in which they are used; the initials given are those of the author(s) of that chapter.

Abnormal regime—the abnormal regime of a cold cathode glow discharge is attained when the applied voltage is raised sufficiently for the potential difference across the cathode dark space (and the current density) to rise with an increase in the applied voltage, i.e. the glow discharge has undergone a transition from the normal regime (q.v.). (LH)

Absorb—see "adsorb".

a.c.—alternating current.

Activated step—a step which requires a fixed amount of energy to be overcome. It is the activation energy. (JCV)

Activity (ion)—the concentration of an ion corrected for deviations from ideal behaviour. The molecular concentration of the ion multiplied by a factor called the activity coefficient. (RPK)

Adhesion—this is said to occur when surfaces in frictional contact weld together at the high spots or asperities. (JCG)

Adsorb—material is *ad*sorbed onto a surface, or *ab*sorbed into the bulk.

Ageing—improvement of mechanical connections (adherence) of electrical or magnetic materials or devices (characteristic data) by heat treatment. (AP & WS)

Air cap—the part of the spray gun which controls the amount of air and direction of its flow in the process of atomization. (JM)

Amorphous materials—those materials that show no atomic ordering, even using electron diffraction techniques. This is in contrast to the ordering shown by single or polycrystalline materials. (DSC)

Anneal (to)—to bake in order to obtain a rearrangement of the sample crystallographic structure. (JCV)

Anodizing—the electrical field-enhanced growth of an oxide on a parent metal immersed in an electrolyte. (DSC)

Anodizing—the formation of a tough, corrosion-resistant oxide coating on certain metals (especially by anodic treatment in a suitable aluminium electrolyte). (HS)

Atomization—the operation of reducing the molten material from the end of the wire into fine particles. (DRM)

Austenite—a solid solution of one or more elements in gamma iron. In steel this is normally a solid solution of carbon in iron but alloy austenites may contain, for example, nickel. (JCG)

Autocatalytic reaction—a reaction which catalyses itself, e.g. electroless plating or reaction of nitric acid on copper. (RPK)

Base part—the subject part to which the surface coating is applied. (HNW)

Beam-lead—free-supporting end of a metal line, cantilevering from a semiconductor chip; to be connected to a conducting part of a substrate. (AP & WS)

B.H.N.—Brinell hardness number. See "hardness".

Blasting—a method of surface roughening by a stream of sharp angular abrasive, forcibly projected against the surface to be metallized. (DRM)

Bleeding—when a prime coat dissolves in a subsequently applied top coat and therefore comes through onto the surface. The prime coat is said to bleed through the top coat and the phenomena is referred to as "bleeding". (LS)

Breech—the closed end of the gun barrel where the fuel gases and powder enter prior to detonation. (RS)

Case depth—the depth of chemical alteration during coating. By convention, this is usually that portion in which the alteration is optically discernible using appropriate etching and microscopic techniques. (RLW)

Cementite—a compound of carbon and iron chemically known as iron carbide and having the formula Fe_3C, although metallographically the term may cover the presence of other carbides present in steel. (JCG)

Chalking—see "surface chalking". (SK)

Chemical polishing—uses pastes which have both an abrasive and an etching effect on a substrate surface. (JCV)

Chemical potential—an energy parameter pertaining to a chemical component. It is the increase of the free energy of a thermodynamic system which is the result of the addition of a small quantity of that component at constant pressure and temperature, divided by this quantity. It is important to note that in the case of complete equilibrium, pressure, temperature and the chemical potentials of the components have uniform values through the system. (MJS)

Chromating—production of a corrosion-resisting yellowish film on metals (especially zinc, cadmium and aluminium) by immersion in acid solutions containing chromium salts. (HS)

Chromizing—a surface treatment at elevated temperature, generally carried out in pack, vapour or salt bath, in which an alloy is formed by the inward diffusion of chromium into the base metal. (ASM Handbook, Vol. 1, 8th edn., Definitions.) (RLW)

Cissing—poor surface wetting of the substrate, manifested by small areas of apparent "misses" with no or little coating. Another term is "fish eyes". (SK)

c.l.a.—centre line average.

Contact angle—the angle formed by a liquid in contact with a solid measured inside the liquid. (AP)

Continuous coating—a process in which a material is continuously coated by another material. (AP)

Cross-hatching—this is an adhesion test. A series of parallel cuts are made through the coating in one direction and a second series at right angles to the first. In strict specification tests, normally eleven cuts are made, $\frac{1}{32}$ in. (0·79 mm) in each direction, forming 100 squares. Adhesive tape is pressed firmly down on the "cut" and sharply pulled off. The number of squares remaining intact gives a measure of the adhesion; for high performance, no loss of adhesion is permitted. (SK)

Cyanine—any of a group of bluish-green crystalline dyes. (LS)

d.c.—direct current.

Detonation phase boundary—the detonation phase boundary is the zone extending from the shock wave to the point at which chemical equilibrium is reached. The thickness of this zone is usually fractions of a millimetre. This detonation phase, referred to as the Chapman-Jouget or C-J phase, is dependent upon temperature, pressure and stoichiometry of the fuel gases used. (RS)

Die-casting—see "zinc alloy die casting".

Dielectric relaxation time—see "relaxation time".

Diffusion coatings—(1) an alloy coating produced at high temperatures by the inward diffusion of the coating material into the base metal. (2) Composite electrodeposited coatings which are subsequently interdiffused by thermal treatment. (ASM Handbook, Vol. 1, 8th edn., Definitions.) (RLW) (See also "pack cementation coating" and "vapour phase coating".)

Diffusion couple—the designation for an interface between two or more dissimilar materials generated under conditions producing diffusion. (RLW)

Dispersant—a substance designed to keep finely divided particles in liquid suspension. (DDG)

Dispersion—the state of finely divided particles in liquid suspension. (DDG)

Dispersion forces—interactions between atoms due to an asymmetrical distribution of their electrons. (AP)

D.P.H.—this stands for "diamond pyramid hardness" and is synonymous with the Vickers hardness number, reference being made to the diamond-shaped indentor used in the Vickers test. (DRM)

Drive-line system—system of metal lines which is necessary to operate (drive) memory devices. It connects the memory elements with the word lines (input) and the sense lines (output). (AP & WS)

Electrocoagulation—the precipitation of charged colloidal particles from the dispersion upon the conductive substrate surface. (AW)

Electrodeposition—deposition of a film of a substance on a substrate surface by means of electrical forces. (AW)

Electrogalvanizing—electroplating with zinc. (HS)

Electrokinetic (ζ, or zeta) potential—the potential drop in the diffuse electrical double layer in an electrolyte near the interface with an electrode. (AW)

Electroless—a process of plating or coating not involving electrodes, i.e. chemical deposition. (MS)

Electroless plating—deposition of a metallic coating by a controlled chemical reduction which is catalysed by the metal or alloy being deposited. (RPK)

Electrolysis—transport of ions through the electrolyte of an electric cell by means of electrical forces. (AW)

Electromotive series—list of metals arranged according to their tendencies to pass into ionic form by losing electrons. (RPK)

Electron gun—the source of electrons in a cathode ray tube. (DDG)

Electronegative metals—the reactive or base metals with strong tendencies to form ions. (RPK)

Electroosmosis—osmotic dehydration of an electrodeposited film by means of electrical forces. (AW)

Electrophoresis—the migration of naturally charged suspended particles in a liquid medium in an electric field. (HS)

Electrophoresis—transport of charged colloidal particles of one phase through another phase by means of electrical forces. (AW)

Electrophotographic industry—that industry which concerns itself with the reproduction of images through the use of uniformly charging a surface and then exposing that surface to light in order to form a latent charge image (LS)

Electropositive metals—metals arranged in the noble end of the electromotive series. They resist the formation of ions and are displaced from their solutions by the electronegative metals. (RPK)

Eutectoid—in a binary alloy, coexistence of two phases appearing simultaneously during cooling. (JJC & CA)

Extraction replica—a replica in which part of the surface is removed with the replicating film. The technique is often used to examine the discontinuous (early) stages of growth of metallic films on water-soluble substrates. The surface is coated with a self-supporting carbon layer and this layer, together with the metal film, is removed by floating off at a water surface. The carbon is picked up on grids and examined by transmission under the electron microscope. (RC)

Fatigue—metals may fail by fatigue due to the imposition of repeated or alternating stresses having a maximum value less than their tensile strength. Failure is initiated by a very small crack which progresses sufficiently to cause fracture. (JCG)

Ferrite—a solid solution of one or more elements in alpha iron, the solute normally being carbon. (JCG)

Field—is an accepted expression in photomicrography. It means a selected area of microscopic dimensions chosen on a polished, or polished and etched, microsection. (JCG)

Film build—the thickness (real or apparent) of the dried paint film on the work surface. (JM)

Flip-chip—interconnection technique in which many contacts between an integrated circuit and a wiring system on a substrate are performed at the same time by bonding. The metal is localized at "bumps" on the surface of the semiconductor chip. (AP & WS)

Flocculate (to)—the coagulation of finely divided particles into particles of a greater mass. (DDG)

Flux—a granular material, used in welding to protect the molten surface from oxidation, to aid in the slow cooling of the weld deposit, and to shield the operator from radiation effects of what would otherwise be an open arc. Although fluxes are usually chemically neutral with respect to the weld metal, on occassion they do have controlled additions of alloying elements for the express purpose of changing the chemical and physical characteristics of the weld deposit. (HNW)

Formvar—trade name for 4g polyvinylformal in 1 litre of ethylene dichloride. (MS)

Free energy G—also known as the Gibbs function or thermodynamic potential. It should be noted that the symbol F is often used particularly in American literature to denote Gibbs free energy. (JCV)

Frit (to)—the assembly of glass parts by means of glass powder which is caused to sinter by heating. (DDG)

Galling—see "scuffing". (JCG)

Gaseous anodization—the electrical field-enhanced growth of an oxide on a parent metal suspended in a low pressure gas discharge. (DSC)

Halftone—a printed illustration in which the lights and shades of the original are represented by small or large dots. (SAP)

Halftone picture—a picture whose densities are formed by dots of various sizes. (SK & PO)

Hardness—the hardness of a material is its resistance to displacement by an indentor under given load, this being determined from the projected area of the indentation surface. In the measurement of coatings, very small impressions (hence very low loads) are necessary to avoid perforation of the coating when taken normal to the surface, or extension of the indent into the substrate, or off the edge when taken end-on in the coating. In conventional load ranges ($\sim$ kg), the hardness number is pretty well independent of the load used, but in the micro range ($\leqslant 100$ g), may vary according to specific indentor characteristics, probably because of difficulty controlling the geometry at the utmost tip of the indentor. (Adapted from RLW)

There are several hardness tests, and corresponding hardness scales, in use. The following table provides an approximate comparison between scales for a uniform steel. Data is taken from Mr. H. N. Watson (see chapter 23) and the A.S.M. (American Society for Metals) Handbook.

VPN or DPN Vickers diamond pyramid hardness (kg/mm^2)	BHN Brinell hardness number 10 mm steel ball, 3000 kg load (kg/mm^2)	Rockwell C 120° diamond cone 150 kg load	Rockwell B $\frac{1}{16}$ in. (1·6 mm) steel ball 100 kg load	Sclero-scope Shore Model C	Moh's Scale
	800	72		100	
	760	70		98	
800	725	67		96	
740	682	65		93	
665	626	62		87	8·0
630	590	59		81	
590	552	56	$\sim$120	76	7·5
540	502	52		70	
480	451	47	$\sim$115	65	7·0
430	401	43		58	
370	351	38	$\sim$110	51	6·5
310	301	33		45	6·0
250	249	25	102	38	5·5
220	224	21	98	33	
210	200	16	95	31	5·0
190	185	12	92	29	
175	170	7	88	27	
160	160	3	85	25	
155	150	1	83	24	4·5
150	145		81	23	
145	140		78	22	
135	130		74	22	

Brinell hardness numbers above 400 are not considered reliable.
The KNOOP scale is approximately equal to the Vickers scale.
The MEYER scale varies as 1–2 times the equivalent Brinell value.

Harmonic oscillator—a particle, mass m, moving along, say, the x-axis subject to a force which is proportional to the distance x of the particle from the equilibrium position $x = 0$ (this force law is due to Hooke). The proportionality constant is $4\pi^2 m v^2$. When the harmonic oscillator is used as a simple model for an atom, the oscillating mass is an outer shell electron (or a collection of outer shell electrons), and the position $x = 0$ is the position of the nucleus. (MJS)

Hiding—the ability of a pigmented coating to mask or cover features or colours over which it is applied, is referred to as hiding. (LS)

Holes; recombination—in a semiconductor like Si a conduction electron e^- can be produced by, for instance, a temperature increase. In such a case we can write a reaction equation: $Si \rightarrow Si^+ + e^-$. The symbol "$Si^+$" means: a Si-atom which has lost a valence electron (this valence electron has become a conduction electron). A hole, h^+, is defined as the difference $Si^+ - Si$. The reaction equation becomes: $0 \rightarrow h^+ + e^-$. The reaction equation $h^+ + e^- \rightarrow 0$ indicates recombination. In order to become a conduction electron, a valence electron must as least overcome the energy difference $E_V - E_C$ (see Fig. 5 and legend). (MJS)

Hydrogen bond—molecular interactions between a hydrogen atom and a polar group. (AP)

i.d.—internal diameter.

impact test—a test to determine the behaviour of materials when subjected to high rates of loading, usually in bending, tension or torsion. The quantity measures is the energy absorbed in breaking the specimen by a single blow, as in the Charpy or Izod tests. (ASM Handbook, Vol. 1, 8th edn., Definitions.) As used by Wachtell, the energy required to deform a test bar under impact to the point where chipping or spalling of the coating first occurs. (RLW)

Imperial units—values of some imperial units in terms of SI units.

Length

1 in.	**25·4** mm
1 ft	**304·8** mm
1 yd	**0·9144** m
1 mile	**1·609 344** km

Area

1 in.^{2}	**645·16** mm^2
1 ft^2	0·092 903 0 m^2
1 yd^2	0·836 127 m^2
1 acre	4046·86 m^2
	(i.e. 0·404 686 ha)
1 mile^2	2·589 99 km^2

Volume

1 in.^{3}	16 387·1 mm^3
1 ft^3	0·028 316 8 m^3
1 UK gal	4·546 09 dm^3

Velocity

1 ft/s	**0·3048** m/s
1 mile/h	**0·447 04** m/s

Mass

1 lb	**0·453 592 37** kg
1 ton	1016·05 kg
	(i.e. 1·016 05 tonne)

Density

1 lb/in.^{3}	2·767 99 × 10^4 kg/m^3
1 lb/ft^3	16·0185 kg/m^3
1 lb/UK gal	0·099 776 3 Mg/m^3
	(i.e. 0·099 776 3 kg/dm^3)

Force

1 pdl	0·138 255 N
1 lbf	4·448 22 N

Pressure

1 lbf/in.^{2}	6·894 76 kPa

Energy (work, heat)

1 ft pdl	0·042 140 1 J
1 ft lbf	1·355 82 J
1 cal	**4·1868** J
1 Btu	1·055 06 kJ

Power

1 hp	745·700 W

Ink distribution system—transfers an ink film of a controlled thickness from the duct to the printing cylinder. (SK & PO)

Ink duct—the reservoir of ink in an ink distribution system. (SK & PO)

Instability constant—the equilibrium constant for the dissociation of a complex particle is called the instability constant of the complex particle. (RPK)

Intermetallics—an intermediate phase in an alloy system, having a narrow range of homogeneity and relatively simple stoichiometric proportions, in which the nature of the atomic binding can vary from metallic to ionic. (ASM Handbook, Vol. 1, 8th edn., Definitions.) (RLW)

Inverted beam-lead—free-supporting end of a metal line, cantilevering from a wiring system on a substrate; to be connected to a metallic pad of a semiconductor chip. (AP & WS)

Lipophilic—affinity to oil high compared with affinity to water, causing easy mixing of oil and a substance or low contact angle of oil on a surface. (SK & PO)

Martensite—a microconstituent in quenched steel having an acicular pattern and consisting of a supersaturated solid solution of carbon in alpha iron. (JCG)

Metal acetylacetonates—metal complex with β-di-ketone of low decomposition temperature. (e.g. $[CH_3COCH = C(CH_3)O]_2Ni$.) (AP & KH)

Metal carbonyl—low boiling point, group V to VIII metal-CO complex. (AP & KH)

Metallizing—the process of spraying metals or intermetallics, in molten or semi-molten condition, to form a spray deposit bonded to the base material. (DRM)

Microprobe—a device permitting chemical analysis on a microscale using X-ray analysis methods. Precisely aligned with an optical microscope, a focused electron beam spot (0.5 μm diam.) produces X-radiation which, properly analysed, permits chemical identification of microconstituents. (RLW)

Molar fraction—number of moles of a particular component divided by the total number of moles. (AP)

Monodisperse—a dispersion characterized by uniform sized particles. (SBS, RB, DAD & JWB)

Nip—the zone of contact between two cylinders or a cylinder and a plate. (SK & PO)

Nital—is a solution of nitric acid in alcohol. 2% nital means a solution of 2% nitric acid in alcohol—i.e., 2 parts nitric acid, 98 parts alcohol. (JCG)

Non-galling—having low coefficients of friction and the ability of resisting galling or seizing when subjected to sliding friction with a mating part. (HNW)

Normal regime—this is a regime of a cold cathode glow discharge in which the cathode is incompletely covered with a glow that increases in area as the total current flowing to the cathode rises; in this regime the cathode current density and the potential difference across the cathode dark space are constant. (LH)

o.d.—outer diameter.

Ohms per square—see "sheet resistance".

Organometallic compound—organic compound containing metal which is directly bonded to carbon. (AP & KH)

Organometallic compound—a compound whose molecule is made of a metal and an organic radical. (JCV)

Overspray—the percentage by weight of the spray material that does not land on the object to be coated. (RPC)

Overspray—sprayed paint which misses the surface to be painted. (JM)

Oxidation—a chemical reaction involving loss of electrons from a reactant. (RPK)

Pack cementation coating—a method of diffusion coating wherein the part to be coated is embedded within a retort, in a powdery or granular compound containing the coating material(s). Although diffusion coating can and does result from solid/solid contact where particles touch the workpiece, the bulk of the material transfer occurs through a gas phase either generated within the compound (pack) by appropriate additives, or percolated through the powdered mass from an external source. (RLW) (See also "diffusion coatings".)

Pearlite—a microconstituent of steel and cast iron consisting of a lamellar aggregate of ferrite and cementite. (JCG)

Perfect solution—a solution in which the enthalpy and the entropy of mixing equal zero. (AP)

Photoconductive material—a material which exhibits low electrical conductivity when in the dark but which becomes conductive when illuminated. (LS)

Photoresist—a varnish, sensitive to ultra-violet light, which is spread onto a surface and dried. After illuminating through a mask, a positive or negative image can be developed, comprising the resistant parts of the film. (AP & WS)

Picking-up—see "scuffing". (JCG)

Pickling—refers to a process used for the removal of scale and other unwanted contaminates from the surfaces of iron and steel (and sometimes non-ferrous metal) parts. It is usually carried out in acid or alkaline solution. (JCG)

Picture information—mathematically defined as $\sum_{i=1}^{m} p_i \log_2 p_i$ (bits/element) in which p_i is the probability of a density level and m the number of density levels. (SK & PO)

Plasma oxidation—the growth of an oxide on a parent metal using ions created by an r.f. discharge. (DSC)

Polyamides—thermoplastic powders, in which the structural units are linked by amide groupings. Typical examples are nylon powders. (SK)

Polyimide—polymer foil, resistant to temperature and organic solvents, prepared from condensation of tetracarbonic acids with organic diamines. An example of a commercial foil is "Kapton" (Du Pont). (AP & WS)

Porous layers—layers that are permeable to fluids. (DSC)

Positive column—the plasma region of a cold cathode glow discharge adjacent to the anode. The positive column has a high electrical conductivity and almost equal concentrations of positive ions and electrons so that only a low voltage gradient is required for electrons to drift to the anode. (LH)

Pneumatic atomization—as shown on Fig. 1, Chapter 18, the working solution is fed into a spray nozzle. A strong gas flow breaks the liquid flow into droplets and throws them towards the substrate. (JCV)

Print density—defined as $\log_{10} (I/I_0)$ in which I is the intensity of the light reflected from a solid or halftone area of a print and I_0 is the intensity of the incident light. (SK & PO)

Printing cylinder/printing plate—contains the image and non-image areas of a print. (SK & PO)

PTFE—polytetrafluoroethylene.

PVC—polyvinylchloride.

Reduction—a chemical reaction involving gain of electrons by a reactant. (RPK)

Relaxation time (or dielectric relaxation time)—is a time constant for the decay of charge density at every point within a material. This time constant is equal to ε/σ, where ε is the dielectric constant of the material and σ is its conductivity. (SBS, RB, DAD & JWB)

Resin—natural or synthetic plastic material obtained by polymerization and used as a binder. (DDG)

Reversible process—a process which is carried out in small steps such that the state of equilibrium is preserved at each step. (MJS)

Rockwell—see "hardness".

Rose bengal—a typical phlthalin dye which fluoresces green when irradiated with ultraviolet light. (LS)

Rough threading—a method of surface roughening which consists of cutting threads with the sides and tops of the threads jagged and torn sufficiently that other roughening is not required. (DRM)

Running-in—the process by which machine parts improve in conformity, surface topography and frictional compatibility during the initial stage of use. (JJC & CA)

Saturation—occurs when the components of reaction diffuse towards the substrate at a rate higher than the reaction rate on the substrate surface. (JCV)

Scuffing—the surface damage occurring due to continuing relative movement of parts after adhesion (welding) of the asperities has occurred. This may also include an element due to the ploughing of the asperities of one component into the surface of the other. Scuffing is also known as "galling" or "picking-up". (JCG)

Seizure—this is the stoppage or stalling of a mechanism when the friction force between them following scuffing has become too great to allow of continued motion. (JCG)

Sessile drop method—used to determine the surface tension of a liquid from the shape of a drop at rest on a flat surface. (AP)

Sheet resistance—this is the electrical resistance of a square coating of a material, measured between two opposite parallel edges of the square. Regardless of the size of the square, since the width equals the length l, then the resistance per square R_s is given by

$$R_s = \frac{\rho l}{a} = \rho \cdot \frac{l}{lt} = \frac{\rho}{t}$$

where ρ is the resistivity, a the cross-sectional area, and t the thickness. Often normalized to 20 μm, the sheet resistance allows immediate comparison between materials, and is widely used in the field. (RGL)

Sintered materials—materials that have been prepared by the fusing of individual particles, usually at high temperature. They can be porous. (DSC)

Solid print—an area covered by a continuous ink layer on the printing plate. (SK & PO)

Solids—that portion of a liquid system which remains as solid film on the surface of the coated article after the volatile portion has evaporated. (LS)

Solids content—the ingredients of a coating composition which remain after drying to constitute the dry film. (JM)

Solvent—a liquid, usually volatile, which dissolves or disperses the film-forming constituents and which evaporates during drying and does not become part of the dried paint film. (JM)

Solvent wash—the washing down of a deposited, but not yet cured, paint film by solvent vapour condensed on the surface of a cavity. (AW)

Specific area—the surface area of a particle of a given metal, produced by atomization of that material from a molten state at a particular temperature and by a particular atomizing gas pressure and velocity. (DRM)

Spray pattern—the wedge-shaped pattern formed by the atomized paint as it leaves the front of the spray gun. (JM)

Spreading—the dynamic aspect of wetting. (AP)

Standard free energy change ΔG_T^0—the free energy change during a reaction occurring at a constant temperature T when the pressures are all fixed at unity. (JCV)

Stock removal—the amount of material to be removed from a surface in order to produce a certain dimension and/or finish. (HNW)

Stoichiometry—the stoichiometry of a compound is the ratio of the anions to cations. In defect lattices, anion or cation vacancies cause deviations from the ideal stoichiometry. For example, in TiC, C lattice sites can be vacant and the stoichiometry would approach 1 in TiC_{1-x} as x approached zero. (RFB)

Stoving—the process of drying and hardening the paint coating by heating in an oven or other apparatus. (JM)

Strike-through—that phenomenon encountered when a dark prime or initial coating causes a discoloration of a subsequently applied top coat of a lighter colour. (LS)

Substrate—a support on which a subsequent thin layer can be grown or deposited. (DSC)

Substrate—the material with the surface upon which a film of another material is deposited. (AW)

Surface chalking—certain resin-based coatings, such as on epoxy resins, will on exterior exposure lose their gloss, accompanied by "chalking"—powdering of the surface film. (SK)

Surface roughening—a group of procedures for producing irregularities on the surface to be metallized. (DRM)

Thermoplastic powder—one based on a resin system such as PVC (polyvinylchloride) which will soften on heating and regain properties on cooling. (SK)

Thermosetting powder—one based on resin and curing agent, the mixture reacting on heating to form a tough cured coating which will not soften on heating. Typical examples are epoxy powders. (SK)

Thixotropy—the property of a liquid of increasing in viscosity with the passage of time. On shaking, the viscosity returns to its original value. (DDG)

Transport-migration—the movement of the ions through the electrolyte associated with the passage of the electric current. (RPK)

Transport number—the proportion of the current carried by the ions of a given kind. (RPK)

Tribology—the science and technology of interacting surfaces in relative motion and of the practices related thereto. (JCG)

Ultrasonic atomization—a piezoelectric crystal is set on the bottom of a 200 cm^3 vessel containing the working solution. When electrically excited, the crystal emits ultrasonic energy which generates a dense mist at the surface of the liquid. A flow of inert gas carries the droplets towards the substrate. (JCV)

Vapour phase coating—a method of coating in which the part to be coated is enveloped by atmosphere which carries, in gaseous form, the material(s) from which the coating will be formed (e.g. nitriding). Vapour phase coatings may or may not be diffused, depending upon the temperature at which they are conducted. (RLW) (See also "diffusion coatings" and "pack cementation coating".)

Volatile compound—compound whose boiling point is not far from room temperature. (JCV)

VPN—Vickers pyramid number. See "hardness".

Wet-on-dry printing/wet-on-wet printing—multicolour printing in which successive ink layers are printed on a dry/wet ink layer. (SK & PO)

Wetting—the ability of a liquid to wet a solid surface. (AP)

Wetting agent—a material which lowers the surface tension of a liquid. (DDG)

Work function—the work function $e\phi$ of a uniform surface of an electronic conductor is the difference between the electrochemical potential (or: the Fermi energy E_F) just inside the conductor, and the electrostatic potential energy just outside the conductor in vacuum. (C. Herring and M. H. Nichols. *Rev. Mod. Phys.*, **21** (1949), 190.) The definition also applies to electronic semiconductors. (MJS)

Wrap-around—the property of a spray system which defines the degree with which the reverse side of the object to be coated receives a satisfactory coating. (RPC)

Yield value—the shear rate at which a thixotropic material starts flowing. (DDG)

Zeta potential—see "electrokinetic potential".

Zinc alloy die-casting—process by which articles are mass produced by injecting molten zinc alloys into steel dies on high speed automatic machines. (HS)

Subject index